International Series in Pure and Applied Mathematics

William Ted Martin, Consulting Editor

VECTOR AND TENSOR ANALYSIS

International Series in Pure and Applied Mathematics

William Ted Martin, CONSULTING EDITOR

Golomb and Shanks: ELEMENTS OF ORDINARY DIFFERENTIAL EQUATIONS (*In Preparation*)

Lass: VECTOR AND TENSOR ANALYSIS

Sneddon: FOURIER TRANSFORMS (*In Preparation*)

VECTOR AND TENSOR ANALYSIS

BY

HARRY LASS

Assistant Professor of Mathematics
University of Illinois

FIRST EDITION

New York Toronto London

McGRAW-HILL BOOK COMPANY, INC.

1950

VECTOR AND TENSOR ANALYSIS

THE MAPLE PRESS COMPANY, YORK, PA.

To My

Mother and Father

PREFACE

This text can be used in a variety of ways. The student totally unfamiliar with vector analysis can peruse Chapters 1, 2, and 4 to gain familiarity with the algebra and calculus of vectors. These chapters cover the ordinary one-semester course in vector analysis. Numerous examples in the fields of differential geometry, electricity, mechanics, hydrodynamics, and elasticity can be found in Chapters 3, 5, 6, and 7, respectively. Those already acquainted with vector analysis who feel that they would like to become better acquainted with the applications of vectors can read the above-mentioned chapters with little difficulty: only a most rudimentary knowledge of these fields is necessary in order that the reader be capable of following their contents, which are fairly complete from an elementary viewpoint. A knowledge of these chapters should enable the reader to further digest the more comprehensive treatises dealing with these subjects, some of which are listed in the reference section. It is hoped that these chapters will give the mathematician a brief introduction to elementary theoretical physics. Finally, the author feels that Chapters 8 and 9 deal sufficiently with tensor analysis and Riemannian geometry to enable the reader to study the theory of relativity with a minimum of effort as far as the mathematics involved is concerned.

In order to cover such a wide range of topics the treatment has necessarily been brief. It is hoped, however, that nothing has been sacrificed in the way of clearness of ideas. The author has attempted to be as rigorous as is possible in a work of this nature. Numerous examples have been worked out fully in the text. The teacher who plans on using this book as a text can surely arrange the topics to suit his needs for a one-, two-, or even three-semester course.

If the book is successful, it is due in no small measure to the composite efforts of those men who have invented and who have applied the vector and tensor analysis. The excellent works listed in the reference section have been of great aid. Finally, I wish to thank Professor Charles de Prima of the California Institute of Technology for his kind interest in the development of this text.

HARRY LASS

URBANA, ILL.
February, 1950

CONTENTS

CHAPTER 1

THE ALGEBRA OF VECTORS

1. Definition of a Vector. Our starting point for the definition of a vector will be the intuitive one encountered in elementary physics. Any directed line segment will be called a vector. The length of the vector will be denoted by the word *magnitude*. Any physical element that has magnitude and direction, and hence can be represented by a vector, will also be designated as a vector. In Chap. 8 we will give a more mathematically rigorous definition of a vector.

Elementary examples of vectors are displacements, velocities, forces, accelerations, etc. Physical concepts, such as speed, temperature, distance, and specific gravity, and arithmetic numbers, such as 2, π, etc., are called *scalars* to distinguish them from vectors. We note that no direction is associated with a scalar.

We shall represent vectors by arrows and use **boldface** type to indicate that we are speaking of a vector. In order to distinguish between scalars and vectors, the student will have to adopt some notation for describing a vector in writing. The student may choose his mode of representing a vector from Fig. 1 or may adopt his own notation.

$$\left[\vec{a},\quad \bar{a},\quad \underline{a},\quad \utilde{a},\quad a\right]$$

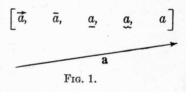

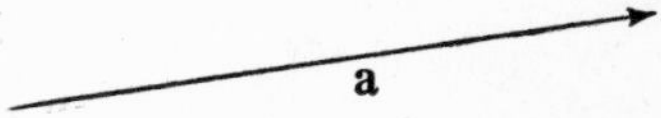

Fig. 1.

To every vector will be associated a real nonnegative number equal to the length of the vector. This number will depend, of course, on the unit chosen to represent a given class of vectors. A vector of length one will be called a unit vector. If a represents the length of the vector $\mathbf{a}$, we shall write $a \equiv |\mathbf{a}|$. If $|\mathbf{a}| = 0$, we define $\mathbf{a}$ as the zero vector.

2. Equality of Vectors. Two vectors will be defined to be equal if, and only if, they are parallel, have the same sense of direction, and the same magnitude. The starting points of the vectors are immaterial. It is the direction and magnitude which are important. Equal vectors, however, may produce different physical effects, as will be seen later. We write **a** = **b** if the vectors are equal (see Fig. 2).

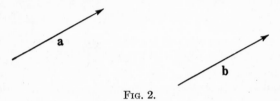

FIG. 2.

3. Multiplication by a Scalar. If we multiply a vector **a** by a real number x, we define the product $x\mathbf{a}$ to be a new vector parallel to **a** whose magnitude has been multiplied by the factor x. Thus 2**a** will be a vector which is twice as long as the vector **a**

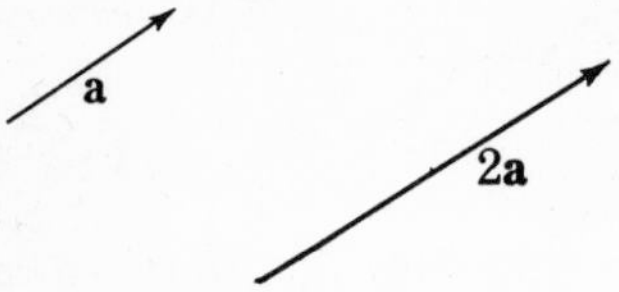

FIG. 3.

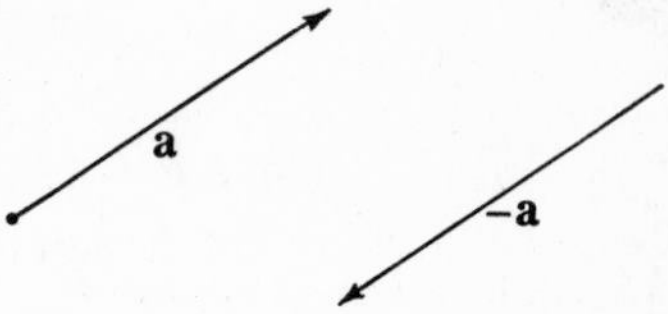

FIG. 4.

and which has the same direction as **a** (see Fig. 3). We define −**a** as the vector obtained from **a** by reversing its direction (see Fig. 4).

We note that

$$x(y\mathbf{a}) = (xy)\mathbf{a} = xy\mathbf{a}$$
$$(x + y)\mathbf{a} = x\mathbf{a} + y\mathbf{a}$$
$$0\mathbf{a} = \mathbf{0} \qquad \text{(zero vector)}$$

It is immediately seen that two vectors are parallel if, and only if, one of them can be written as a scalar multiple of the other.

4. Addition of Vectors. Let us suppose we have two vectors given, say **a** and **b**. We form a third vector by constructing a triangle with **a** and **b** forming two sides of the triangle, **b** adjoined to **a** (see Fig. 5). The vector starting from the origin of **a** and ending at the arrow of **b** is defined as the vector sum **a** + **b**.

We see that $\mathbf{a} + \mathbf{0} = \mathbf{a}$, and if $\mathbf{a} = \mathbf{b}$, $\mathbf{c} = \mathbf{d}$, then

$$\mathbf{a} + \mathbf{c} = \mathbf{b} + \mathbf{d}$$

From Euclidean geometry we note that

$$\mathbf{a} + \mathbf{b} = \mathbf{b} + \mathbf{a} \tag{1}$$
$$(\mathbf{a} + \mathbf{b}) + \mathbf{c} = \mathbf{a} + (\mathbf{b} + \mathbf{c}) \tag{2}$$
$$x(\mathbf{a} + \mathbf{b}) = x\mathbf{a} + x\mathbf{b} \tag{3}$$

(1) is called the commutative law of vector addition; (2) is called the associative law of vector addition; (3) is the distributive law for multiplication by a scalar. The reader should have no trouble proving these three results geometrically.

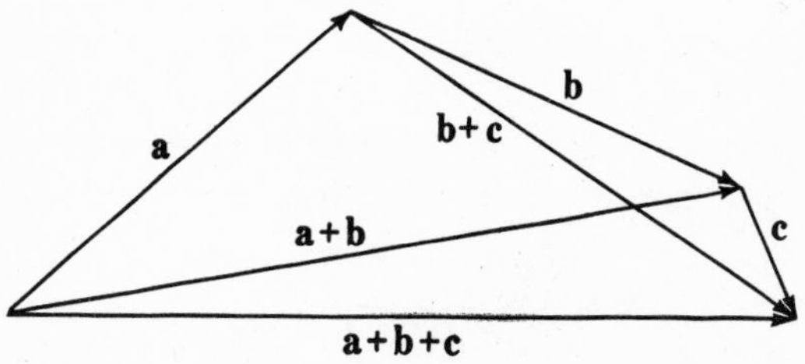

FIG. 5.

5. Subtraction of Vectors. Given the two vectors **a** and **b**, we can ask ourselves the following question: What vector **c** must be added to **b** to give **a**? The vector **c** is defined to be the vector **a** − **b**. We can obtain the desired result by two methods. First, construct −**b** and then add this vector to **a**, or second, let **b** and **a** have a common origin and construct the third side of the triangle. The two possible directions will give **a** − **b** and **b** − **a** (see Fig. 6). Thus **a** − **b** = **a** + (−**b**).

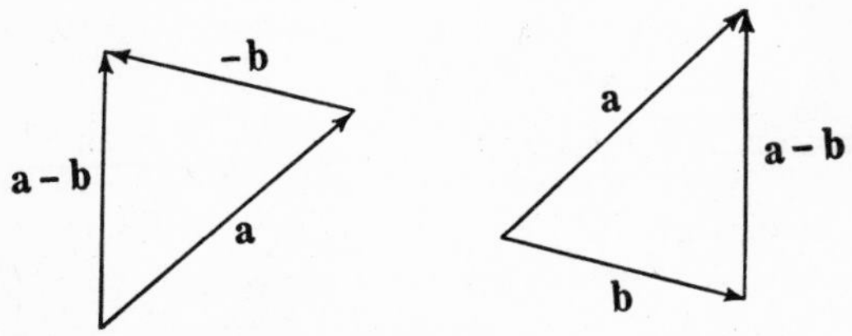

FIG. 6.

6. Linear Functions. Let us consider all vectors in the two-dimensional Euclidean plane. We choose a basis for this system of vectors by considering any two nonparallel, nonzero vectors. Call them **a** and **b**. Any third vector **c** can be written as a linear combination or function of **a** and **b**,

$$\mathbf{c} = x\mathbf{a} + y\mathbf{b} \tag{4}$$

The proof of (4) is by construction (see Fig. 7).

Let us now consider the following problem: Let **a** and **b** have a common origin, O, and let **c** be any vector starting from O whose end point lies on the line joining the ends of **a** and **b** (see Fig. 8).

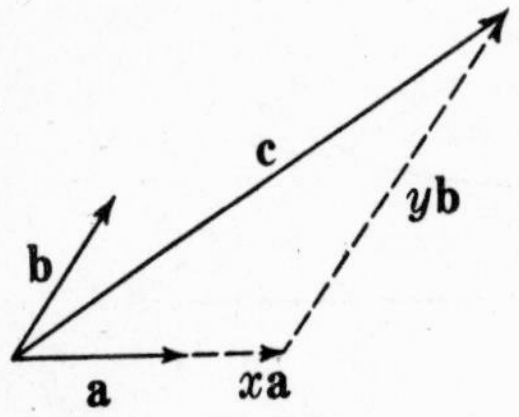

FIG. 7.

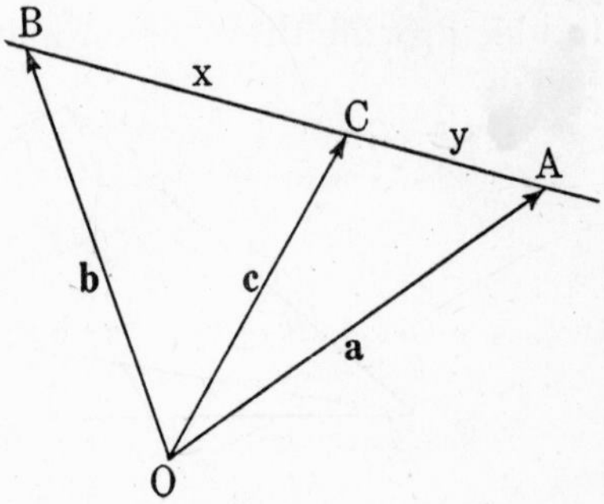

FIG. 8.

Let C divide BA in the ratio $x:y$ where $x + y = 1$. In particular, if C is the mid-point of BA, then $x = y = \frac{1}{2}$. Now

$$\begin{aligned}\mathbf{c} &= \overrightarrow{OB} + \overrightarrow{BC} \\ &= \mathbf{b} + x(\mathbf{a} - \mathbf{b}) \\ &= x\mathbf{a} + (1 - x)\mathbf{b}\end{aligned}$$

so that

$$\boxed{\mathbf{c} = x\mathbf{a} + y\mathbf{b}} \tag{5}$$

Now conversely, assume $\mathbf{c} = x\mathbf{a} + y\mathbf{b}$, $x + y = 1$. Then

$$\mathbf{c} = x\mathbf{a} + (1 - x)\mathbf{b} = x(\mathbf{a} - \mathbf{b}) + \mathbf{b}$$

We now note that **c** is a vector that is obtained by adding to **b** the vector $x(\mathbf{a} - \mathbf{b})$, this latter vector being parallel to the vector $\mathbf{a} - \mathbf{b}$. This immediately implies that the end point of **c** lies on the line joining A to B. We can rewrite (5) as

$$\begin{aligned}\mathbf{c} - x\mathbf{a} - y\mathbf{b} &= \mathbf{0} \\ 1 - x - y &= 0\end{aligned} \tag{6}$$

We have proved our first important theorem. A necessary and sufficient condition that the end points of any three vectors with common origin be on a straight line is that real constants l, m, n exist such that

$$\begin{aligned} l\mathbf{a} + m\mathbf{b} + n\mathbf{c} &= \mathbf{0} \\ l + m + n &= 0 \end{aligned} \tag{7}$$

with $l^2 + m^2 + n^2 \neq 0$.

We shall, however, find (5) more useful for solving problems.

Example 1. Let us prove that the medians of a triangle meet at a point P which divides each median in the ratio 1:2.

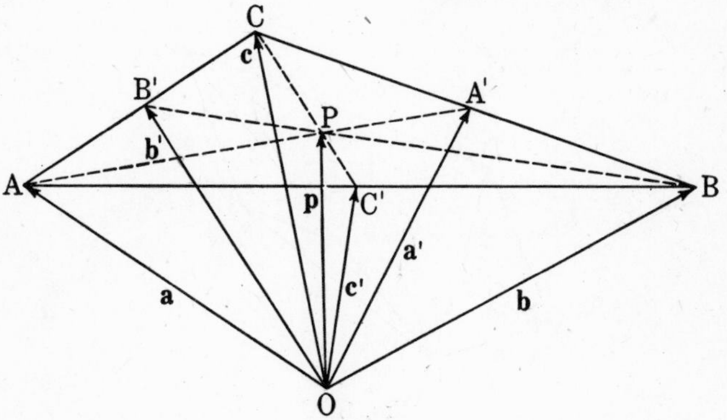

FIG. 9.

Let ABC be the given triangle and let A', B', C' be the midpoints. Choose O anywhere in space and construct the vectors from O to A, B, C, A', B', C', calling them $\mathbf{a}$, $\mathbf{b}$, $\mathbf{c}$, $\mathbf{a}'$, $\mathbf{b}'$, $\mathbf{c}'$ (see Fig. 9). From (5) we have

$$\begin{aligned} \mathbf{a}' &= \tfrac{1}{2}\mathbf{b} + \tfrac{1}{2}\mathbf{c} \\ \mathbf{b}' &= \tfrac{1}{2}\mathbf{a} + \tfrac{1}{2}\mathbf{c} \end{aligned} \tag{8}$$

Now P (the intersection of two of the medians) lies on the line joining A and A' and on the line joining B and B'. We shall thus find it expedient to find a relationship between the four vectors $\mathbf{a}$, $\mathbf{b}$, $\mathbf{a}'$, $\mathbf{b}'$ associated with A, B, A', B'. From (8) we eliminate the vector $\mathbf{c}$ and obtain

$$2\mathbf{a}' + \mathbf{a} = 2\mathbf{b}' + \mathbf{b}$$

or

$$\tfrac{2}{3}\mathbf{a}' + \tfrac{1}{3}\mathbf{a} = \tfrac{2}{3}\mathbf{b}' + \tfrac{1}{3}\mathbf{b} \tag{9}$$

But from (5), $\frac{2}{3}\mathbf{a}' + \frac{1}{3}\mathbf{a}$ represents a vector whose origin is at O and whose end point lies on the line joining A to A'. Similarly, $\frac{2}{3}\mathbf{b}' + \frac{1}{3}\mathbf{b}$ represents a vector whose origin is at O and whose end point lies on the line joining B to B'. There can only be one vector having both these properties, and this is the vector $\mathbf{p} = \overrightarrow{OP}$. Hence $\mathbf{p} = \frac{2}{3}\mathbf{a}' + \frac{1}{3}\mathbf{a} = \frac{2}{3}\mathbf{b}' + \frac{1}{3}\mathbf{b}$. Note that P divides AA' and BB' in the ratios 2:1. Had we considered the median CC' in connection with AA', we would have obtained that $\mathbf{p} = \frac{2}{3}\mathbf{c}' + \frac{1}{3}\mathbf{c}$, and this completes the proof of the theorem.

Example 2. To prove that the diagonals of a parallelogram bisect each other. Let $ABCD$ be the parallelogram and O any

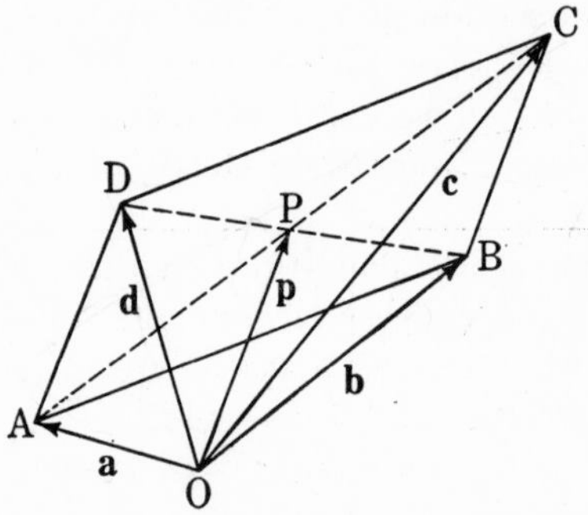

FIG. 10.

point in space (see Fig. 10). The equation $\mathbf{d} - \mathbf{a} = \mathbf{c} - \mathbf{b}$ implies that $ABCD$ is a parallelogram. Hence

$$\tfrac{1}{2}\mathbf{a} + \tfrac{1}{2}\mathbf{c} = \tfrac{1}{2}\mathbf{b} + \tfrac{1}{2}\mathbf{d} = \mathbf{p}$$

so that P bisects AC and BD.

Problems

1. Interpret $\dfrac{\mathbf{a}}{|\mathbf{a}|}$.

2. Give a geometric proof of (3).

3. $\mathbf{a}, \mathbf{b}, \mathbf{c}$ are consecutive vectors forming a triangle. What is the vector sum $\mathbf{a} + \mathbf{b} + \mathbf{c}$? Generalize this result for any closed polygon.

4. Vectors are drawn from the center of a regular polygon to its vertices. From symmetry considerations show that the vector sum is zero.

5. $\mathbf{a}$ and $\mathbf{b}$ are consecutive vectors of a parallelogram. Express the diagonal vectors in terms of $\mathbf{a}$ and $\mathbf{b}$.

6. $\mathbf{a}, \mathbf{b}, \mathbf{c}, \mathbf{d}$ are consecutive vector sides of a quadrilateral. Show that a necessary and sufficient condition that the figure be a parallelogram is that $\mathbf{a} + \mathbf{c} = \mathbf{0}$ and show that this implies $\mathbf{b} + \mathbf{d} = \mathbf{0}$.

7. Show graphically that $|\mathbf{a}| + |\mathbf{b}| \geqq |\mathbf{a} + \mathbf{b}|$. From this show that $|\mathbf{a} - \mathbf{b}| \geqq |\mathbf{a}| - |\mathbf{b}|$.

8. $\mathbf{a}, \mathbf{b}, \mathbf{c}, \mathbf{d}$ are vectors from O to A, B, C, D. If

$$\mathbf{b} - \mathbf{a} = 2(\mathbf{d} - \mathbf{c})$$

show that the intersection point of the two lines joining A and D and B and C trisects these lines.

9. $\mathbf{a}, \mathbf{b}, \mathbf{c}, \mathbf{d}$ are four vectors with a common origin. Find a necessary and sufficient condition that their end points lie in a plane.

10. What is the vector condition that the end points of the vectors of Prob. 9 form the vertices of a parallelogram?

11. Show that the mid-points of the lines which join the mid-points of the opposite sides of a quadrilateral coincide. The four sides of the quadrilateral are not necessarily coplanar.

12. Show that the line which joins one vertex of a parallelogram to the mid-point of an opposite side trisects the diagonal.

13. A line from a vertex of a triangle trisects the opposite side. It intersects a similar line issuing from another vertex. In what ratio do these lines intersect one another?

14. A line from a vertex of a triangle bisects the opposite side. It is trisected by a similar line issuing from another vertex. How does this latter line intersect the opposite side?

15. Show that the bisectors of a triangle meet in a point.

16. Show that if two triangles in space are so situated that the three points of intersection of corresponding sides lie on a line, then the lines joining the corresponding vertices pass through a common point, and conversely. This is Desargues's theorem.

17. $\mathbf{b} = (\sin t)\mathbf{a}$ is a variable vector which always remains parallel to the fixed vector $\mathbf{a}$. What is $\dfrac{d\mathbf{b}}{dt}$? Explain geometrically the meaning of $\dfrac{d\mathbf{b}}{dt}$.

18. Let $\mathbf{v}_1$ be the velocity of A relative to B and let $\mathbf{v}_2$ be the velocity of B relative to C. What is the velocity of A relative to C? Of C relative to A? Are these results obvious?

19. Let **a, b** be constant vectors and let **c** be defined by the equation

$$\mathbf{c} = (\cos t)\mathbf{a} + (\sin t)\mathbf{b}$$

When is **c** parallel to **a**? Parallel to **b**? Can **c** ever be parallel to **a** + **b**? Perpendicular to **a** + **b**? Find $\frac{d\mathbf{c}}{dt}, \frac{d^2\mathbf{c}}{dt^2}$. If **a** and **b** are unit orthogonal vectors with common origin, describe the positions of **c** and show that **c** is perpendicular to $\frac{d\mathbf{c}}{dt}$.

20. If **a** and **b** are not parallel, show that $m\mathbf{a} + n\mathbf{b} = k\mathbf{a} + j\mathbf{b}$ implies $m = k$, $n = j$.

21. *Theorem of Ceva.* A necessary and sufficient condition that the lines which join three points, one on each side of a triangle, to the opposite vertices be concurrent is that the product of the algebraic ratios in which the three points divide the sides be -1.

22. *Theorem of Menelaus.* Three points, one on each side of a triangle ABC, are collinear if and only if the product of the algebraic ratios in which they divide the sides BC, CA, AB is unity.

7. Coordinate Systems. For a considerable portion of the text we shall deal with the Euclidean space of three dimensions. This is the ordinary space encountered by students of analytic geometry and the calculus. We choose a right-handed coordinate system. If we rotate the x axis into the y axis, a right-hand screw will advance along the positive z axis.

We let **i, j, k** be the three unit vectors along the positive x, y, and z axes, respectively. The vectors **i, j, k** form a very simple and elegant basis for our three-dimensional Euclidean space. From Fig. 11 we observe that

$$\mathbf{r} = x\mathbf{i} + y\mathbf{j} + z\mathbf{k} \tag{10}$$

The numbers x, y, z are called the components of the vector **r**. Note that they represent the projections of the vector **r** on the x, y, and z axes. **r** is called the position vector of the point P

and will be used quite frequently in what follows. The most general space-time vector that we shall encounter will be of the form

$$\mathbf{u} = \mathbf{u}(x, y, z, t) = \alpha(x, y, z, t)\mathbf{i} + \beta(x, y, z, t)\mathbf{j} + \gamma(x, y, z, t)\mathbf{k} \quad (11)$$

It is of the utmost importance that the student understand the meaning of (11). To be more specific, let us consider a fluid in motion. At any time t the particle which happens to be at the

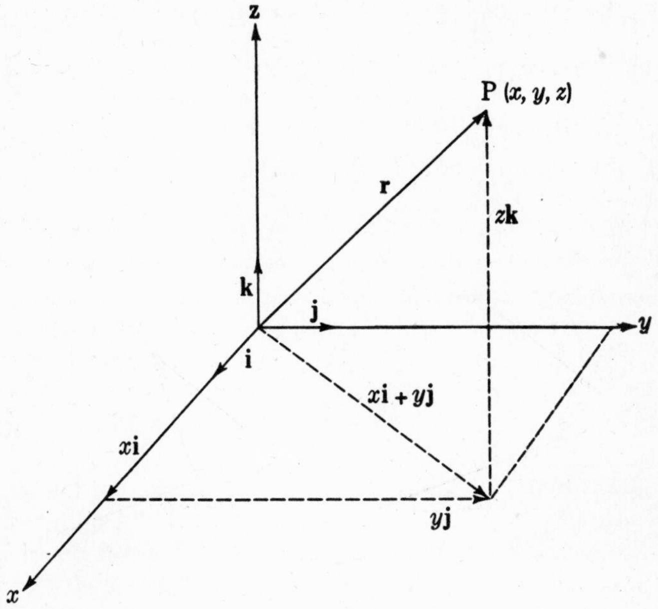

FIG. 11.

point $P(x, y, z)$ will have a velocity which depends on the coordinates x, y, z and on the time t. As time goes on, various particles arrive at $P(x, y, z)$ and have the velocities $\mathbf{u}(x, y, z, t)$ with components along the x, y, z axes given by $\alpha(x, y, z, t)$, $\beta(x, y, z, t)$, $\gamma(x, y, z, t)$.

Whenever we have a vector of the type (11), we say that we have a *vector field*. An elementary example would be the vector $\mathbf{u} = y\mathbf{i} - x\mathbf{j}$. This vector field is time-independent and so is called a *steady field*. At the point $P(1, -2, 3)$ it has the value $-2\mathbf{i} - \mathbf{j}$. Another example would be $\mathbf{u} = 3xze^t\mathbf{i} - xyzt\mathbf{j} + 5x\mathbf{k}$. We shall have more to say about this type of vector in later chapters and will, for the present, be interested only in constant vectors (*uniform fields*).

A moment's reflection shows that if $\mathbf{a} = a_1\mathbf{i} + a_2\mathbf{j} + a_3\mathbf{k}$, $\mathbf{b} = b_1\mathbf{i} + b_2\mathbf{j} + b_3\mathbf{k}$, then

$$\begin{aligned} \mathbf{a} + \mathbf{b} &= (a_1 + b_1)\mathbf{i} + (a_2 + b_2)\mathbf{j} + (a_3 + b_3)\mathbf{k} \\ x\mathbf{a} + y\mathbf{b} &= (xa_1 + yb_1)\mathbf{i} + (xa_2 + yb_2)\mathbf{j} + (xa_3 + yb_3)\mathbf{k} \end{aligned} \tag{12}$$

8. Scalar, or Dot, Product. We define the scalar or dot product of two vectors by the identity

$$\mathbf{a} \cdot \mathbf{b} \equiv |\mathbf{a}||\mathbf{b}| \cos \theta \tag{13}$$

where θ is the angle between the two vectors when drawn from a common origin. It makes no difference whether we choose θ or $-\theta$ since $\cos \theta = \cos(-\theta)$. This definition of the scalar product arose in physics and will play a dominant role in the development of the text.

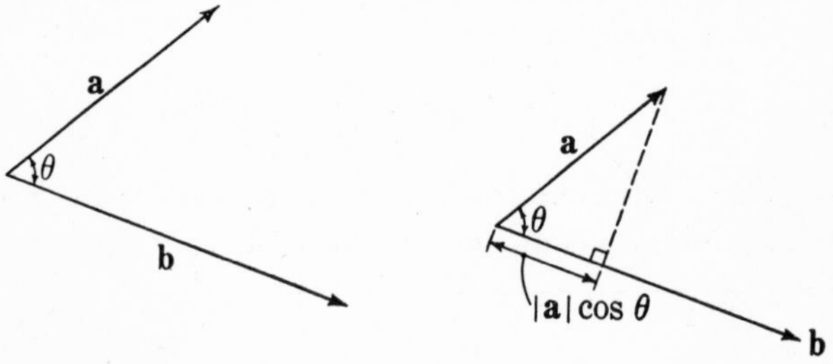

FIG. 12.

From (13) we at once verify that

$$\mathbf{a} \cdot \mathbf{b} = \mathbf{b} \cdot \mathbf{a} \tag{14}$$

$$\mathbf{a} \cdot \mathbf{a} = |\mathbf{a}|^2 = \mathbf{a}^2 \tag{15}$$

If $\mathbf{a}$ is perpendicular to $\mathbf{b}$, then

$$\mathbf{a} \cdot \mathbf{b} = 0 \tag{16}$$

However, if $\mathbf{a} \cdot \mathbf{b} = 0$, then either (1) $\mathbf{a} = \mathbf{0}$, (2) $\mathbf{b} = \mathbf{0}$, or (3) $\mathbf{a}$ is perpendicular to $\mathbf{b}$.

Now $\mathbf{a} \cdot \mathbf{b}$ is equal to the projection of $\mathbf{a}$ onto $\mathbf{b}$ multiplied by the length of $\mathbf{b}$ (see Fig. 12).

$$\mathbf{a} \cdot \mathbf{b} = (\text{proj } \mathbf{a})_b\, |\mathbf{b}| = (\text{proj } \mathbf{b})_a\, |\mathbf{a}| \tag{17}$$

With this in mind we proceed to prove the distributive law, which states that

$$\mathbf{a} \cdot (\mathbf{b} + \mathbf{c}) = \mathbf{a} \cdot \mathbf{b} + \mathbf{a} \cdot \mathbf{c} \tag{18}$$

From Fig. 13 it is apparent that

$$\begin{aligned} \mathbf{a} \cdot (\mathbf{b} + \mathbf{c}) &= [\text{proj } (\mathbf{b} + \mathbf{c})]_a \, |\mathbf{a}| \\ &= (\text{proj } \mathbf{b})_a \, |\mathbf{a}| + (\text{proj } \mathbf{c})_a \, |\mathbf{a}| \\ &= \mathbf{a} \cdot \mathbf{b} + \mathbf{a} \cdot \mathbf{c} \end{aligned}$$

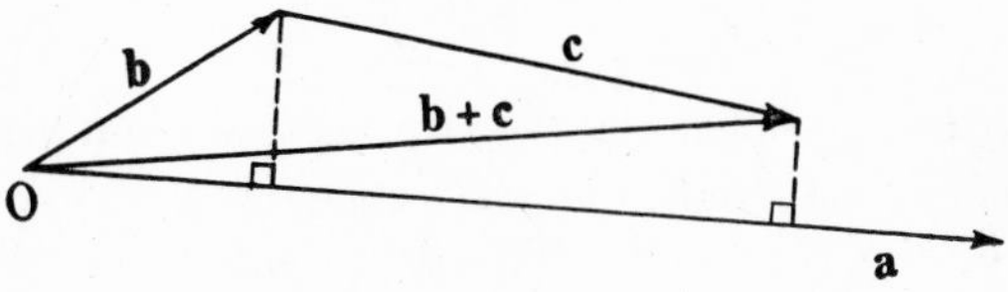

FIG. 13.

since the projection of the sum is the sum of the projections. Let the reader now prove that

$$(\mathbf{a} + \mathbf{b}) \cdot (\mathbf{c} + \mathbf{d}) = \mathbf{a} \cdot \mathbf{c} + \mathbf{a} \cdot \mathbf{d} + \mathbf{b} \cdot \mathbf{c} + \mathbf{b} \cdot \mathbf{d}$$

Example 3. To prove that the median to the base of an isosceles triangle is perpendicular to the base (see Fig. 14). From (5) we see that

$$\mathbf{m} = \tfrac{1}{2}\mathbf{a} + \tfrac{1}{2}\mathbf{b}$$

so that

$$\mathbf{m} \cdot (\mathbf{b} - \mathbf{a}) = \tfrac{1}{2}(\mathbf{b}^2 - \mathbf{a}^2) = 0$$

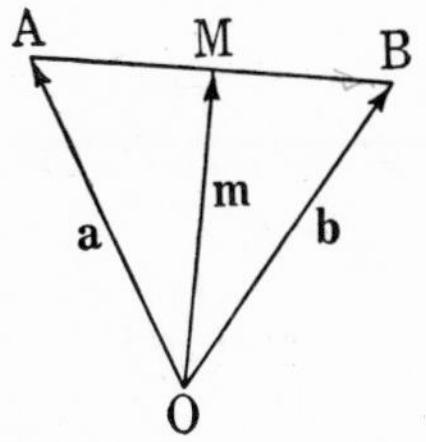

FIG. 14.

which proves that OM is perpendicular to AB.

Example 4. To prove that an angle inscribed in a semicircle is a right angle (see Fig. 15).

$$\overrightarrow{BC} = \mathbf{a} + \mathbf{c}$$
$$\overrightarrow{AC} = \mathbf{c} - \mathbf{a}$$
$$\begin{aligned} BC \cdot \overrightarrow{AC} &= (\mathbf{a} + \mathbf{c}) \cdot (\mathbf{c} - \mathbf{a}) \\ &= \mathbf{c}^2 - \mathbf{a}^2 = 0 \end{aligned}$$

so that ∡BCA is a right angle.

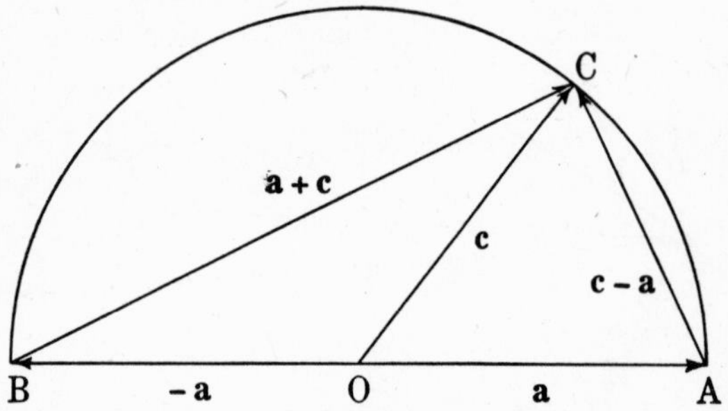

FIG. 15.

Example 5. *Cosine Law of Trigonometry.*

$$\begin{aligned} \mathbf{c} &= \mathbf{b} - \mathbf{a} \\ \mathbf{c} \cdot \mathbf{c} &= (\mathbf{b} - \mathbf{a}) \cdot (\mathbf{b} - \mathbf{a}) \\ &= \mathbf{b}^2 + \mathbf{a}^2 - 2\mathbf{a} \cdot \mathbf{b} \end{aligned}$$

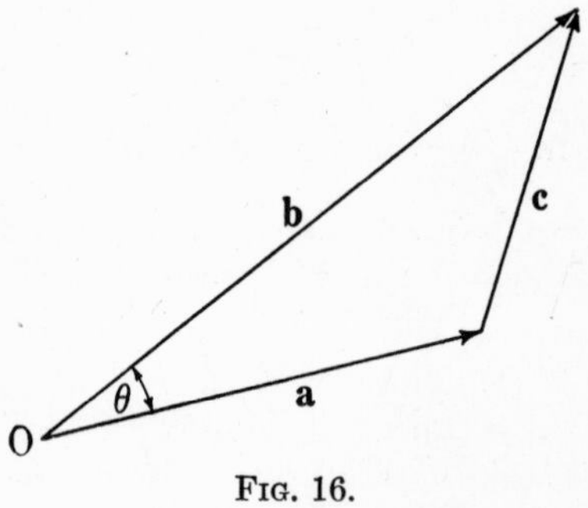

FIG. 16.

so that

$$c^2 = b^2 + a^2 - 2ab \cos \theta$$

Example 6

$$\begin{aligned} \mathbf{i} \cdot \mathbf{i} &= \mathbf{j} \cdot \mathbf{j} = \mathbf{k} \cdot \mathbf{k} = 1 \\ \mathbf{i} \cdot \mathbf{j} &= \mathbf{j} \cdot \mathbf{k} = \mathbf{k} \cdot \mathbf{i} = 0 \end{aligned}$$

$\overrightarrow{\text{Tensorial}}$ $a_j b_j = a_1 b_1 + a_2 b_2 + a_3 b_3$

Single order Tensor.

Hence if $\mathbf{a} = a_1\mathbf{i} + a_2\mathbf{j} + a_3\mathbf{k}$, $\mathbf{b} = b_1\mathbf{i} + b_2\mathbf{j} + b_3\mathbf{k}$, then

$$\mathbf{a} \cdot \mathbf{b} = a_1b_1 + a_2b_2 + a_3b_3 \tag{19}$$

Formula (19) is of the utmost importance.   Notice that $\mathbf{0} \cdot \mathbf{a} = 0$.

Example 7.   Cauchy's Inequality

$$(\mathbf{a} \cdot \mathbf{b})(\mathbf{a} \cdot \mathbf{b}) = |\mathbf{a}|^2 |\mathbf{b}|^2 \cos^2 \theta \leqq |\mathbf{a}|^2 |\mathbf{b}|^2$$

so that from (19)

$$(a_1b_1 + a_2b_2 + a_3b_3)^2 \leqq (a_1^2 + a_2^2 + a_3^2)(b_1^2 + b_2^2 + b_3^2)$$

In general

$$\sum_{\alpha=1}^{n} a_\alpha b_\alpha \leqq \left(\sum_{\alpha=1}^{n} a_\alpha^2 \right)^{\frac{1}{2}} \left(\sum_{\alpha=1}^{n} b_\alpha^2 \right)^{\frac{1}{2}} \tag{20}$$

Example 8.   Let $\mathbf{i}'$ be a unit vector making angles α, β, γ with the x, y, z axes.   The projections of $\mathbf{i}'$ on the x, y, z axes are $\cos \alpha$, $\cos \beta$, $\cos \gamma$, so that

$$\begin{aligned} \mathbf{i}' &= \cos \alpha\, \mathbf{i} + \cos \beta\, \mathbf{j} + \cos \gamma\, \mathbf{k} \\ &= p_1\mathbf{i} + q_1\mathbf{j} + r_1\mathbf{k} \end{aligned} \tag{21}$$

Notice that $p_1^2 + q_1^2 + r_1^2 = 1$. p_1, q_1, r_1 are called the direction cosines of the vector $\mathbf{i}'$.   Similarly, let $\mathbf{j}'$ and $\mathbf{k}'$ be unit vectors with direction cosines p_2, q_2, r_2 and p_3, q_3, r_3.   Thus

$$\begin{aligned} \mathbf{j}' &= p_2\mathbf{i} + q_2\mathbf{j} + r_2\mathbf{k} \\ \mathbf{k}' &= p_3\mathbf{i} + q_3\mathbf{j} + r_3\mathbf{k} \end{aligned} \tag{22}$$

We also impose the condition that $\mathbf{i}', \mathbf{j}', \mathbf{k}'$ be mutually orthogonal, so that the x', y', z' axes form a coordinate system similar to the x-y-z coordinate system with common origin O (see Fig. 17).

We have $\mathbf{r} = \mathbf{r}'$ so that $x\mathbf{i} + y\mathbf{j} + z\mathbf{k} = x'\mathbf{i}' + y'\mathbf{j}' + z'\mathbf{k}'$, where x, y, z are the coordinates of a point P as measured in the x-y-z coordinate system and x', y', z' are the coordinates of the same point P as measured in the x'-y'-z' coordinate system. Making use of (21) and (22) and equating components, we find that

$$\begin{aligned} x &= p_1x' + p_2y' + p_3z' \\ y &= q_1x' + q_2y' + q_3z' \\ z &= r_1x' + r_2y' + r_3z' \end{aligned} \tag{23}$$

We now find it more convenient to rename the x-y-z coordinate system. Let $x = x^1$, $y = x^2$, $z = x^3$, where the superscripts do not designate powers but are just labels which enable us to differentiate between the various axes. Similarly, let $x' = \bar{x}^1$, $y' = \bar{x}^2$,

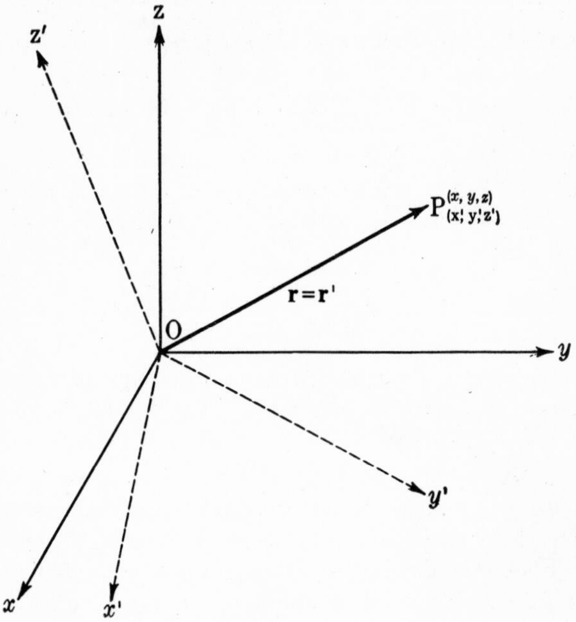

FIG. 17.

$z' = \bar{x}^3$. Now let $a_\beta{}^\alpha$ represent the cosine of the angle between the x^α and $\bar{x}^\beta$ axes. We can write (23) as

$$x^\alpha = \sum_{\beta=1}^{3} a_\beta{}^\alpha \bar{x}^\beta, \qquad \alpha = 1, 2, 3 \tag{24}$$

By making use of the fact that $\mathbf{i}' \cdot \mathbf{j}' = \mathbf{j}' \cdot \mathbf{k}' = \mathbf{k}' \cdot \mathbf{i}' = 0$, we can prove that

$$\bar{x}^\alpha = \sum_{\beta=1}^{3} A_\beta{}^\alpha x^\beta, \qquad \alpha = 1, 2, 3 \tag{25}$$

where $A_\beta{}^\alpha = a_\alpha{}^\beta$. We leave this as an exercise for the reader. Let us notice that differentiating (24) yields

$$\frac{\partial x^\alpha}{\partial \bar{x}^\sigma} = a_\sigma{}^\alpha, \qquad \alpha, \sigma = 1, 2, 3 \tag{26}$$

Example 9. The vector $\mathbf{a} = a^1\mathbf{i} + a^2\mathbf{j} + a^3\mathbf{k}$ may be represented by the number triple (a^1, a^2, a^3). Hence, without appealing to geometry we could develop an algebraic theory of vectors. If $\mathbf{b} = (b^1, b^2, b^3)$, then $\mathbf{a} + \mathbf{b}$ is defined by the number triple $(a^1 + b^1, a^2 + b^2, a^3 + b^3)$, and $x\mathbf{a} = x(a^1, a^2, a^3)$ is defined by the number triple (xa^1, xa^2, xa^3). From this the reader can prove that

$$(a^1, a^2, a^3) = a^1(1, 0, 0) + a^2(0, 1, 0) + a^3(0, 0, 1)$$

The triples (1, 0, 0), (0, 1, 0), (0, 0, 1) form a basis for our linear vector space, that is, the space of number triples. We note that the determinant formed from these triples, namely,

$$\begin{vmatrix} 1 & 0 & 0 \\ 0 & 1 & 0 \\ 0 & 0 & 1 \end{vmatrix} \equiv 1$$

does not vanish. Any three triples whose determinant does not vanish can be used to form a basis. Let the reader prove this result. We can define the scalar product (inner product) of two triples by the law $(\mathbf{a} \cdot \mathbf{b}) = a^1b^1 + a^2b^2 + a^3b^3$.

9. Applications of the Scalar Product to Space Geometry

(*a*) We define a plane as the locus of lines passing through a fixed point perpendicular to a fixed direction. Let the fixed point be $P_0(x_0, y_0, z_0)$ and let the fixed direction be given by the vector $\mathbf{N} = A\mathbf{i} + B\mathbf{j} + C\mathbf{k}$. Let $\mathbf{r}$ be the position vector to any point $P(x, y, z)$ on the plane (Fig. 18). Now $\overrightarrow{P_0P} = \mathbf{r} - \mathbf{r}_0$ is perpendicular to $\mathbf{N}$ so that

$$(\mathbf{r} - \mathbf{r}_0) \cdot (A\mathbf{i} + B\mathbf{j} + C\mathbf{k}) = 0$$

or

$$[(x - x_0)\mathbf{i} + (y - y_0)\mathbf{j} + (z - z_0)\mathbf{k}] \cdot (A\mathbf{i} + B\mathbf{j} + C\mathbf{k}) = 0$$

and

$$A(x - x_0) + B(y - y_0) + C(z - z_0) = 0 \tag{27}$$

This is the equation of the plane. The point $P_0(x_0, y_0, z_0)$ obviously lies in the plane since its coordinates satisfy (27). Equation (27) is linear in x, y, z.

(*b*) Consider the surface $Ax + By + Cz + D = 0$. Let $P(x_0, y_0, z_0)$ be any point on the surface. Of necessity,

$$Ax_0 + By_0 + Cz_0 + D = 0.$$

Subtracting we have

$$A(x - x_0) + B(y - y_0) + C(z - z_0) = 0 \tag{28}$$

Now consider the two vectors $A\mathbf{i} + B\mathbf{j} + C\mathbf{k}$ and

$$(x - x_0)\mathbf{i} + (y - y_0)\mathbf{j} + (z - z_0)\mathbf{k}$$

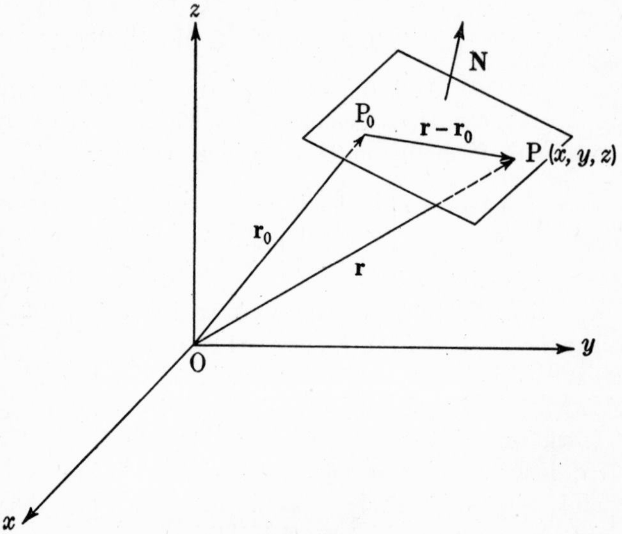

FIG. 18.

Equation (28) shows that these two vectors are perpendicular. Hence the constant vector $A\mathbf{i} + B\mathbf{j} + C\mathbf{k}$ is normal to the surface at every point so that the surface is a plane.

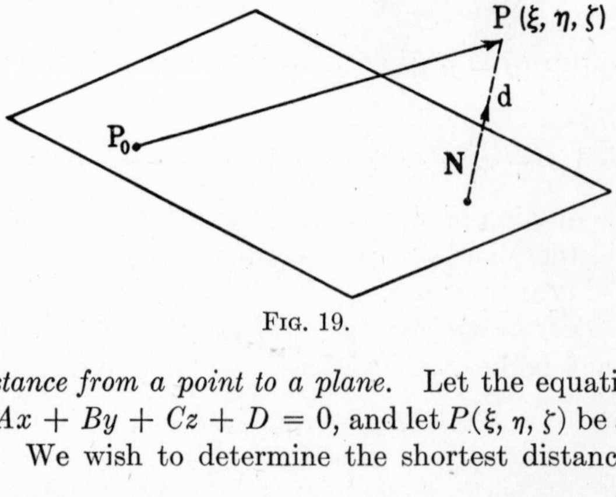

FIG. 19.

(c) *Distance from a point to a plane.* Let the equation of the plane be $Ax + By + Cz + D = 0$, and let $P(\xi, \eta, \zeta)$ be any point in space. We wish to determine the shortest distance from P

to the plane. Choose any point P_0 lying in the plane. It is apparent that the shortest distance will be the projection of $\overrightarrow{P_0P}$ on $\mathbf{N}$, where $\mathbf{N}$ is a unit vector normal to the plane (see Fig. 19). Now

$$d = |\overrightarrow{P_0P} \cdot \mathbf{N}|$$

$$= \left| \frac{A\xi + B\eta + C\zeta + D}{(A^2 + B^2 + C^2)^{\frac{1}{2}}} \right| \qquad (29)$$

where use has been made of the fact that

$$Ax_0 + By_0 + Cz_0 + D = 0.$$

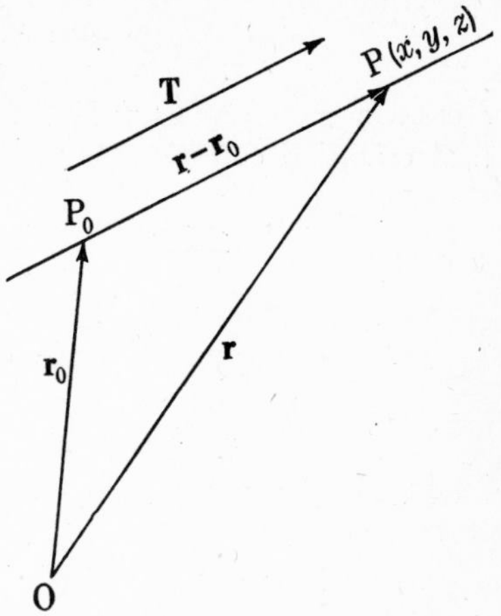

FIG. 20.

(d) *Equation of a straight line through the point $P_0(x_0, y_0, z_0)$ parallel to the vector* $\mathbf{T} = l\mathbf{i} + m\mathbf{j} + n\mathbf{k}$. From Fig. 20 it is

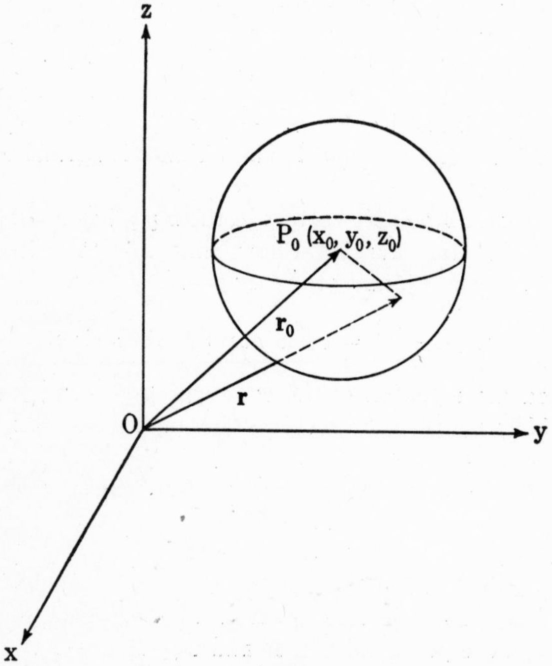

FIG. 21.

apparent that $\mathbf{r} - \mathbf{r}_0$ is parallel to $\mathbf{T}$ so that $\mathbf{r} - \mathbf{r}_0 = \lambda\mathbf{T}$,

$$-\infty < \lambda < +\infty$$

Hence $(x - x_0)\mathbf{i} + (y - y_0)\mathbf{j} + (z - z_0)\mathbf{k} = \lambda(l\mathbf{i} + m\mathbf{j} + n\mathbf{k})$, so that equating components yields

$$\frac{x - x_0}{l} = \frac{y - y_0}{m} = \frac{z - z_0}{n} = \lambda \tag{30}$$

By allowing λ to vary from $-\infty$ to $+\infty$ we generate every point on the line.

(e) *Equation of a sphere with center at* $P_0(x_0, y_0, z_0)$ *and radius* a. In Fig. 21 obviously $(\mathbf{r} - \mathbf{r}_0) \cdot (\mathbf{r} - \mathbf{r}_0) = a^2$, or

$$(x - x_0)^2 + (y - y_0)^2 + (z - z_0)^2 = a^2$$

Problems

1. Add and subtract the vectors $\mathbf{a} = 2\mathbf{i} - 3\mathbf{j} + 5\mathbf{k}$,

$$\mathbf{b} = -2\mathbf{i} + 2\mathbf{j} + 2\mathbf{k}$$

Show that the vectors are perpendicular.

2. Find the cosine of the angle between the two vectors $\mathbf{a} = 2\mathbf{i} - 3\mathbf{j} + \mathbf{k}$ and $\mathbf{b} = 3\mathbf{i} - \mathbf{j} - 2\mathbf{k}$.

3. If $\mathbf{c}$ is normal to $\mathbf{a}$ and $\mathbf{b}$, show that $\mathbf{c}$ is normal to $\mathbf{a} + \mathbf{b}$, $\mathbf{a} - \mathbf{b}$.

4. Let $\mathbf{a}$ and $\mathbf{b}$ be unit vectors in the x-y plane making angles α and β with the x axis. Show that $\mathbf{a} = \cos\alpha\,\mathbf{i} + \sin\alpha\,\mathbf{j}$, $\mathbf{b} = \cos\beta\,\mathbf{i} + \sin\beta\,\mathbf{j}$, and prove that

$$\cos(\alpha - \beta) = \cos\alpha\cos\beta + \sin\alpha\sin\beta$$

5. Find the equation of the cone whose generators make an angle of 30° with the unit vector which makes equal angles with the x, y, and z axes.

6. The position vectors of the foci of an ellipse are $\mathbf{c}$ and $-\mathbf{c}$, and the length of the major axis is $2a$. Show that the equation of the ellipse is $a^4 - a^2(\mathbf{r}^2 + \mathbf{c}^2) + (\mathbf{c} \cdot \mathbf{r})^2 = 0$.

7. Prove that the altitudes of a triangle are concurrent.

8. Find the shortest distance from the point $A(1, 0, 1)$ to the line through the points $B(2, 3, 4)$ and $C(-1, 1, -2)$.

9. Let $\mathbf{a} = 2\mathbf{i} - \mathbf{j} + \mathbf{k}$, $\mathbf{b} = \mathbf{i} - 3\mathbf{j} - 5\mathbf{k}$. Find a vector $\mathbf{c}$ so that $\mathbf{a}$, $\mathbf{b}$, $\mathbf{c}$ form the sides of a right triangle.

10. Let $\mathbf{r}$ be the position vector of a point $P(x, y, z)$, and let $\mathbf{a}$ be a constant vector. Interpret the equation $(\mathbf{r} - \mathbf{a}) \cdot \mathbf{a} = 0$.

11. Given $\mathbf{a} = 2\mathbf{i} - 3\mathbf{j} + \mathbf{k}$, $\mathbf{b} = 3\mathbf{j} - 4\mathbf{k}$, find the projection of $\mathbf{a}$ along $\mathbf{b}$.

12. Show that the line joining the end points of the vectors $\mathbf{a} = 2\mathbf{i} - \mathbf{j} - \mathbf{k}$, $\mathbf{b} = -\mathbf{i} + 3\mathbf{j} - \mathbf{k}$ with common origin at O is parallel to the x-y plane, and find its length.

13. Prove that the sum of the squares of the diagonals of a parallelogram is equal to the sum of the squares of its sides.

14. Let $\mathbf{a} = \mathbf{a}_1 + \mathbf{a}_2$ where $\mathbf{a}_1 \cdot \mathbf{b} = 0$ and $\mathbf{a}_2$ is parallel to $\mathbf{b}$. Show that $\mathbf{a}_2 = \dfrac{(\mathbf{a} \cdot \mathbf{b})}{|\mathbf{b}|} \mathbf{b}$, $\mathbf{a}_1 = \mathbf{a} - \dfrac{(\mathbf{a} \cdot \mathbf{b})}{|\mathbf{b}|} \mathbf{b}$.

15. Derive (25).

16. Verify (26).

17. Find a vector perpendicular to the vectors $\mathbf{a} = \mathbf{i} - \mathbf{j} + \mathbf{k}$, $\mathbf{b} = 2\mathbf{i} + 3\mathbf{j} - \mathbf{k}$.

18. Let $\mathbf{a} = f(t)\mathbf{i} + g(t)\mathbf{j} + h(t)\mathbf{k}$, and define

$$\frac{d\mathbf{a}}{dt} = f'(t)\mathbf{i} + g'(t)\mathbf{j} + h'(t)\mathbf{k}$$

Show that $\dfrac{d\mathbf{a}}{dt} \cdot \mathbf{a} = |\mathbf{a}| \dfrac{d|\mathbf{a}|}{dt}$.

19. Find the angle between the plane $Ax + By + Cz + D = 0$ and the plane $ax + by + cz + d = 0$.

20. $\mathbf{a}$, $\mathbf{b}$, $\mathbf{c}$ are coplanar. If $\mathbf{a}$ is not parallel to $\mathbf{b}$, show that

$$\mathbf{c} = \frac{\begin{vmatrix} \mathbf{c} \cdot \mathbf{a} & \mathbf{a} \cdot \mathbf{b} \\ \mathbf{c} \cdot \mathbf{b} & \mathbf{b} \cdot \mathbf{b} \end{vmatrix} \mathbf{a} + \begin{vmatrix} \mathbf{a} \cdot \mathbf{a} & \mathbf{c} \cdot \mathbf{a} \\ \mathbf{a} \cdot \mathbf{b} & \mathbf{c} \cdot \mathbf{b} \end{vmatrix} \mathbf{b}}{\begin{vmatrix} \mathbf{a} \cdot \mathbf{a} & \mathbf{a} \cdot \mathbf{b} \\ \mathbf{a} \cdot \mathbf{b} & \mathbf{b} \cdot \mathbf{b} \end{vmatrix}}$$

21. For the $a_\beta{}^\alpha$, $A_\beta{}^\alpha$ defined by (24) and (25), show that

$$\sum_{\gamma=1}^{3} a_\gamma{}^\alpha A_\beta{}^\gamma = \delta_\beta{}^\alpha$$

where $\delta_\beta{}^\alpha = 1$ if $\alpha = \beta$, $\delta_\beta{}^\alpha = 0$ if $\alpha \neq \beta$.

22. If B^1, B^2, B^3 are the components of a vector $\mathbf{B}$, that is, $\mathbf{B} = B^1\mathbf{i} + B^2\mathbf{j} + B^3\mathbf{k}$ (see Example 8 in regard to the super-

scripts), show that for a rotation of axes the components of the vector **B** become $(\bar{B}^1, \bar{B}^2, \bar{B}^3)$ where $\bar{B}^\alpha = \sum_{\beta=1}^{3} a_\beta{}^\alpha B^\beta$, $\alpha = 1, 2, 3$, and $\mathbf{B} = \bar{B}^1\mathbf{i}' + \bar{B}^2\mathbf{j}' + \bar{B}^3\mathbf{k}'$. Read Example 8 carefully.

23. Show that for a rotation of axes,

$$B^1C^1 + B^2C^2 + B^3C^3 = \bar{B}^1\bar{C}^1 + \bar{B}^2\bar{C}^2 + \bar{B}^3\bar{C}^3$$

This shows the invariance of the scalar product for rotations of axes. The invariance here refers to both the numerical invariance of the scalar product and the formal invariance,

$$\sum_{\alpha=1}^{3} B^\alpha C^\alpha \equiv \sum_{\alpha=1}^{3} \bar{B}^\alpha \bar{C}^\alpha$$

24. Prove the statements made in Example 9.

25. Generalize the statements of Example 9 for n-tuples

$$(a^1, a^2, \ldots, a^n)$$

10. Vector, or Cross, Product. Given any two nonparallel vectors **a** and **b**, we may construct a third vector **c** as follows: When translated so that they have a common origin, the two vectors **a**, **b** form two sides of a parallelogram. We define **c** to be perpendicular to the plane of this parallelogram with magnitude equal to the area of the parallelogram. We choose that normal obtained by the motion of a right-hand screw when **a** is rotated into **b** (angle of rotation less than 180°) (see Fig. 22). A cross is placed between the vectors **a** and **b** to denote the vector $\mathbf{c} = \mathbf{a} \times \mathbf{b}$. The vector **c** is called the cross, or vector, product of **a** and **b** and is given by

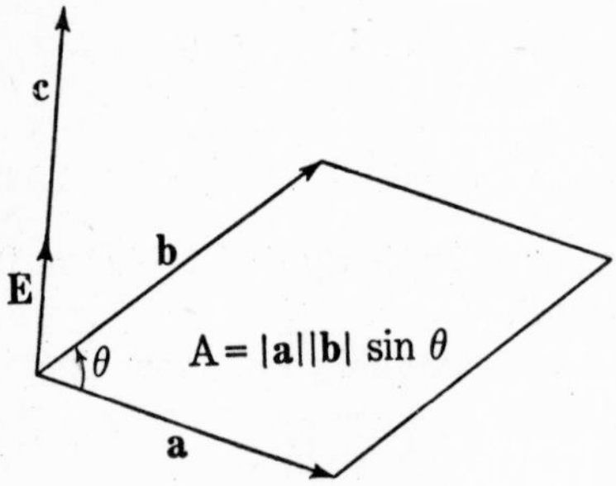

FIG. 22.

$$\mathbf{c} = \mathbf{a} \times \mathbf{b} = |\mathbf{a}||\mathbf{b}| \sin\theta \, \mathbf{E} \tag{31}$$

where $|\mathbf{E}| = 1$. The area of the parallelogram is

$$A = |\mathbf{a}||\mathbf{b}| \sin\theta$$

The cross product will occur frequently in mechanics and electricity, but for the present we discuss its algebraic behavior. It is obvious that $\mathbf{a} \times \mathbf{b} = -\mathbf{b} \times \mathbf{a}$, so that vector multiplication is not commutative. If $\mathbf{a}$ and $\mathbf{b}$ are parallel, $\mathbf{a} \times \mathbf{b} = \mathbf{0}$. In particular, $\mathbf{a} \times \mathbf{a} = \mathbf{0}$.

11. The Distributive Law for the Vector Product. We desire to prove that $\mathbf{a} \times (\mathbf{b} + \mathbf{c}) = \mathbf{a} \times \mathbf{b} + \mathbf{a} \times \mathbf{c}$. Let

$$\mathbf{u} \equiv \mathbf{a} \times (\mathbf{b} + \mathbf{c}) - \mathbf{a} \times \mathbf{b} - \mathbf{a} \times \mathbf{c}$$

and form the scalar product of this vector with an arbitrary vector $\mathbf{v}$. We obtain

$$\mathbf{v} \cdot \mathbf{u} = \mathbf{v} \cdot [\mathbf{a} \times (\mathbf{b} + \mathbf{c})] - \mathbf{v} \cdot (\mathbf{a} \times \mathbf{b}) - \mathbf{v} \cdot (\mathbf{a} \times \mathbf{c})$$

In Sec. 13 we shall show that $\mathbf{a} \cdot (\mathbf{b} \times \mathbf{c}) = (\mathbf{a} \times \mathbf{b}) \cdot \mathbf{c}$. Hence

$$\begin{aligned} \mathbf{v} \cdot \mathbf{u} &= (\mathbf{v} \times \mathbf{a}) \cdot (\mathbf{b} + \mathbf{c}) - (\mathbf{v} \times \mathbf{a}) \cdot \mathbf{b} - (\mathbf{v} \times \mathbf{a}) \cdot \mathbf{c} \\ &= (\mathbf{v} \times \mathbf{a}) \cdot \mathbf{b} + (\mathbf{v} \times \mathbf{a}) \cdot \mathbf{c} - (\mathbf{v} \times \mathbf{a}) \cdot \mathbf{b} - (\mathbf{v} \times \mathbf{a}) \cdot \mathbf{c} \equiv 0 \end{aligned}$$

This implies either that $\mathbf{u} = \mathbf{0}$ or that $\mathbf{v}$ is perpendicular to $\mathbf{u}$. Since $\mathbf{v}$ is arbitrary, we can choose it not perpendicular to $\mathbf{u}$. Hence $\mathbf{u} = \mathbf{0}$ and

$$\mathbf{a} \times (\mathbf{b} + \mathbf{c}) = \mathbf{a} \times \mathbf{b} + \mathbf{a} \times \mathbf{c} \tag{32}$$

This proof is by Professor Morgan Ward of the California Institute of Technology.

12. Examples of the Vector Product

Example 10

$$\mathbf{i} \times \mathbf{i} = \mathbf{j} \times \mathbf{j} = \mathbf{k} \times \mathbf{k} = \mathbf{0}$$
$$\mathbf{i} \times \mathbf{j} = \mathbf{k}, \qquad \mathbf{j} \times \mathbf{k} = \mathbf{i}, \qquad \mathbf{k} \times \mathbf{i} = \mathbf{j}$$

For the vectors $\mathbf{a} = a_1\mathbf{i} + a_2\mathbf{j} + a_3\mathbf{k}$, $\mathbf{b} = b_1\mathbf{i} + b_2\mathbf{j} + b_3\mathbf{k}$ we obtain $\mathbf{a} \times \mathbf{b} = (a_2b_3 - a_3b_2)\mathbf{i} + (a_3b_1 - a_1b_3)\mathbf{j} + (a_1b_2 - a_2b_1)\mathbf{k}$ by making use of the distributive law of Sec. 11. Symbolically we have

$$\mathbf{a} \times \mathbf{b} = \begin{vmatrix} \mathbf{i} & \mathbf{j} & \mathbf{k} \\ a_1 & a_2 & a_3 \\ b_1 & b_2 & b_3 \end{vmatrix} \tag{33}$$

where (33) is to be expanded by the ordinary method of determinants.

Example 11. $\mathbf{a} = 2\mathbf{i} - 3\mathbf{j} + 5\mathbf{k}$, $\mathbf{b} = -\mathbf{i} + 2\mathbf{j} - 3\mathbf{k}$, so that

$$\mathbf{a} \times \mathbf{b} = \begin{vmatrix} \mathbf{i} & \mathbf{j} & \mathbf{k} \\ 2 & -3 & 5 \\ -1 & 2 & -3 \end{vmatrix} = -\mathbf{i} + \mathbf{j} + \mathbf{k}$$

Example 12. *Sine law of trigonometry*

$$\mathbf{c} = \mathbf{b} - \mathbf{a}$$
$$\mathbf{c} \times \mathbf{c} = \mathbf{c} \times (\mathbf{b} - \mathbf{a})$$
$$\mathbf{0} = \mathbf{c} \times \mathbf{b} - \mathbf{c} \times \mathbf{a}$$

or

$$\mathbf{c} \times \mathbf{a} = \mathbf{c} \times \mathbf{b}$$

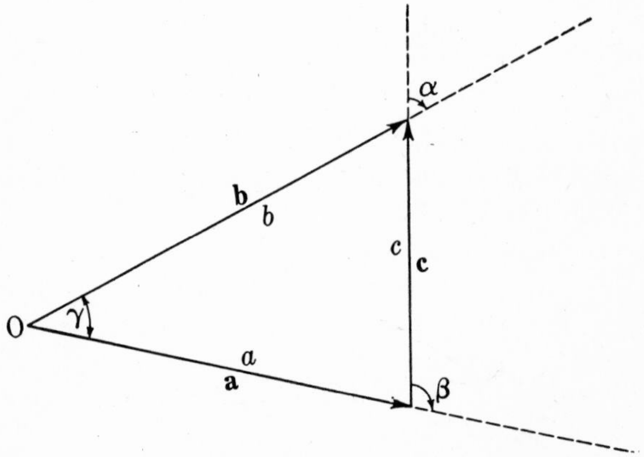

FIG. 23.

However, if two vectors are equal, their magnitudes are equal so that

$$|\mathbf{c}||\mathbf{a}| \sin \beta = |\mathbf{c}||\mathbf{b}| \sin \alpha$$

and

$$\frac{a}{\sin \alpha} = \frac{b}{\sin \beta} = \frac{c}{\sin \gamma}$$

Example 13. *Rotation of a Particle.* Assume that a particle is rotating about a fixed line L with angular speed ω. We assume that its distance from L remains constant. Let us define the angular velocity of the particle as the vector $\boldsymbol{\omega}$, whose direction is along L and whose length is ω. We choose the direction of $\boldsymbol{\omega}$ in the usual sense of a right-hand screw advance (see Fig. 24).

It is our aim to prove that the velocity vector **v** can be represented by $\boldsymbol{\omega} \times \mathbf{r}$, where **r** is the position vector of P from any origin taken on the line L. Let the reader show that **v** and $\boldsymbol{\omega} \times \mathbf{r}$ are parallel. Now $|\boldsymbol{\omega} \times \mathbf{r}| = \omega a =$ speed of P, so that

$$\mathbf{v} = \boldsymbol{\omega} \times \mathbf{r} \tag{34}$$

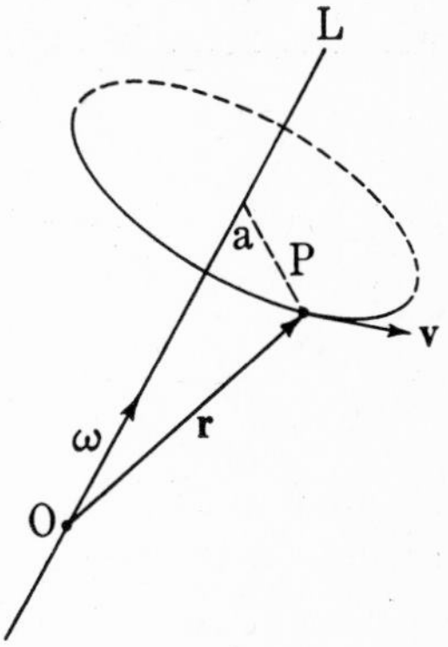

FIG. 24.

13. The Triple Scalar Product. Let us consider the scalar $\mathbf{a} \cdot (\mathbf{b} \times \mathbf{c})$. This scalar represents the volume of the parallelepiped formed by the coterminous sides **a**, **b**, **c**, since

$$\begin{aligned} \mathbf{a} \cdot (\mathbf{b} \times \mathbf{c}) &= |\mathbf{a}||\mathbf{b}||\mathbf{c}| \sin \theta \cos \alpha \\ &= hA = \text{volume} \end{aligned} \tag{35}$$

A being the area of the parallelogram with sides **b** and **c**, the altitude of the parallelepiped being denoted by h (see Fig. 25).

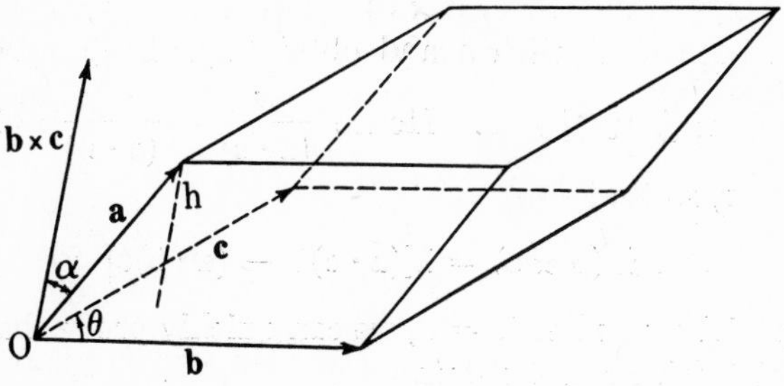

FIG. 25.

Now

$$\begin{aligned} \mathbf{a} \cdot (\mathbf{b} \times \mathbf{c}) &= (a_1\mathbf{i} + a_2\mathbf{j} + a_3\mathbf{k}) \cdot \begin{vmatrix} \mathbf{i} & \mathbf{j} & \mathbf{k} \\ b_1 & b_2 & b_3 \\ c_1 & c_2 & c_3 \end{vmatrix} \\ &= a_1(b_2c_3 - b_3c_2) + a_2(b_3c_1 - b_1c_3) + a_3(b_1c_2 - b_2c_1) \end{aligned}$$

so that

$$\mathbf{a} \cdot (\mathbf{b} \times \mathbf{c}) = \begin{vmatrix} a_1 & a_2 & a_3 \\ b_1 & b_2 & b_3 \\ c_1 & c_2 & c_3 \end{vmatrix} \tag{36}$$

Notice that $\mathbf{a} \cdot (\mathbf{b} \times \mathbf{c}) = (\mathbf{a} \times \mathbf{b}) \cdot \mathbf{c}$ since both terms represent the volume of the parallelepiped. It is also very easy to show that the determinant of (36) represents $(\mathbf{a} \times \mathbf{b}) \cdot \mathbf{c}$. We usually write $\mathbf{a} \cdot (\mathbf{b} \times \mathbf{c}) \equiv (\mathbf{abc})$ since there can be no confusion as to where the dot and cross belong. We note that

$$(\mathbf{abc}) = (\mathbf{cab}) = (\mathbf{bca})$$

and that $(\mathbf{abc}) = -(\mathbf{bac}) = -(\mathbf{cba}) = -(\mathbf{acb})$. These results follow from elementary theorems on determinants. We are thus allowed to interchange the dot and the cross when working with the triple scalar product. This result was used to prove (32). If the three vectors $\mathbf{a}, \mathbf{b}, \mathbf{c}$ are coplanar, no volume exists, and we at once have $(\mathbf{abc}) = 0$. In particular, if two of the three vectors are equal, the triple scalar product vanishes.

14. The Triple Vector Product. The triple vector product $\mathbf{a} \times (\mathbf{b} \times \mathbf{c})$ plays an important role in the development of vector analysis and in its applications. The result is a vector since it is the vector product of $\mathbf{a}$ and $(\mathbf{b} \times \mathbf{c})$. This vector is therefore perpendicular to $\mathbf{b} \times \mathbf{c}$ so that it lies in the plane of $\mathbf{b}$ and $\mathbf{c}$. If $\mathbf{b}$ is not parallel to $\mathbf{c}$, $\mathbf{a} \times (\mathbf{b} \times \mathbf{c}) = x\mathbf{b} + y\mathbf{c}$, from Sec. 6. Now dot both sides with $\mathbf{a}$ and obtain $x(\mathbf{a} \cdot \mathbf{b}) + y(\mathbf{a} \cdot \mathbf{c}) = 0$, since $\mathbf{a} \cdot [\mathbf{a} \times (\mathbf{b} \times \mathbf{c})] = 0$. Hence $\dfrac{x}{(\mathbf{a} \cdot \mathbf{c})} = \dfrac{-y}{(\mathbf{a} \cdot \mathbf{b})} = \lambda$, where λ is a scalar, so that

$$\mathbf{a} \times (\mathbf{b} \times \mathbf{c}) = \lambda[(\mathbf{a} \cdot \mathbf{c})\mathbf{b} - (\mathbf{a} \cdot \mathbf{b})\mathbf{c}] \tag{37}$$

In the special case when $\mathbf{b} = \mathbf{a}$, we can quickly prove that $\lambda \equiv 1$. We dot (37) with $\mathbf{c}$ and obtain

$$\mathbf{c} \cdot [\mathbf{a} \times (\mathbf{a} \times \mathbf{c})] = \lambda[(\mathbf{a} \cdot \mathbf{c})^2 - \mathbf{a}^2\mathbf{c}^2]$$

or

$$-(\mathbf{a} \times \mathbf{c})^2 = \lambda[(\mathbf{a} \cdot \mathbf{c})^2 - \mathbf{a}^2\mathbf{c}^2]$$

by an interchange of dot and cross. Hence

$$-\mathbf{a}^2\mathbf{c}^2 \sin^2 \theta = \lambda(\mathbf{a}^2\mathbf{c}^2 \cos^2 \theta - \mathbf{a}^2\mathbf{c}^2) = -\lambda\mathbf{a}^2\mathbf{c}^2 \sin^2 \theta$$

so that $\lambda = 1$. Hence $\mathbf{a} \times (\mathbf{a} \times \mathbf{c}) = (\mathbf{a} \cdot \mathbf{c})\mathbf{a} - (\mathbf{a} \cdot \mathbf{a})\mathbf{c}$. From this it immediately follows that $(\mathbf{a} \times \mathbf{b}) \times \mathbf{b} = (\mathbf{a} \cdot \mathbf{b})\mathbf{b} - (\mathbf{b} \cdot \mathbf{b})\mathbf{a}$. Now we prove that $\lambda \equiv 1$ for the general case. We dot (37) with $\mathbf{b}$ and obtain

$$\lambda[(\mathbf{a}\cdot\mathbf{c})(\mathbf{b}\cdot\mathbf{b}) - (\mathbf{a}\cdot\mathbf{b})(\mathbf{c}\cdot\mathbf{b})] = \mathbf{b}\cdot[\mathbf{a}\times(\mathbf{b}\times\mathbf{c})]$$
$$= -(\mathbf{a}\times\mathbf{b})\cdot(\mathbf{b}\times\mathbf{c})$$
$$= -[(\mathbf{a}\times\mathbf{b})\times\mathbf{b}]\cdot\mathbf{c}$$

Now $(\mathbf{a}\times\mathbf{b})\times\mathbf{b} = (\mathbf{a}\cdot\mathbf{b})\mathbf{b} - b^2\mathbf{a}$, so that

$$-[(\mathbf{a}\times\mathbf{b})\times\mathbf{b}]\cdot\mathbf{c} = b^2(\mathbf{a}\cdot\mathbf{c}) - (\mathbf{a}\cdot\mathbf{b})(\mathbf{b}\cdot\mathbf{c})$$

implying $\lambda = 1$. Thus

$$\mathbf{a}\times(\mathbf{b}\times\mathbf{c}) = (\mathbf{a}\cdot\mathbf{c})\mathbf{b} - (\mathbf{a}\cdot\mathbf{b})\mathbf{c} \tag{38}$$

We leave it to the reader to show that

$$(\mathbf{a}\times\mathbf{b})\times\mathbf{c} = (\mathbf{a}\cdot\mathbf{c})\mathbf{b} - (\mathbf{b}\cdot\mathbf{c})\mathbf{a}$$

Notice that $\mathbf{a}\times(\mathbf{b}\times\mathbf{c}) \neq (\mathbf{a}\times\mathbf{b})\times\mathbf{c}$. If $\mathbf{b}$ is parallel to $\mathbf{c}$, (38) reduces to the identity $\mathbf{0} = \mathbf{0}$, so that (38) holds for any three vectors. The expansion (38) of $\mathbf{a}\times(\mathbf{b}\times\mathbf{c})$ is often referred to as the rule of the middle factor.

More complicated products are simplified by use of the triple products. For example, we can expand $(\mathbf{a}\times\mathbf{b})\times(\mathbf{c}\times\mathbf{d})$ by considering $(\mathbf{a}\times\mathbf{b})$ as a single vector and applying (38).

$$(\mathbf{a}\times\mathbf{b})\times(\mathbf{c}\times\mathbf{d}) = (\mathbf{a}\times\mathbf{b}\cdot\mathbf{d})\mathbf{c} - (\mathbf{a}\times\mathbf{b}\cdot\mathbf{c})\mathbf{d}$$
$$= (\mathbf{abd})\mathbf{c} - (\mathbf{abc})\mathbf{d} \tag{39}$$

Also

$$(\mathbf{a}\times\mathbf{b})\cdot(\mathbf{c}\times\mathbf{d}) = \mathbf{a}\cdot\mathbf{b}\times(\mathbf{c}\times\mathbf{d})$$
$$= \mathbf{a}\cdot[(\mathbf{b}\cdot\mathbf{d})\mathbf{c} - (\mathbf{b}\cdot\mathbf{c})\mathbf{d}]$$
$$= (\mathbf{b}\cdot\mathbf{d})(\mathbf{a}\cdot\mathbf{c}) - (\mathbf{b}\cdot\mathbf{c})(\mathbf{a}\cdot\mathbf{d}) \tag{40}$$

15. Applications to the Spherical Trigonometry. Consider the spherical triangle ABC (sides are arcs of great circles) (see Fig. 26). Let the sphere be of radius 1. Now from (40) we see that

$$(\mathbf{a}\times\mathbf{b})\cdot(\mathbf{a}\times\mathbf{c}) = (\mathbf{b}\cdot\mathbf{c}) - (\mathbf{a}\cdot\mathbf{b})(\mathbf{a}\cdot\mathbf{c})$$

The angle between $\mathbf{a}\times\mathbf{b}$ and $\mathbf{a}\times\mathbf{c}$ is the same as the dihedral angle A between the planes OAC and OAB, since $\mathbf{a}\times\mathbf{b}$ is perpendicular to the plane of OAB and since $\mathbf{a}\times\mathbf{c}$ is perpendicular to the plane of OAC. Hence

$$\sin\gamma\sin\beta\cos A = \cos\alpha - \cos\gamma\cos\beta.$$

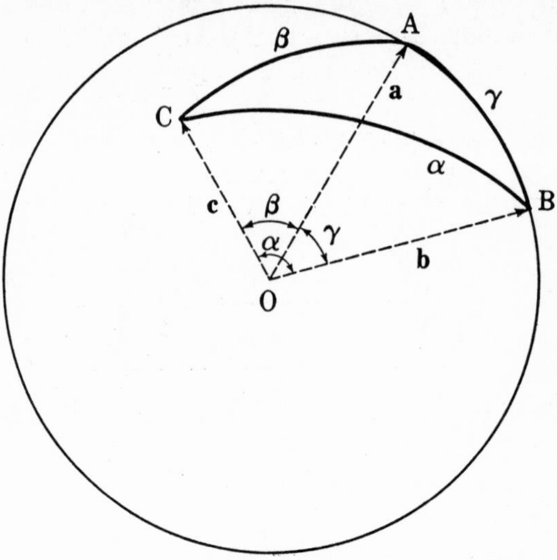

FIG. 26.

Problems

1. Show by two methods that the vectors $\mathbf{a} = 2\mathbf{i} - 3\mathbf{j} - \mathbf{k}$, $\mathbf{b} = -6\mathbf{i} + 9\mathbf{j} + 3\mathbf{k}$ are parallel.

2. Find a unit vector perpendicular to the vectors

$$\mathbf{a} = \mathbf{i} - \mathbf{j} + \mathbf{k}$$
$$\mathbf{b} = \mathbf{i} + \mathbf{j} - \mathbf{k}$$

3. A particle has an angular speed of 2 radians per second, and its axis of rotation passes through the points $P(0, 1, 2)$,

$$Q(1, 3, -2)$$

Find the velocity of the particle when it is located at the point $R(3, 6, 4)$.

4. Find the equation of the plane passing through the end points of the vectors $\mathbf{a} = a_1\mathbf{i} + a_2\mathbf{j} + a_3\mathbf{k}$, $\mathbf{b} = b_1\mathbf{i} + b_2\mathbf{j} + b_3\mathbf{k}$, $\mathbf{c} = c_1\mathbf{i} + c_2\mathbf{j} + c_3\mathbf{k}$, all three vectors with origin at $P(0, 0, 0)$.

5. Show that $(\mathbf{a} \times \mathbf{b}) \times (\mathbf{c} \times \mathbf{d}) = (\mathbf{acd})\mathbf{b} - (\mathbf{bcd})\mathbf{a}$.

6. Prove that $\mathbf{d} \times (\mathbf{a} \times \mathbf{b}) \cdot (\mathbf{a} \times \mathbf{c}) = (\mathbf{abc})(\mathbf{a} \cdot \mathbf{d})$.

7. If $\mathbf{a} + \mathbf{b} + \mathbf{c} = \mathbf{0}$, prove that $\mathbf{a} \times \mathbf{b} = \mathbf{b} \times \mathbf{c} = \mathbf{c} \times \mathbf{a}$, and interpret this result trigonometrically.

8. Four vectors have directions which are outward perpendiculars to the four faces of a tetrahedron, and their lengths are equal to the areas of the faces they represent. Show that the sum of these four vectors is the zero vector.

9. Prove that $(\mathbf{a} \times \mathbf{b}) \cdot (\mathbf{b} \times \mathbf{c}) \times (\mathbf{c} \times \mathbf{a}) = (\mathbf{abc})^2$.

10. If **a, b, c** are not coplanar, show that

$$\mathbf{d} = \frac{(\mathbf{dbc})}{(\mathbf{abc})}\mathbf{a} + \frac{(\mathbf{adc})}{(\mathbf{abc})}\mathbf{b} + \frac{(\mathbf{abd})}{(\mathbf{abc})}\mathbf{c}$$

for any vector **d.**

11. If **a, b, c** are not coplanar, show that

$$\mathbf{d} = \frac{(\mathbf{c} \cdot \mathbf{d})}{(\mathbf{abc})}\mathbf{a} \times \mathbf{b} + \frac{(\mathbf{a} \cdot \mathbf{d})}{(\mathbf{abc})}\mathbf{b} \times \mathbf{c} + \frac{(\mathbf{b} \cdot \mathbf{d})}{(\mathbf{abc})}\mathbf{c} \times \mathbf{a}$$

for any vector **d.**

12. Prove that

$$\mathbf{a} \times (\mathbf{b} \times \mathbf{c}) + \mathbf{b} \times (\mathbf{c} \times \mathbf{a}) + \mathbf{c} \times (\mathbf{a} \times \mathbf{b}) = \mathbf{0}$$

13. The four vectors **a, b, c, d** are coplanar. Show that $(\mathbf{a} \times \mathbf{b}) \times (\mathbf{c} \times \mathbf{d}) = \mathbf{0}$.

14. By considering the expansion for $(\mathbf{a} \times \mathbf{b}) \times (\mathbf{a} \times \mathbf{c})$, derive a spherical trigonometric identity (see Sec. 15).

15. Show that

$$\begin{aligned}(\mathbf{a} \times \mathbf{b}) \cdot (\mathbf{c} \times \mathbf{d}) \times (\mathbf{e} \times \mathbf{f}) &= (\mathbf{abd})(\mathbf{cef}) - (\mathbf{abc})(\mathbf{def}) \\ &= (\mathbf{abe})(\mathbf{fcd}) - (\mathbf{abf})(\mathbf{ecd}) \\ &= (\mathbf{cda})(\mathbf{bef}) - (\mathbf{cdb})(\mathbf{aef})\end{aligned}$$

16. Find an expression for the shortest distance from the end point of the vector $\mathbf{r}_1$ to the plane passing through the end points of the vectors $\mathbf{r}_2$, $\mathbf{r}_3$, $\mathbf{r}_4$. All four vectors have their origin at $P(0, 0, 0)$.

17. Consider the system of equations

$$\begin{aligned}a_1x + b_1y + c_1z &= d_1 \\ a_2x + b_2y + c_2z &= d_2 \\ a_3x + b_3y + c_3z &= d_3\end{aligned} \tag{41}$$

Let $\mathbf{a} = a_1\mathbf{i} + a_2\mathbf{j} + a_3\mathbf{k}$, etc. Write (41) as a single vector equation, and assuming $(\mathbf{abc}) \neq 0$, solve for x, y, z.

18. Find the shortest distance between two straight lines in space.

19. Show directly that $\mathbf{a} \times (\mathbf{b} \times \mathbf{c}) = (\mathbf{a} \cdot \mathbf{c})\mathbf{b} - (\mathbf{a} \cdot \mathbf{b})\mathbf{c}$, where $\mathbf{a}, \mathbf{b}, \mathbf{c}$ take on the values $\mathbf{i}, \mathbf{j}, \mathbf{k}$ in all possible ways. Now show that (38) is linear in $\mathbf{a}, \mathbf{b}, \mathbf{c}$, that is,

$$\mathbf{a} \times [\mathbf{b} \times (\alpha \mathbf{c} + \beta \mathbf{d})] = \alpha \mathbf{a} \times (\mathbf{b} \times \mathbf{c}) + \beta \mathbf{a} \times (\mathbf{b} \times \mathbf{d})$$

etc. Since any vector is a linear combination of $\mathbf{i}, \mathbf{j}, \mathbf{k}$, explain why (38) holds for all vectors.

20. If $\mathbf{a}$ and $\mathbf{b}$ lie in a plane normal to a plane containing $\mathbf{c}$ and $\mathbf{d}$, show that $(\mathbf{a} \times \mathbf{b}) \cdot (\mathbf{c} \times \mathbf{d}) = 0$.

21. If (a^1, a^2, a^3), (b^1, b^2, b^3) are the components of the vectors $\mathbf{a}, \mathbf{b}$, show that three of the nine numbers $c^{\alpha\beta}$ obtained by considering $c^{\alpha\beta} \equiv a^\alpha b^\beta - a^\beta b^\alpha$, $\alpha, \beta = 1, 2, 3$, represent the components of $\mathbf{a} \times \mathbf{b}$, that three others represent $\mathbf{b} \times \mathbf{a}$, while the remaining three vanish. Represent $c^{\alpha\beta}$ as a matrix and show that

$$c^{\alpha\beta} = -c^{\beta\alpha}$$

By considering a rotation of axes (see Example 8), show that the $\bar{c}^{\alpha\beta}$ in the new coordinate system are related to the $c^{\alpha\beta}$ in the old coordinate system by the equations $\bar{c}^{\alpha\beta} = \sum_{\tau=1}^{3} \sum_{\sigma=1}^{3} a_\sigma{}^\alpha a_\tau{}^\beta c^{\sigma\tau}$, $\alpha, \beta = 1, 2, 3$. Show that $\bar{c}^{\alpha\beta} = -\bar{c}^{\beta\alpha}$.

The numbers $c^1 = a^2b^3 - a^3b^2$, $c^2 = a^3b^1 - a^1b^3$,

$$c^3 = a^1b^2 - a^2b^1$$

are the components of $\mathbf{a} \times \mathbf{b}$. Show that $\bar{c}^\alpha \neq \sum_{\beta=1}^{3} a_\beta{}^\alpha c^\beta$ under a rotation of axes (see Prob. 22, Sec. 9). It is for this reason that the vector product is not considered to be a vector in the tensor analysis.

22. We can construct $\mathbf{a} \times \mathbf{b}$ by three geometrical constructions. We first construct a vector normal to $\mathbf{b}$ lying in the plane of $\mathbf{a}$ and $\mathbf{b}$. We project $\mathbf{a}$ onto this vector, and finally we rotate this new vector through an angle of 90° about the axis parallel to $\mathbf{b}$, magnifying this newly constructed vector by the factor $|\mathbf{b}|$. The final result yields $\mathbf{a} \times \mathbf{b}$. Use this to prove that

$$\mathbf{a} \times (\mathbf{b} + \mathbf{c}) = \mathbf{a} \times \mathbf{b} + \mathbf{a} \times \mathbf{c}$$

CHAPTER 2

DIFFERENTIAL VECTOR CALCULUS

16. Differentiation of Vectors. Let us consider the vector field

$$\mathbf{u} = \alpha(x, y, z, t)\mathbf{i} + \beta(x, y, z, t)\mathbf{j} + \gamma(x, y, z, t)\mathbf{k} \tag{42}$$

At any point $P(x, y, z)$ and at any time t, (42) defines a vector. If we keep P fixed, the vector $\mathbf{u}$ can still change because of the time dependence of its components α, β, and γ. If we keep the time fixed, we note that the vector at the point $P(x, y, z)$ will, in general, be different from that at the point

$$Q(x + dx, y + dy, z + dz)$$

Now, in the calculus, the student has learned how to find the change in a single function of x, y, z, t. What difficulties do we encounter in the case of a vector? Actually none, since we easily note that $\mathbf{u}$ will change if and only if its components change. Thus a change in $\alpha(x, y, z, t)$ produces a change in $\mathbf{u}$ in the x direction, and similarly changes in β and γ produce changes in $\mathbf{u}$ in the y and z directions, respectively. We are thus led to the following definition:

$$d\mathbf{u} = d\alpha\,\mathbf{i} + d\beta\,\mathbf{j} + d\gamma\,\mathbf{k} \tag{43}$$

$$\begin{aligned} d\mathbf{u} = \left(\frac{\partial\alpha}{\partial x}\,dx + \frac{\partial\alpha}{\partial y}\,dy + \frac{\partial\alpha}{\partial z}\,dz + \frac{\partial\alpha}{\partial t}\,dt\right)\mathbf{i} \\ + \left(\frac{\partial\beta}{\partial x}\,dx + \frac{\partial\beta}{\partial y}\,dy + \frac{\partial\beta}{\partial z}\,dz + \frac{\partial\beta}{\partial t}\,dt\right)\mathbf{j} \\ + \left(\frac{\partial\gamma}{\partial x}\,dx + \frac{\partial\gamma}{\partial y}\,dy + \frac{\partial\gamma}{\partial z}\,dz + \frac{\partial\gamma}{\partial t}\,dt\right)\mathbf{k} \end{aligned}$$

For example, let $\mathbf{r} = x\mathbf{i} + y\mathbf{j} + z\mathbf{k}$ be the position vector of a moving particle $P(x, y, z)$ in three-space. Then

$$d\mathbf{r} = dx\,\mathbf{i} + dy\,\mathbf{j} + dz\,\mathbf{k}$$

and

$$\mathbf{v} = \frac{d\mathbf{r}}{dt} = \frac{dx}{dt}\mathbf{i} + \frac{dy}{dt}\mathbf{j} + \frac{dz}{dt}\mathbf{k} \tag{44}$$

$$\mathbf{a} = \frac{d^2\mathbf{r}}{dt^2} = \frac{d^2x}{dt^2}\mathbf{i} + \frac{d^2y}{dt^2}\mathbf{j} + \frac{d^2z}{dt^2}\mathbf{k} \tag{45}$$

Equations (44) and (45) are, by definition, the velocity and acceleration of the particle. We have assumed that the vectors **i, j, k** remain fixed in space.

If the vector **u** depends on a single variable t, we can define

$$\frac{d\mathbf{u}}{dt} = \lim_{\Delta t \to 0} \frac{\mathbf{u}(t + \Delta t) - \mathbf{u}(t)}{\Delta t} \tag{46}$$

(see Fig. 27). It is easy to verify that (46) is equivalent to (43).

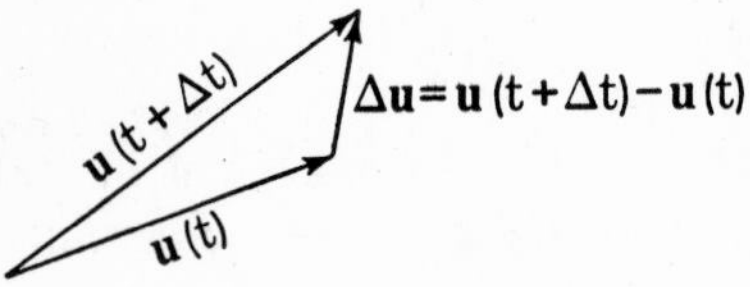

FIG. 27.

Example 14. Consider a particle P moving on a circle of radius r with constant angular speed $\omega = \frac{d\theta}{dt}$ (Fig. 28). We note that

$$\mathbf{r} = r\cos\theta\,\mathbf{i} + r\sin\theta\,\mathbf{j}$$

so that

$$\mathbf{v} = \frac{d\mathbf{r}}{dt} = (-r\sin\theta\,\mathbf{i} + r\cos\theta\,\mathbf{j})\frac{d\theta}{dt}$$

and

$$\mathbf{a} = \frac{d\mathbf{v}}{dt} = \frac{d^2\mathbf{r}}{dt^2} = (-r\cos\theta\,\mathbf{i} - r\sin\theta\,\mathbf{j})\left(\frac{d\theta}{dt}\right)^2$$

Therefore the acceleration is

$$\mathbf{a} = -\omega^2\mathbf{r} \tag{47}$$

The point P has an acceleration toward the origin of constant magnitude $\omega^2 r$. This acceleration is due to the fact that the velocity vector is changing direction at a constant rate; it is called the *centripetal* acceleration.

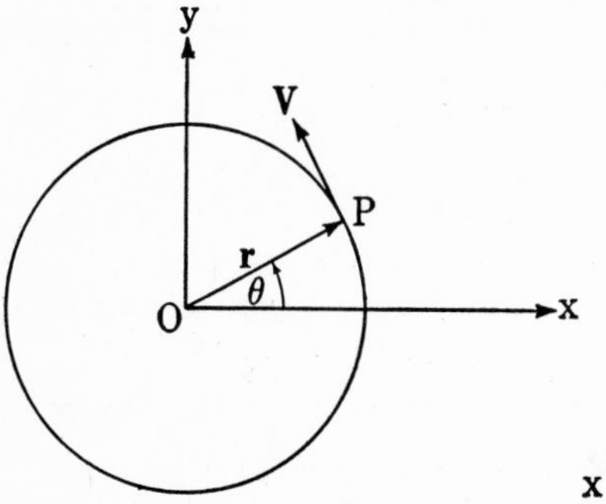

FIG. 28.

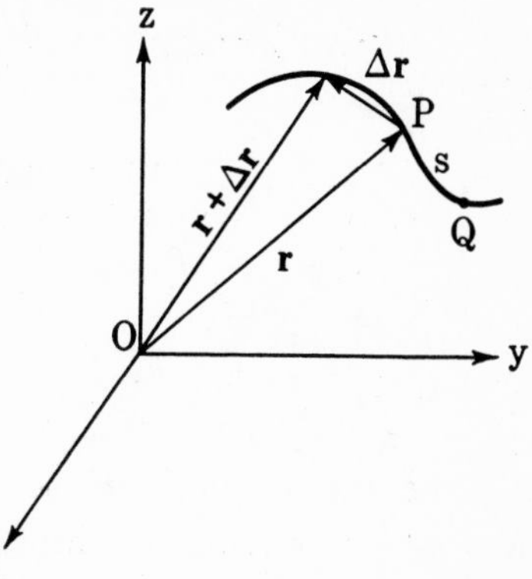

FIG. 29.

Example 15. Let P be any point on the space curve (Fig. 29)

$$x = x(s)$$
$$y = y(s)$$
$$z = z(s)$$

where s is arc length measured from some fixed point Q. Now

$$\mathbf{r} = x(s)\mathbf{i} + y(s)\mathbf{j} + z(s)\mathbf{k} \tag{48}$$

so that

$$\frac{d\mathbf{r}}{ds} = \frac{dx}{ds}\mathbf{i} + \frac{dy}{ds}\mathbf{j} + \frac{dz}{ds}\mathbf{k} \tag{49}$$

and

$$\frac{d\mathbf{r}}{ds} \cdot \frac{d\mathbf{r}}{ds} = \left(\frac{dx}{ds}\right)^2 + \left(\frac{dy}{ds}\right)^2 + \left(\frac{dz}{ds}\right)^2$$
$$= \frac{dx^2 + dy^2 + dz^2}{ds^2} \equiv 1$$

from the calculus. Hence $\dfrac{d\mathbf{r}}{ds}$ is a unit vector. As $\Delta s \to 0$, the position of $\dfrac{\Delta \mathbf{r}}{\Delta s}$ approaches the tangent line at P. Hence (49) represents the unit tangent vector to the space curve (48).

17. Differentiation Rules. Consider

$$\varphi(t) = \mathbf{u}(t) \cdot \mathbf{v}(t)$$
$$\varphi(t + \Delta t) - \varphi(t) = \mathbf{u}(t + \Delta t) \cdot \mathbf{v}(t + \Delta t) - \mathbf{u}(t) \cdot \mathbf{v}(t)$$

Now

$$\mathbf{u}(t + \Delta t) = \mathbf{u}(t) + \Delta\mathbf{u}$$
$$\mathbf{v}(t + \Delta t) = \mathbf{v}(t) + \Delta\mathbf{v}$$

(see Fig. 27), so that

$$\frac{\varphi(t + \Delta t) - \varphi(t)}{\Delta t} = \mathbf{u} \cdot \frac{\Delta\mathbf{v}}{\Delta t} + \frac{\Delta\mathbf{u}}{\Delta t} \cdot \mathbf{v} + \frac{\Delta\mathbf{u}}{\Delta t} \cdot \Delta\mathbf{v}$$

and passing to the limit, we obtain

$$\frac{d(\mathbf{u} \cdot \mathbf{v})}{dt} = \mathbf{u} \cdot \frac{d\mathbf{v}}{dt} + \frac{d\mathbf{u}}{dt} \cdot \mathbf{v} \tag{50}$$

Similarly

$$\frac{d(\mathbf{u} \times \mathbf{v})}{dt} = \mathbf{u} \times \frac{d\mathbf{v}}{dt} + \frac{d\mathbf{u}}{dt} \times \mathbf{v} \tag{51}$$

$$\frac{d(f\mathbf{u})}{dt} = f\frac{d\mathbf{u}}{dt} + \frac{df}{dt}\mathbf{u} \tag{52}$$

Notice how these formulas conform to the rules of the calculus.

Example 16. Let $\mathbf{u}(t)$ be a vector of constant magnitude. Therefore

$$\mathbf{u} \cdot \mathbf{u} = u^2 = \text{constant}$$

By differentiating we obtain

$$\mathbf{u} \cdot \frac{d\mathbf{u}}{dt} + \frac{d\mathbf{u}}{dt} \cdot \mathbf{u} = 0$$
$$\mathbf{u} \cdot \frac{d\mathbf{u}}{dt} = 0$$

Hence either $\frac{d\mathbf{u}}{dt} = 0$ or $\frac{d\mathbf{u}}{dt}$ is perpendicular to $\mathbf{u}$. This is an important result and should be fully understood by the student. The reader should give a geometric proof of this theorem.

Example 17. In all cases $\mathbf{u} \cdot \mathbf{u} = u^2$ where u is the length of $\mathbf{u}$. Differentiation yields $2\mathbf{u} \cdot \frac{d\mathbf{u}}{dt} = 2u \frac{du}{dt}$ and

$$\mathbf{u} \cdot \frac{d\mathbf{u}}{dt} = u \frac{du}{dt} \tag{53}$$

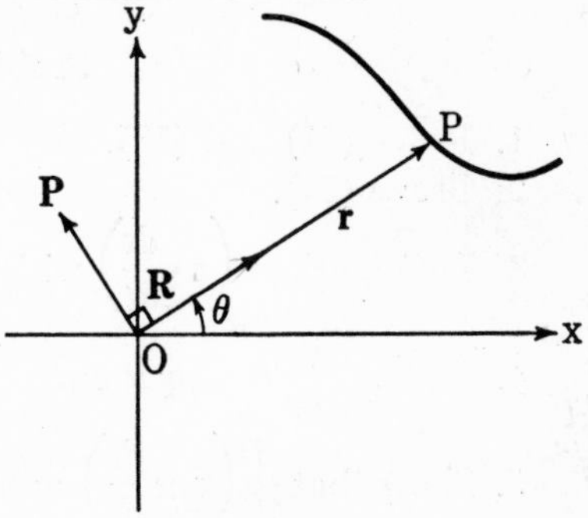

FIG. 30.

This result is not trivial, for $|d\mathbf{u}| \neq du$.

Example 18. *Motion in a Plane.* Now $\mathbf{r} = r\mathbf{R}$, where $\mathbf{R}$ is a unit vector (see Fig. 30). Hence

$$\mathbf{v} = \frac{d\mathbf{r}}{dt} = \frac{dr}{dt}\mathbf{R} + r\frac{d\mathbf{R}}{dt}$$

Now $\frac{d\mathbf{R}}{dt}$ is perpendicular to $\mathbf{R}$ (see Example 15). Also $\left|\frac{d\mathbf{R}}{dt}\right| = \frac{d\theta}{dt}$ since $\mathbf{R}$ is a unit vector. We can easily verify this by differentiating $\mathbf{R} = \cos\theta\,\mathbf{i} + \sin\theta\,\mathbf{j}$. Hence $\mathbf{v} = \frac{dr}{dt}\mathbf{R} + r\frac{d\theta}{dt}\mathbf{P}$, where

$\mathbf{P}$ is a unit vector perpendicular to $\mathbf{R}$. Differentiating again we obtain

$$\mathbf{a} = \frac{d\mathbf{v}}{dt} = \frac{d^2r}{dt^2}\mathbf{R} + \frac{dr}{dt}\frac{d\mathbf{R}}{dt} + \frac{dr}{dt}\frac{d\theta}{dt}\mathbf{P} + r\frac{d^2\theta}{dt^2}\mathbf{P} + r\frac{d\theta}{dt}\frac{d\mathbf{P}}{dt}$$

or

$$\mathbf{a} = \frac{d^2r}{dt^2}\mathbf{R} + 2\frac{dr}{dt}\frac{d\theta}{dt}\mathbf{P} + r\frac{d^2\theta}{dt^2}\mathbf{P} - r\left(\frac{d\theta}{dt}\right)^2\mathbf{R}$$

since

$$\frac{d\mathbf{P}}{dt} = -\frac{d\theta}{dt}\mathbf{R} \tag{54}$$

Thus

$$\mathbf{a} = \left[\frac{d^2r}{dt^2} - r\left(\frac{d\theta}{dt}\right)^2\right]\mathbf{R} + \frac{1}{r}\frac{d}{dt}\left(r^2\frac{d\theta}{dt}\right)\mathbf{P} \tag{55}$$

Problems

1. Prove (51) and (52).
2. Prove (54).
3. Differentiate $\left(\mathbf{r} \cdot \frac{d\mathbf{r}}{dt}\right)$ with respect to t.
4. Expand $\frac{d}{dt}[\mathbf{p} \times (\mathbf{q} \times \mathbf{r})]$.
5. Show that $\frac{d}{dt}\left(\mathbf{r} \times \frac{d\mathbf{r}}{dt}\right) = \mathbf{r} \times \frac{d^2\mathbf{r}}{dt^2}$.
6. Find the first and second derivatives of $\left(\mathbf{r}\,\frac{d\mathbf{r}}{dt}\,\frac{d^2\mathbf{r}}{dt^2}\right)$.
7. $\mathbf{r} = \mathbf{a} \cos \omega t + \mathbf{b} \sin \omega t$; $\mathbf{a}$, $\mathbf{b}$, ω are constants. Prove that $\mathbf{r} \times \frac{d\mathbf{r}}{dt} = \omega\mathbf{a} \times \mathbf{b}$ and $\frac{d^2\mathbf{r}}{dt^2} + \omega^2\mathbf{r} = 0$.
8. If $\mathbf{r} \times \frac{d\mathbf{r}}{dt} \equiv 0$, show that $\mathbf{r}$ has a constant direction.
9. $\mathbf{R}$ is a unit vector in the direction $\mathbf{r}$. Show that

$$\mathbf{R} \times d\mathbf{R} = \frac{\mathbf{r} \times d\mathbf{r}}{r^2}$$

10. If $\frac{d\mathbf{a}}{dt} = \boldsymbol{\omega} \times \mathbf{a}$, $\frac{d\mathbf{b}}{dt} = \boldsymbol{\omega} \times \mathbf{b}$, show that

$$\frac{d}{dt}(\mathbf{a} \times \mathbf{b}) = \boldsymbol{\omega} \times (\mathbf{a} \times \mathbf{b})$$

11. If $\mathbf{r} = \mathbf{a}e^{\omega t} + \mathbf{b}e^{\omega t}$, show that $\frac{d^2\mathbf{r}}{dt^2} - \omega^2\mathbf{r} = 0$. $\mathbf{a}$, $\mathbf{b}$ are constant vectors.
12. Find a vector $\mathbf{u}$ which satisfies $\frac{d^2\mathbf{u}}{dt^2} = \mathbf{a}t + \mathbf{b}$. $\mathbf{a}$, $\mathbf{b}$ are constant vectors. Is $\mathbf{u}$ unique?
13. Show that $\frac{d}{dt}\left(\frac{\mathbf{r}}{r}\right) = \frac{1}{r}\frac{d\mathbf{r}}{dt} - \frac{1}{r^2}\frac{dr}{dt}\mathbf{r}$.

14. Let $\mathbf{r}_0$ be the position vector to a fixed point P in space, and let $\mathbf{r}$ be the position vector to a variable point Q lying on a space curve $\mathbf{r} = \mathbf{r}(s)$. Show that if the distance $\overline{PQ}$ is a minimum, then $\mathbf{r} - \mathbf{r}_0$ is perpendicular to the tangent at Q. Show also that

$$\mathbf{r} \cdot \frac{d^2\mathbf{r}}{ds^2} + \left(\frac{d\mathbf{r}}{ds}\right)^2 > \mathbf{r}_0 \cdot \frac{d^2\mathbf{r}}{ds^2}$$

15. If $\mathbf{u} = \alpha(x, y, z, t)\mathbf{i} + \beta(x, y, z, t)\mathbf{j} + \alpha(x, y, z, t)\mathbf{k}$, show that $\dfrac{d\mathbf{u}}{dt} = \dfrac{\partial \mathbf{u}}{\partial t} + \dfrac{\partial \mathbf{u}}{\partial x}\dfrac{dx}{dt} + \dfrac{\partial \mathbf{u}}{\partial y}\dfrac{dy}{dt} + \dfrac{\partial \mathbf{u}}{\partial z}\dfrac{dz}{dt}$.

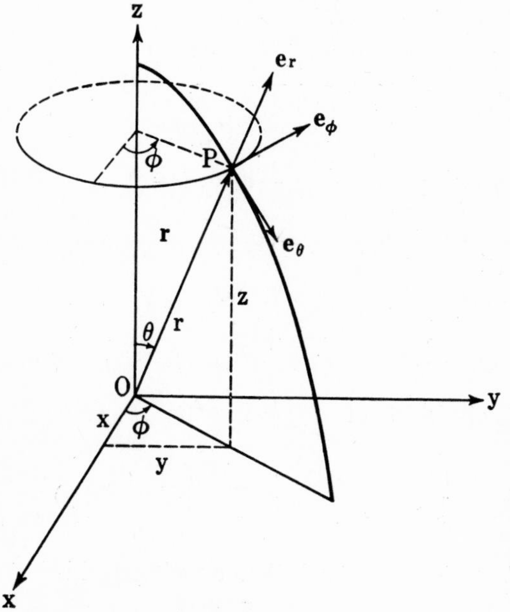

FIG. 31.

16. The transformation between rectangular coordinates and spherical coordinates is given by

$$x = r \sin \theta \cos \varphi$$
$$y = r \sin \theta \sin \varphi$$
$$z = r \cos \theta$$

where θ is the colatitude, φ is the longitudinal or azimuthal angle, and r is the magnitude of the position vector $\mathbf{r}$ from the origin to the particle in question. Find the components of the velocity and acceleration of the particle along the unit orthogonal vectors $\mathbf{e}_r$, $\mathbf{e}_\theta$, $\mathbf{e}_\varphi$ (see Fig. 31).

17. Consider the differential equation

$$\text{(i)} \qquad \frac{d^2\mathbf{u}}{dt^2} + 2A\frac{d\mathbf{u}}{dt} + B\mathbf{u} = 0$$

where A, B are constants. Assume a solution of the form $\mathbf{u}(t) = e^{wt}\mathbf{C}$, where $\mathbf{C}$ is a constant vector, and show that $\mathbf{u}(t) = \mathbf{C}_1e^{w_1t} + \mathbf{C}_2e^{w_2t}$ is a solution of (i), w_1, w_2 being roots of $w^2 + 2Aw + B = 0$. Consider the cases for which $A^2 - B < 0$, $A^2 - B = 0$, $A^2 - B > 0$.

18. Find the vector $\mathbf{u}$ which satisfies

$$\frac{d^3\mathbf{u}}{dt^3} - \frac{d^2\mathbf{u}}{dt^2} - 2\frac{d\mathbf{u}}{dt} = \mathbf{0}$$

such that $\mathbf{u} = \mathbf{i}$, $\dfrac{d\mathbf{u}}{dt} = \mathbf{j}$, $\dfrac{d^2\mathbf{u}}{dt^2} = \mathbf{k}$ for $t = 0$.

19. If $\mathbf{u}_1$ is a solution of

$$\frac{d^3\mathbf{u}}{dt^3} + A\frac{d^2\mathbf{u}}{dt^2} + B\frac{d\mathbf{u}}{dt} + C\mathbf{u} = 0$$

and if $\mathbf{u}_2$ is a solution of

$$\text{(ii)} \qquad \frac{d^3\mathbf{u}}{dt^3} + A\frac{d^2\mathbf{u}}{dt^2} + B\frac{d\mathbf{u}}{dt} + C\mathbf{u} = \mathbf{F}(t)$$

show that $\mathbf{u}_1 + \mathbf{u}_2$ is a solution of (ii) provided A, B, C are independent of $\mathbf{u}$. Why is this necessary?

20. A particle moving in the plane of (r, θ) has no transverse acceleration, that is, $\dfrac{1}{r}\dfrac{d}{dt}\left(r^2\dfrac{d\theta}{dt}\right) = 0$. Show that the radius vector from the origin to the particle sweeps out equal areas in equal intervals of time.

18. The Gradient. Let $\varphi(x, y, z)$ be any continuous differentiable space function. From the calculus

$$d\varphi = \frac{\partial\varphi}{\partial x}dx + \frac{\partial\varphi}{\partial y}dy + \frac{\partial\varphi}{\partial z}dz \tag{56}$$

Now let $\mathbf{r}$ be the position vector to the point $P(x, y, z)$.

$$\mathbf{r} = x\mathbf{i} + y\mathbf{j} + z\mathbf{k}$$

If we move to the point $Q(x + dx, y + dy, z + dz)$ (Fig. 32),

$$d\mathbf{r} = dx\,\mathbf{i} + dy\,\mathbf{j} + dz\,\mathbf{k}$$

Now notice that (56) contains the terms dx, dy, dz and the terms $\frac{\partial\varphi}{\partial x}, \frac{\partial\varphi}{\partial y}, \frac{\partial\varphi}{\partial z}$. We define a new vector formed from φ by taking its three partial derivatives. Let del $\varphi \equiv \nabla\varphi$ be defined by

$$\nabla\varphi = \frac{\partial\varphi}{\partial x}\mathbf{i} + \frac{\partial\varphi}{\partial y}\mathbf{j} + \frac{\partial\varphi}{\partial z}\mathbf{k} \tag{57}$$

We immediately see that

$$d\varphi = d\mathbf{r} \cdot \nabla\varphi \tag{58}$$

We shall now give a geometrical interpretation of $\nabla\varphi$. At the point $P(x_0, y_0, z_0)$, φ has the value $\varphi(x_0, y_0, z_0)$ so that

$$\varphi(x, y, z) = \varphi(x_0, y_0, z_0)$$

represents a surface which obviously contains the point

$$P(x_0, y_0, z_0)$$

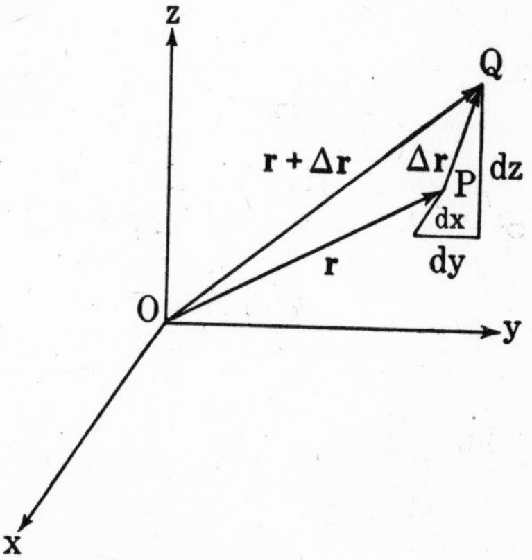

FIG. 32.

As long as we move along this surface, φ has the constant value $\varphi(x_0, y_0, z_0)$ and $d\varphi = 0$. Consequently, from (58),

$$d\mathbf{r} \cdot \nabla\varphi = 0 \tag{59}$$

Now $\nabla\varphi$ is a vector which is at once completely determined after φ has been differentiated, and Eq. (59) states that $\nabla\varphi$ is perpendicular to $d\mathbf{r}$ as long as $d\mathbf{r}$ represents a change from P to Q, where Q remains on the surface φ = constant. Thus $\nabla\varphi$ is normal to all the possible tangents to the surface at P so that $\nabla\varphi$ must necessarily be normal to the surface $\varphi(x, y, z)$ = constant (see

Fig. 33). Let us now return to $d\varphi = d\mathbf{r} \cdot \nabla\varphi$. The vector $\nabla\varphi$ is fixed at any point $P(x, y, z)$, so that $d\varphi$ (the change in φ) will depend to a great extent on $d\mathbf{r}$. Certainly $d\varphi$ will be a maximum when $d\mathbf{r}$ is parallel to $\nabla\varphi$, since $d\mathbf{r} \cdot \nabla\varphi = |d\mathbf{r}||\nabla\varphi| \cos \theta$, and $\cos \theta$ is a maximum for $\theta = 0°$. Thus $\nabla\varphi$ is in the direction of maximum increase of $\varphi(x, y, z)$. Let $|d\mathbf{r}| = ds$ so that

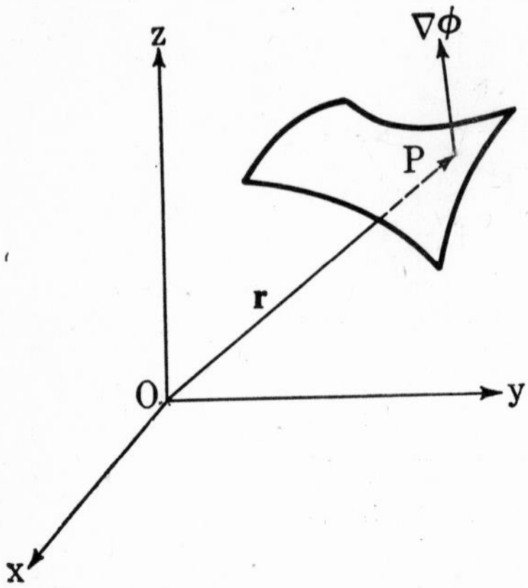

FIG. 33.

$$\frac{d\varphi}{ds} = \mathbf{u} \cdot \nabla\varphi \tag{60}$$

where $\mathbf{u}$ is a unit vector in the direction $d\mathbf{r}$. Hence the change of φ in any direction is the projection of $\nabla\varphi$ on the unit vector having this direction.

Example 19. To find a unit vector normal to the surface $x^2 + y^2 - z = 1$ at the point $P(1, 1, 1)$. Here

$$\begin{aligned} \varphi(x, y, z) &= x^2 + y^2 - z \\ \nabla\varphi &= 2x\mathbf{i} + 2y\mathbf{j} - \mathbf{k} \\ &= 2\mathbf{i} + 2\mathbf{j} - \mathbf{k} \text{ at } P(1, 1, 1) \end{aligned}$$

Thus

$$\mathbf{N} = \frac{2\mathbf{i} + 2\mathbf{j} - \mathbf{k}}{3}$$

Example 20. We find ∇r if $r = (x^2 + y^2 + z^2)^{\frac{1}{2}}$. The surface r = constant is a sphere. Hence ∇r is normal to the sphere and so is parallel to the position vector $\mathbf{r}$. Thus $\nabla r = k\mathbf{r}$. Now

$$dr = d\mathbf{r} \cdot \nabla r = k\, d\mathbf{r} \cdot \mathbf{r} = kr\, dr \qquad \text{from (53)}$$

Therefore

$$k = \frac{1}{r} \qquad \text{and} \qquad \nabla r = \frac{\mathbf{r}}{r} = \mathbf{R} \tag{61}$$

Example 21

$$\nabla f(u) = f'(u)\, \nabla u, \qquad u = u(x, y, z)$$

Proof:

$$\begin{aligned}\nabla f(u) &= \frac{\partial f}{\partial x}\,\mathbf{i} + \frac{\partial f}{\partial y}\,\mathbf{j} + \frac{\partial f}{\partial z}\,\mathbf{k} \\ &= f'(u)\,\frac{\partial u}{\partial x}\,\mathbf{i} + f'(u)\,\frac{\partial u}{\partial y}\,\mathbf{j} + f'(u)\,\frac{\partial u}{\partial z}\,\mathbf{k} \\ &= f'(u)\,\nabla u\end{aligned}$$

Example 22

$$\begin{aligned}\nabla f(u_1, u_2, \ldots, u_n) &= \frac{\partial f}{\partial x}\,\mathbf{i} + \frac{\partial f}{\partial y}\,\mathbf{j} + \frac{\partial f}{\partial z}\,\mathbf{k} \\ &= \sum_{\alpha=1}^{n} \left(\frac{\partial f}{\partial u_\alpha}\frac{\partial u_\alpha}{\partial x}\,\mathbf{i} + \frac{\partial f}{\partial u_\alpha}\frac{\partial u_\alpha}{\partial y}\,\mathbf{j} + \frac{\partial f}{\partial u_\alpha}\frac{\partial u_\alpha}{\partial z}\,\mathbf{k} \right) \\ &= \sum_{\alpha=1}^{n} \frac{\partial f}{\partial u_\alpha}\,\nabla u_\alpha \qquad (62)\end{aligned}$$

Example 23. Consider the ellipse given by $r_1 + r_2 =$ constant (see Fig. 34). Now $\nabla(r_1 + r_2)$ is normal to the ellipse. Let

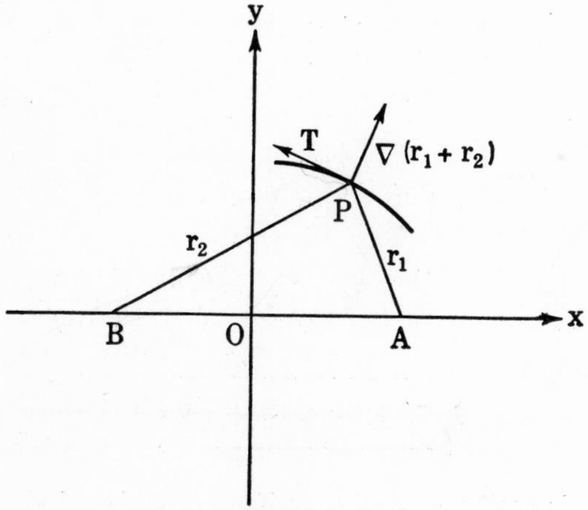

FIG. 34.

$\mathbf{T}$ be a unit tangent to the ellipse. Thus $\nabla(r_1 + r_2) \cdot \mathbf{T} = 0$, and

$$\nabla r_1 \cdot \mathbf{T} = -\nabla r_2 \cdot \mathbf{T} \qquad (63)$$

But from Example ~~18~~ 20, ∇r_1 is a unit vector parallel to the vector $\overrightarrow{AP}$, and ∇r_2 is a unit vector parallel to the vector $\overrightarrow{BP}$. Equation

(63) shows that $\overrightarrow{AP}$ and $\overrightarrow{BP}$ make equal angles with the tangent to the ellipse.

19. The Vector Operator ∇. We define

$$\nabla \equiv \mathbf{i}\frac{\partial}{\partial x} + \mathbf{j}\frac{\partial}{\partial y} + \mathbf{k}\frac{\partial}{\partial z} \tag{64}$$

Notice that ∇ is an operator, just as $\frac{d}{dx}$ is an operator in the differential calculus. Thus

$$\begin{aligned} \nabla\varphi &= \left(\mathbf{i}\frac{\partial}{\partial x} + \mathbf{j}\frac{\partial}{\partial y} + \mathbf{k}\frac{\partial}{\partial z}\right)\varphi \\ &= \mathbf{i}\frac{\partial\varphi}{\partial x} + \mathbf{j}\frac{\partial\varphi}{\partial y} + \mathbf{k}\frac{\partial\varphi}{\partial z} \end{aligned}$$

We call ∇ (del), a vector operator because of its components $\frac{\partial}{\partial x}, \frac{\partial}{\partial y}, \frac{\partial}{\partial z}$. It will help us in the future to keep in mind that ∇ acts both as a differential operator and as a vector.

Example 24

$$\begin{aligned} \nabla(uv) &= \mathbf{i}\frac{\partial(uv)}{\partial x} + \mathbf{j}\frac{\partial(uv)}{\partial y} + \mathbf{k}\frac{\partial(uv)}{\partial z} \\ &= \left(\mathbf{i}\frac{\partial v}{\partial x} + \mathbf{j}\frac{\partial v}{\partial y} + \mathbf{k}\frac{\partial v}{\partial z}\right)u + \left(\mathbf{i}\frac{\partial u}{\partial x} + \mathbf{j}\frac{\partial u}{\partial y} + \mathbf{k}\frac{\partial u}{\partial z}\right)v \end{aligned}$$

$$\nabla(uv) = u\,\nabla v + v\,\nabla u \tag{65}$$

This result is easily remembered if we keep in mind that ∇ is a differential operator, so that we can apply the ordinary rules of calculus.

Problems

1. Find the equation of the tangent plane to the surface $xy - z = 1$ at the point (2, 1, 1).

2. Show that $\nabla(\mathbf{a} \cdot \mathbf{r}) = \mathbf{a}$, where $\mathbf{a}$ is a constant vector and $\mathbf{r}$ is the position vector.

3. If $r = (x^2 + y^2 + z^2)^{\frac{1}{2}}$, find ∇r^n by explicit use of (57).

4. If $\varphi = (\mathbf{r} \times \mathbf{a}) \cdot (\mathbf{r} \times \mathbf{b})$, show that

$$\nabla\varphi = \mathbf{b} \times (\mathbf{r} \times \mathbf{a}) + \mathbf{a} \times (\mathbf{r} \times \mathbf{b})$$

when $\mathbf{a}$ and $\mathbf{b}$ are constant vectors.

5. Let $\varphi = x^2 + y^2$. Find $\nabla\varphi$ and show that it is the maximum change of φ.

6. Find the cosine of the angle between the surfaces

$$x^2y + z = 3$$

and $x \log z - y^2 = -4$ at the point of intersection $P(-1, 2, 1)$.

7. What is the value of $\nabla\varphi(x, y, z)$ at a point that makes φ a maximum?

8. The surfaces $\varphi(x, y, z) =$ constant and $\psi(x, y, z) =$ constant are normal along a curve of intersection. What is the value of $\nabla\varphi \cdot \nabla\psi$ along this curve?

9. What is the direction for the maximum change of the space function $\varphi(x, y, z) = x \sin z - y \cos z$ at the origin?

10. Expand $\nabla(u/v)$ where $u = u(x, y, z)$, $v = v(x, y, z)$.

11. Let r and x be the distances from the focus and directrix to any point on a parabola. We know that $r = x$. Show that $(\mathbf{R} - \mathbf{i}) \cdot \mathbf{T} = 0$, where $\mathbf{T}$ is a unit tangent vector to the parabola, and interpret this equation.

12. Show that the ellipse $r_1 + r_2 = c_1$ and the hyperbola $r_1 - r_2 = c_2$ intersect at right angles when they have the same foci.

13. If $\nabla\varphi$ is always parallel to the position vector $\mathbf{r}$, show that $\varphi = \varphi(r)$, $r^2 = x^2 + y^2 + z^2$.

14. Find the change of $\varphi = xyz$ in the direction normal to the surface $yx^2 + xy^2 + z^2y = 3$ at the point $P(1, 1, 1)$.

15. If $f = f(x^1, x^2, x^3)$ (see Example 8), and if

$$x^\alpha = x^\alpha(y^1, y^2, y^3),$$

$\alpha = 1, 2, 3$, show that

$$\frac{\partial f}{\partial y^\alpha} = \sum_{\beta=1}^{3} \frac{\partial f}{\partial x^\beta} \frac{\partial x^\beta}{\partial y^\alpha} \qquad \alpha = 1, 2, 3$$

Using the fact that $\dfrac{\partial x^\alpha}{\partial x^\beta} = \sum_{\sigma=1}^{3} \dfrac{\partial x^\alpha}{\partial y^\sigma}\dfrac{\partial y^\sigma}{\partial x^\beta} \equiv \begin{matrix} 1 \text{ if } \alpha = \beta \\ 0 \text{ if } \alpha \neq \beta \end{matrix}$, show that $\dfrac{\partial f}{\partial x^\alpha} = \sum_{\beta=1}^{3} \dfrac{\partial f}{\partial y^\beta}\dfrac{\partial y^\beta}{\partial x^\alpha}$, $\alpha = 1, 2, 3$.

16. Apply the results of Prob. 15 above to the transformation

$$\begin{aligned} x &= r \cos \theta \\ y &= r \sin \theta \\ z &= z \end{aligned}$$

and show that $\dfrac{\partial f}{\partial r}, \dfrac{1}{r}\dfrac{\partial f}{\partial \theta}, \dfrac{\partial f}{\partial z}$ are the components of ∇f along the three mutually orthogonal unit vectors $\mathbf{e}_r$, $\mathbf{e}_\theta$, $\mathbf{e}_z$ which occur in cylindrical coordinates.

17. If $\varphi = \varphi(x, y, z, t)$, show that $\dfrac{d\varphi}{dt} = \dfrac{\partial \varphi}{\partial t} + \dfrac{d\mathbf{r}}{dt} \cdot \nabla\varphi$.

18. If $\mathbf{r} = x\mathbf{i} + y\mathbf{j} + z\mathbf{k}$, find $\mathbf{r} \cdot \nabla\varphi$ for

$$\varphi = xye^{yz/x^2} + \log\left(\frac{y}{x} + \frac{x}{z} + \frac{z}{y}\right)$$

19. If $\mathbf{u} = \mathbf{u}(x, y, z, t)$ show that $\dfrac{d\mathbf{u}}{dt} = \dfrac{\partial \mathbf{u}}{\partial t} + \left(\dfrac{d\mathbf{r}}{dt} \cdot \nabla\right)\mathbf{u}$.

20. If $\mathbf{u}(tx, ty, tz) = t^n\mathbf{u}(x, y, z)$, show that $(\mathbf{r} \cdot \nabla)\mathbf{u} = n\mathbf{u}$.

20. The Divergence of a Vector. Let us consider the motion of a fluid of density $\rho(x, y, z)$. We assume that the velocity field is given by $\mathbf{f} = u(x, y, z)\mathbf{i} + v(x, y, z)\mathbf{j} + w(x, y, z)\mathbf{k}$. This type of motion is called *steady* motion because of the explicit independence of ρ and $\mathbf{f}$ on the time. We now concentrate on the flow through a small parallelepiped $ABCDEFGH$ (Fig. 35), of dimensions dx, dy, dz.

Let us first calculate the amount of fluid passing through the face $ABCD$ per unit time. The x and z components of the velocity $\mathbf{f}$ contribute nothing to the flow through $ABCD$. The mass of fluid entering $ABCD$ per unit time is given by $\rho v\, dx\, dz$. The mass of fluid leaving the face $EFGH$ per unit time is

$$\left[\rho v + \frac{\partial(\rho v)}{\partial y}\, dy\right] dx\, dz$$

The loss of mass per unit time is thus seen to be equal to

$$\frac{\partial(\rho v)}{\partial y}\,dx\,dy\,dz$$

If we also take into consideration the other two faces, we find that the total loss of mass per unit time is

$$\left[\frac{\partial(\rho u)}{\partial x}+\frac{\partial(\rho v)}{\partial y}+\frac{\partial(\rho w)}{\partial z}\right]dx\,dy\,dz$$

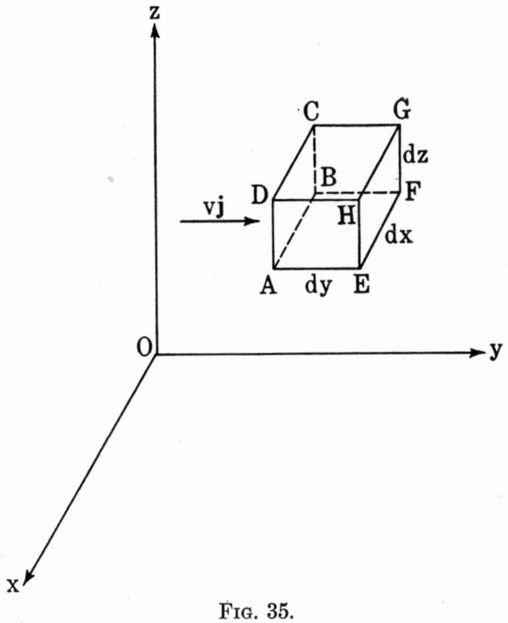

FIG. 35.

so that

$$\frac{\partial(\rho u)}{\partial x}+\frac{\partial(\rho v)}{\partial y}+\frac{\partial(\rho w)}{\partial z} \tag{66}$$

represents the loss of mass per unit time per unit volume. This quantity is called the divergence of the vector $\rho\mathbf{f}$. We see at once that

$$\nabla\cdot(\rho\mathbf{f}) = \operatorname{div}(\rho\mathbf{f}) = \frac{\partial(\rho u)}{\partial x}+\frac{\partial(\rho v)}{\partial y}+\frac{\partial(\rho w)}{\partial z} = \frac{1}{V}\frac{dM}{dt} \tag{67}$$

since **i, j, k** are constant vectors. M and V are the mass and volume of the fluid.

The divergence of any vector **f** is defined as $\nabla \cdot \mathbf{f}$. We now calculate the divergence of $\varphi(x, y, z)\mathbf{f}$.

$$\begin{aligned}\nabla \cdot (\varphi \mathbf{f}) &= \frac{\partial(\varphi u)}{\partial x} + \frac{\partial(\varphi v)}{\partial y} + \frac{\partial(\varphi w)}{\partial z} \\ &= \varphi\left(\frac{\partial u}{\partial x} + \frac{\partial v}{\partial y} + \frac{\partial w}{\partial z}\right) + \left(u\,\frac{\partial \varphi}{\partial x} + v\,\frac{\partial \varphi}{\partial y} + w\,\frac{\partial \varphi}{\partial z}\right)\end{aligned}$$

$$\nabla \cdot (\varphi \mathbf{f}) = \varphi\, \nabla \cdot \mathbf{f} + \mathbf{f} \cdot \nabla \varphi \tag{68}$$

We remember this result easily enough if we consider ∇ as a vector differential operator. Thus, when operating on $\varphi\mathbf{f}$, we first keep φ fixed and let ∇ operate on **f**, and then we keep **f** fixed and let ∇ operate on φ($\nabla \cdot \varphi$ is nonsense), and since **f** and $\nabla\varphi$ are vectors we complete their multiplication by taking their dot product.

Example 25. Compute $\nabla \cdot \mathbf{f}$ if $\mathbf{f} = \mathbf{r}/r^3$ (inverse-square force).

$$\begin{aligned}\nabla \cdot (r^{-3}\mathbf{r}) &= r^{-3}\nabla \cdot \mathbf{r} + \mathbf{r} \cdot \nabla r^{-3} \\ &= 3r^{-3} + \mathbf{r} \cdot (-3r^{-4}\,\nabla r) \\ &= 3r^{-3} - 3r^{-5}\mathbf{r} \cdot \mathbf{r} = 3r^{-3} - 3r^{-3} = 0\end{aligned}$$

$$\nabla \cdot (r^{-3}\mathbf{r}) = 0 \tag{69}$$

This is an important result. The divergence of an inverse-square force is zero. We note that

$$\nabla \cdot \mathbf{r} = \frac{\partial x}{\partial x} + \frac{\partial y}{\partial y} + \frac{\partial z}{\partial z} = 3$$

Example 26. What is the divergence of a gradient?

$$\begin{aligned}\nabla \cdot (\nabla \varphi) &= \nabla \cdot \left(\frac{\partial \varphi}{\partial x}\,\mathbf{i} + \frac{\partial \varphi}{\partial y}\,\mathbf{j} + \frac{\partial \varphi}{\partial z}\,\mathbf{k}\right) \\ &= \frac{\partial^2 \varphi}{\partial x^2} + \frac{\partial^2 \varphi}{\partial y^2} + \frac{\partial^2 \varphi}{\partial z^2}\end{aligned}$$

This important quantity is called the Laplacian of φ.

$$\text{Lap } \varphi = \nabla \cdot (\nabla\varphi) = \nabla^2\varphi = \frac{\partial^2\varphi}{\partial x^2} + \frac{\partial^2\varphi}{\partial y^2} + \frac{\partial^2\varphi}{\partial z^2} \tag{70}$$

21. The Curl of a Vector. We postpone the physical meaning of the curl and define

$$\text{curl } \mathbf{f} = \nabla \times \mathbf{f} = \begin{vmatrix} \mathbf{i} & \mathbf{j} & \mathbf{k} \\ \dfrac{\partial}{\partial x} & \dfrac{\partial}{\partial y} & \dfrac{\partial}{\partial z} \\ u & v & w \end{vmatrix}$$

$$\nabla \times \mathbf{f} = \mathbf{i}\left(\frac{\partial w}{\partial y} - \frac{\partial v}{\partial z}\right) + \mathbf{j}\left(\frac{\partial u}{\partial z} - \frac{\partial w}{\partial x}\right) + \mathbf{k}\left(\frac{\partial v}{\partial x} - \frac{\partial u}{\partial y}\right) \tag{71}$$

Example 27

$$\nabla \times \mathbf{r} = \begin{vmatrix} \mathbf{i} & \mathbf{j} & \mathbf{k} \\ \dfrac{\partial}{\partial x} & \dfrac{\partial}{\partial y} & \dfrac{\partial}{\partial z} \\ x & y & z \end{vmatrix} = 0$$

Example 28

$$\nabla \times (\varphi\mathbf{f}) = \begin{vmatrix} \mathbf{i} & \mathbf{j} & \mathbf{k} \\ \dfrac{\partial}{\partial x} & \dfrac{\partial}{\partial y} & \dfrac{\partial}{\partial z} \\ \varphi u & \varphi v & \varphi w \end{vmatrix}$$

$$= \mathbf{i}\left[\frac{\partial(\varphi w)}{\partial y} - \frac{\partial(\varphi v)}{\partial z}\right] + \mathbf{j}\left[\frac{\partial(\varphi u)}{\partial z} - \frac{\partial(\varphi w)}{\partial x}\right] + \mathbf{k}\left[\frac{\partial(\varphi v)}{\partial x} - \frac{\partial(\varphi u)}{\partial y}\right]$$

$$\nabla \times (\varphi\mathbf{f}) = \varphi\left[\mathbf{i}\left(\frac{\partial w}{\partial y} - \frac{\partial v}{\partial z}\right) + \mathbf{j}\left(\frac{\partial u}{\partial z} - \frac{\partial w}{\partial x}\right) + \mathbf{k}\left(\frac{\partial v}{\partial x} - \frac{\partial u}{\partial y}\right)\right] + \begin{vmatrix} \mathbf{i} & \mathbf{j} & \mathbf{k} \\ \dfrac{\partial\varphi}{\partial x} & \dfrac{\partial\varphi}{\partial y} & \dfrac{\partial\varphi}{\partial z} \\ u & v & w \end{vmatrix}$$

$$\nabla \times (\varphi\mathbf{f}) = \varphi\,\nabla \times \mathbf{f} + \nabla\varphi \times \mathbf{f} \tag{72}$$

This result is easily obtained by considering ∇ as a vector differential operator.

Example 29. To show that the curl of a gradient is zero.

$$\nabla \times (\nabla\varphi) = \begin{vmatrix} \mathbf{i} & \mathbf{j} & \mathbf{k} \\ \dfrac{\partial}{\partial x} & \dfrac{\partial}{\partial y} & \dfrac{\partial}{\partial z} \\ \dfrac{\partial\varphi}{\partial x} & \dfrac{\partial\varphi}{\partial y} & \dfrac{\partial\varphi}{\partial z} \end{vmatrix} = \mathbf{i}\left(\frac{\partial^2\varphi}{\partial y\,\partial z} - \frac{\partial^2\varphi}{\partial z\,\partial y}\right)$$

$$+ \mathbf{j}\left(\frac{\partial^2\varphi}{\partial z\,\partial x} - \frac{\partial^2\varphi}{\partial x\,\partial z}\right) + \mathbf{k}\left(\frac{\partial^2\varphi}{\partial x\,\partial y} - \frac{\partial^2\varphi}{\partial y\,\partial x}\right)$$

Hence

$$\boxed{\nabla \times \nabla\varphi = 0} \tag{73}$$

provided φ has continuous second derivatives.

Example 30. To show that the divergence of a curl is zero.

$$\nabla \cdot (\nabla \times \mathbf{f}) = \frac{\partial}{\partial x}\left(\frac{\partial w}{\partial y} - \frac{\partial v}{\partial z}\right) + \frac{\partial}{\partial y}\left(\frac{\partial u}{\partial z} - \frac{\partial w}{\partial x}\right) + \frac{\partial}{\partial z}\left(\frac{\partial v}{\partial x} - \frac{\partial u}{\partial y}\right)$$

$$= \frac{\partial^2 u}{\partial y\,\partial z} - \frac{\partial^2 u}{\partial z\,\partial y} + \frac{\partial^2 v}{\partial z\,\partial x} - \frac{\partial^2 v}{\partial x\,\partial z} + \frac{\partial^2 w}{\partial x\,\partial y} - \frac{\partial^2 w}{\partial y\,\partial x}$$

Thus

$$\boxed{\nabla \cdot (\nabla \times \mathbf{f}) = 0} \tag{74}$$

Example 31. What does $(\mathbf{u} \cdot \nabla)\mathbf{v}$ mean? We first dot $\mathbf{u}$ with ∇. This yields the scalar differential operator

$$u_x \frac{\partial}{\partial x} + u_y \frac{\partial}{\partial y} + u_z \frac{\partial}{\partial z}$$

Then we operate on $\mathbf{v}$ obtaining

$$(\mathbf{u} \cdot \nabla)\mathbf{v} = u_x \frac{\partial \mathbf{v}}{\partial x} + u_y \frac{\partial \mathbf{v}}{\partial y} + u_z \frac{\partial \mathbf{v}}{\partial z}$$

Thus

$$d\mathbf{f} = \frac{\partial \mathbf{f}}{\partial x}\,dx + \frac{\partial \mathbf{f}}{\partial y}\,dy + \frac{\partial \mathbf{f}}{\partial z}\,dz$$

$$= dx\,\frac{\partial \mathbf{f}}{\partial x} + dy\,\frac{\partial \mathbf{f}}{\partial y} + dz\,\frac{\partial \mathbf{f}}{\partial z}$$

and

$$df = (dr \cdot \nabla)f \tag{75}$$

since $d\mathbf{r} = dx\,\mathbf{i} + dy\,\mathbf{j} + dz\,\mathbf{k}$.
If $\mathbf{f} = \mathbf{f}(x, y, z, t)$,

$$d\mathbf{f} = (d\mathbf{r} \cdot \nabla)\mathbf{f} + \frac{\partial \mathbf{f}}{\partial t}\,dt \tag{76}$$

Example 32

$$\begin{aligned}(\mathbf{v} \cdot \nabla)\mathbf{r} &= v_x \frac{\partial \mathbf{r}}{\partial x} + v_y \frac{\partial \mathbf{r}}{\partial y} + v_z \frac{\partial \mathbf{r}}{\partial z} \\ &= v_x\mathbf{i} + v_y\mathbf{j} + v_z\mathbf{k}\end{aligned}$$

$$(\mathbf{v} \cdot \nabla)\mathbf{r} = \mathbf{v} \tag{77}$$

where $\mathbf{r}$ is the position vector $x\mathbf{i} + y\mathbf{j} + z\mathbf{k}$.

Example 33. Let us expand $\nabla(\mathbf{u} \cdot \mathbf{v})$. Now

$$\mathbf{u} \times (\nabla \times \mathbf{v}) = \nabla_v(\mathbf{u} \cdot \mathbf{v}) - (\mathbf{u} \cdot \nabla)\mathbf{v}$$

Here we have applied the rule of the middle factor, noting also that ∇ operates only on $\mathbf{v}$. $\nabla_v(\mathbf{u} \cdot \mathbf{v})$ means that we keep the components of $\mathbf{u}$ fixed and differentiate only the components of $\mathbf{v}$.

Similarly, $\mathbf{v} \times (\nabla \times \mathbf{u}) = \nabla_u(\mathbf{u} \cdot \mathbf{v}) - (\mathbf{v} \cdot \nabla)\mathbf{u}$. Adding, we obtain

$$\begin{aligned}\nabla_u(\mathbf{u} \cdot \mathbf{v}) + \nabla_v(\mathbf{u} \cdot \mathbf{v}) = \mathbf{u} \times (\nabla \times \mathbf{v}) &+ \mathbf{v} \times (\nabla \times \mathbf{u}) \\ &+ (\mathbf{u} \cdot \nabla)\mathbf{v} + (\mathbf{v} \cdot \nabla)\mathbf{u}\end{aligned}$$

and

$$\nabla(\mathbf{u} \cdot \mathbf{v}) = \mathbf{u} \times (\nabla \times \mathbf{v}) + \mathbf{v} \times (\nabla \times \mathbf{u}) + (\mathbf{u} \cdot \nabla)\mathbf{v} + (\mathbf{v} \cdot \nabla)\mathbf{u} \tag{78}$$

Example 34

$$\nabla \times (\mathbf{u} \times \mathbf{v}) = (\mathbf{v} \cdot \nabla)\mathbf{u} - \mathbf{v}(\nabla \cdot \mathbf{u}) + \mathbf{u}(\nabla \cdot \mathbf{v}) - (\mathbf{u} \cdot \nabla)\mathbf{v} \tag{79}$$

Example 35

$$\nabla \cdot (\mathbf{u} \times \mathbf{v}) = \nabla_u \cdot (\mathbf{u} \times \mathbf{v}) + \nabla_v \cdot (\mathbf{u} \times \mathbf{v})$$

$$\boxed{\nabla \cdot (\mathbf{u} \times \mathbf{v}) = (\nabla \times \mathbf{u}) \cdot \mathbf{v} - (\nabla \times \mathbf{v}) \cdot \mathbf{u}} \tag{80}$$

Example 36

$$\boxed{\nabla \times (\nabla \times \mathbf{v}) = \nabla(\nabla \cdot \mathbf{v}) - \nabla^2 \mathbf{v}} \tag{81}$$

Example 37. Let $\mathbf{A} = \nabla \times (\varphi \mathbf{i})$ where $\nabla^2\varphi = 0$. We now compute $\mathbf{A} \cdot \nabla \times \mathbf{A}$. Since $\mathbf{A} = \nabla\varphi \times \mathbf{i}$, we obtain

$$\nabla \times \mathbf{A} = \nabla \times (\nabla\varphi \times \mathbf{i}) = (\mathbf{i} \cdot \nabla)\, \nabla\varphi - \mathbf{i}\, \nabla^2\varphi = \nabla \frac{\partial \varphi}{\partial x}$$

from (7), Sec. 22. Thus

$$\mathbf{A} \cdot \nabla \times \mathbf{A} = \begin{vmatrix} \mathbf{i} & \mathbf{j} & \mathbf{k} \\ \dfrac{\partial \varphi}{\partial x} & \dfrac{\partial \varphi}{\partial y} & \dfrac{\partial \varphi}{\partial z} \\ 1 & 0 & 0 \end{vmatrix} \cdot \left(\mathbf{i}\, \frac{\partial^2 \varphi}{\partial x^2} + \mathbf{j}\, \frac{\partial^2 \varphi}{\partial y\, \partial x} + \mathbf{k}\, \frac{\partial^2 \varphi}{\partial z\, \partial x} \right)$$

$$= \frac{\partial \varphi}{\partial z} \frac{\partial^2 \varphi}{\partial y\, \partial x} - \frac{\partial \varphi}{\partial y} \frac{\partial^2 \varphi}{\partial z\, \partial x}$$

If also $\varphi = X(x)Y(y)Z(z)$, we can immediately conclude that $\mathbf{A} \cdot \nabla \times \mathbf{A} = 0$.

22. Recapitulation. We relist the above results:

1. $\nabla(uv) = u\, \nabla v + v\, \nabla u$
2. $\nabla \cdot (\varphi \mathbf{v}) = \varphi\, \nabla \cdot \mathbf{v} + \nabla\varphi \cdot \mathbf{v}$
3. $\nabla \times (\varphi \mathbf{v}) = \varphi\, \nabla \times \mathbf{v} + \nabla\varphi \times \mathbf{v}$
4. $\nabla \times (\nabla\varphi) = 0$
5. $\nabla \cdot (\nabla \times \mathbf{v}) = 0$
6. $\nabla \cdot (\mathbf{u} \times \mathbf{v}) = (\nabla \times \mathbf{u}) \cdot \mathbf{v} - (\nabla \times \mathbf{v}) \cdot \mathbf{u}$
7. $\nabla \times (\mathbf{u} \times \mathbf{v}) = (\mathbf{v} \cdot \nabla)\mathbf{u} - \mathbf{v}(\nabla \cdot \mathbf{u}) + \mathbf{u}(\nabla \cdot \mathbf{v}) - (\mathbf{u} \cdot \nabla)\mathbf{v}$
8. $\nabla \times (\nabla \times \mathbf{v}) = \nabla(\nabla \cdot \mathbf{v}) - \nabla^2 \mathbf{v}$
9. $\nabla(\mathbf{u} \cdot \mathbf{v}) = \mathbf{u} \times (\nabla \times \mathbf{v}) + \mathbf{v} \times (\nabla \times \mathbf{u}) + (\mathbf{u} \cdot \nabla)\mathbf{v} + (\mathbf{v} \cdot \nabla)\mathbf{u}$
10. $(\mathbf{v} \cdot \nabla)\mathbf{r} = \mathbf{v}$
11. $\nabla \cdot \mathbf{r} = 3$
12. $\nabla \times \mathbf{r} = 0$
13. $d\mathbf{f} = (d\mathbf{r} \cdot \nabla)\mathbf{f} + \dfrac{\partial \mathbf{f}}{\partial t}\, dt$

14. $d\varphi = d\mathbf{r} \cdot \nabla\varphi + \dfrac{\partial\varphi}{\partial t}\, dt$

15. $\nabla \cdot (r^{-3}\mathbf{r}) = 0$

Problems

1. Show that $\nabla^2(1/r) = 0$ where $r = (x^2 + y^2 + z^2)^{\frac{1}{2}}$.

2. Compute $\nabla^2 r$, $\nabla^2 r^2$, $\nabla^2(1/r^2)$ where $r = (x^2 + y^2 + z^2)^{\frac{1}{2}}$.

3. Expand $\nabla(\mathbf{uvw})$.

4. Find the divergence and curl of $(x\mathbf{i} - y\mathbf{j})/(x + y)$; of $x \cos z\, \mathbf{i} + y \log x\, \mathbf{j} - z^2\mathbf{k}$.

5. If $\mathbf{a} = \alpha x\mathbf{i} + \beta y\mathbf{j} + \gamma z\mathbf{k}$, show that $\nabla(\mathbf{a} \cdot \mathbf{r}) = 2\mathbf{a}$.

6. Show that $\nabla \times [f(r)\mathbf{r}] = 0$ when $r = (x^2 + y^2 + z^2)^{\frac{1}{2}}$ and $\mathbf{r} = x\mathbf{i} + y\mathbf{j} + z\mathbf{k}$.

7. Let $u = u(x, y, z)$, $v = v(x, y, z)$. Suppose u and v satisfy an equation of the form $f(u, v) = 0$. Show that $\nabla u \times \nabla v = 0$.

8. Assume $\nabla u \times \nabla v = 0$ and assume that we move on the surface $u(x, y, z) =$ constant. Show that v remains constant and hence $v = f(u)$ or $F(u, v) = 0$.

9. Prove that a necessary and sufficient condition that u, v, w satisfy an equation $f(u, v, w) = 0$ is that $\nabla u \cdot \nabla v \times \nabla w = 0$, or

$$\begin{vmatrix} \dfrac{\partial u}{\partial x} & \dfrac{\partial u}{\partial y} & \dfrac{\partial u}{\partial z} \\ \dfrac{\partial v}{\partial x} & \dfrac{\partial v}{\partial y} & \dfrac{\partial v}{\partial z} \\ \dfrac{\partial w}{\partial x} & \dfrac{\partial w}{\partial y} & \dfrac{\partial w}{\partial z} \end{vmatrix} = 0$$

This determinant is called the Jacobian of (u, v, w) with respect to (x, y, z), written $J[(u, v, w)/(x, y, z)]$.

10. If $\mathbf{w}$ is a constant vector, prove that $\nabla \times (\mathbf{w} \times \mathbf{r}) = 2\mathbf{w}$, where $\mathbf{r} = x\mathbf{i} + y\mathbf{j} + z\mathbf{k}$.

11. If $\rho\mathbf{f} = \nabla p$, prove that $\mathbf{f} \cdot \nabla \times \mathbf{f} = 0$. ρ not constant

12. Prove that $(\mathbf{v} \cdot \nabla)\mathbf{v} = \frac{1}{2}\nabla \mathbf{v}^2 - (\mathbf{v} \times \nabla)\mathbf{v}$.

13. If $\mathbf{A}$ is a constant unit vector, show that

$$\mathbf{A} \cdot [\nabla(\mathbf{v} \cdot \mathbf{A}) - \nabla \times (\mathbf{v} \times \mathbf{A})] = \nabla \cdot \mathbf{v}$$

14. If f^1, f^2, f^3 are the components of the vector $\mathbf{f}$ in one set of rectangular axes and $\bar{f}^1, \bar{f}^2, \bar{f}^3$ are the components of $\mathbf{f}$ after a

rotation of axes (see Example 8), show that

$$\sum_{\alpha=1}^{3} \frac{\partial f^\alpha}{\partial x^\alpha} \equiv \sum_{\alpha=1}^{3} \frac{\partial \bar{f}^\alpha}{\partial \bar{x}^\alpha}$$

so that $\nabla \cdot \mathbf{f}$ is a scalar invariant under a rotation of axes. Also see Prob. 21, Sec. 9.

15. Prove (79), (80), (81).

16. Let $\mathbf{f} = f_1\mathbf{i} + f_2\mathbf{j} + f_3\mathbf{k}$ and consider nine quantities

$$g_{ij} = \frac{\partial f_i}{\partial x^j} - \frac{\partial f_j}{\partial x^i}, \qquad i, j = 1, 2, 3$$

Show that $g_{ij} = -g_{ji}$ and that three of the nine quantities yield the three components of $\nabla \times \mathbf{f}$. Use this result to show that $\nabla \times (\varphi \mathbf{f}) \equiv \varphi \nabla \times \mathbf{f} + \nabla\varphi \times \mathbf{f}$.

23. Curvilinear Coordinates. Often the mathematician, physicist, or engineer finds it convenient to use a coordinate system other than the familiar rectangular cartesian coordinate system. If he is dealing with spheres, he will probably find it expedient to describe the position of a point in space by the spherical coordinates r, θ, φ (see Fig. 31). Let us note the following: The sphere $x^2 + y^2 + z^2 = r^2$, the cone $z/(x^2 + y^2 + z^2)^{\frac{1}{2}} = \cos\theta$, and the plane $y/x = \tan\varphi$ pass through the point $P(r, \theta, \varphi)$. We may consider the transformations

$$r = (x^2 + y^2 + z^2)^{\frac{1}{2}}$$

$$\theta = \cos^{-1} \frac{z}{(x^2 + y^2 + z^2)^{\frac{1}{2}}}$$

$$\varphi = \tan^{-1} \frac{y}{x}$$

as a change of coordinates from the x-y-z coordinate system to the r-θ-φ coordinate system. The surfaces

$$r = (x^2 + y^2 + z^2)^{\frac{1}{2}} = c_1$$

$\theta = \cos^{-1}[z/(x^2 + y^2 + z^2)^{\frac{1}{2}}] = c_2$, $\varphi = \tan^{-1} y/x = c_3$ are respectively, a sphere, cone, and plane. Through any point P in space, except the origin, there will pass exactly one surface of each type, the coordinates of the point P determining the constants c_1, c_2, c_3.

The intersection of the sphere and the cone is a circle, the circle of latitude, having $\mathbf{e}_\varphi$ as its unit tangent vector at P. This circle is called the φ-curve since r and θ remain constant on this curve so that only the coordinate φ changes as we move along this curve. The intersection of the sphere and the plane yields the θ-curve, the circle of longitude, while the intersection of the cone and plane yields the straight line from the origin through P, the r curve. $\mathbf{e}_\theta$ and $\mathbf{e}_r$ are the unit tangent vectors to the θ- and r curves, respectively. The three unit vectors at P, $\mathbf{e}_r$, $\mathbf{e}_\theta$, $\mathbf{e}_\varphi$, are mutually perpendicular to each other and can be considered as forming a basis for a coordinate system in the neighborhood of P. Unlike **i, j, k,** they are not fixed, for as we move from point to point their directions change. Thus we may expect to find more complicated formulas for the gradient, divergence, curl, and Laplacian when dealing with spherical coordinates.

Since $\nabla\varphi$ is perpendicular to the plane $\varphi =$ constant, we must have $\nabla\varphi$ parallel to $\mathbf{e}_\varphi$. Hence $\mathbf{e}_\varphi = h_3 \nabla\varphi$, where h_3 is the scalar factor of proportionality between $\mathbf{e}_\varphi$ and $\nabla\varphi$. If $d\mathbf{r}_3$ is a vector tangent to the φ-curve, of length $ds_3 = |d\mathbf{r}_3|$, we have from (58) $d\varphi = d\mathbf{r}_3 \cdot \nabla\varphi = d\mathbf{r}_3 \cdot \dfrac{\mathbf{e}_\varphi}{h_3} = \dfrac{ds_3}{h_3}$ so that $ds_3 = h_3\, d\varphi$. Hence h_3 is that quantity which must be multiplied into the differential change of coordinate φ, namely, $d\varphi$, to yield arc length along the φ-curve. Thus $\mathbf{e}_\varphi = r \sin\theta\, \nabla\varphi$, while similarly $\mathbf{e}_r = \nabla r$ and $\mathbf{e}_\theta = r\,\nabla\theta$. We note that $\mathbf{e}_r = \mathbf{e}_\theta \times \mathbf{e}_\varphi = r^2 \sin\theta\, \nabla\theta \times \nabla\varphi$,

$$\mathbf{e}_\theta = \mathbf{e}_\varphi \times \mathbf{e}_r = r \sin\theta\, \nabla\varphi \times \nabla r$$
$$\mathbf{e}_\varphi = \mathbf{e}_r \times \mathbf{e}_\theta = r\,\nabla r \times \nabla\theta$$

Any vector at P may be represented as $\mathbf{f} = f_1\mathbf{e}_r + f_2\mathbf{e}_\theta + f_3\mathbf{e}_\varphi$. The scalars f_1, f_2, f_3 can be functions of r, θ, φ. We may also represent $\mathbf{f}$ as $\mathbf{f} = f_1 \nabla r + f_2 r\,\nabla\theta + f_3 r \sin\theta\,\nabla\varphi$ and also by $\mathbf{f} = f_1 r^2 \sin\theta\,\nabla\theta \times \nabla\varphi + f_2 r \sin\theta \nabla\varphi \times \nabla\mathbf{r} + f_3 r\,\nabla r \times \nabla\theta$. We also note that the triple scalar product $\nabla r \cdot \nabla\theta \times \nabla\varphi$ is equal to $(r^2 \sin\theta)^{-1}$ and that $dV \equiv ds_1\, ds_2\, ds_3 = r^2 \sin\theta\, dr\, d\theta\, d\varphi$.

Spherical coordinates are special cases of orthogonal curvilinear coordinate systems so that we will proceed to discuss these more general coordinate systems in order to obtain expressions for the gradient, divergence, curl, and Laplacian.

Let us make a change of coordinates from the x-y-z system to a u_1-u_2-u_3 system as given by the equations

$$\begin{aligned} u_1 &= u_1(x, y, z) \\ u_2 &= u_2(x, y, z) \\ u_3 &= u_3(x, y, z) \end{aligned} \tag{82}$$

We assume that the Jacobian $J[(u_1, u_2, u_3)/(x, y, z)] \neq 0$ so that the transformation (82) is one to one in the neighborhood of a point. A point in space is determined when x, y, z are known and hence when u_1, u_2, u_3 are known. By considering

$$u_1(x, y, z) = c_1$$

$u_2(x, y, z) = c_2$, $u_3(x, y, z) = c_3$, we obtain a family of surfaces. Through a point $P(x_0, y_0, z_0)$ will pass the three surfaces

$$u_1(x, y, z) = u_1(x_0, y_0, z_0)$$

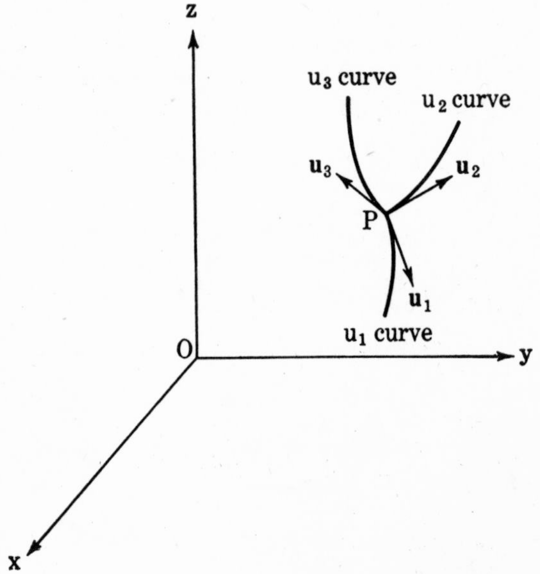

FIG. 36.

$u_2(x, y, z) = u_2(x_0, y_0, z_0)$, and $u_3(x, y, z) = u_3(x_0, y_0, z_0)$. Let us assume that the three surfaces intersect one another orthogonally. The surfaces will intersect in pairs, yielding three curves which intersect orthogonally at the point $P(x_0, y_0, z_0)$. The curve of intersection of the surfaces $u_1 = c_1$ and $u_2 = c_2$ we shall call the u_3 curve, since along this curve only the variable u_3 is allowed to change. Let $\mathbf{u}_1$, $\mathbf{u}_2$, $\mathbf{u}_3$ be three unit vectors issuing from P tangent to the u_1, u_2, u_3 curves, respectively (see Fig. 36).

Now ∇u_3 is perpendicular to the surface

$$u_3(x, y, z) = u_3(x_0, y_0, z_0)$$

so that ∇u_3 is parallel to the unit vector $\mathbf{u}_3$. Hence $\mathbf{u}_3 = h_3 \nabla u_3$ where h_3 is the scalar factor of proportionality between $\mathbf{u}_3$ and ∇u_3. Now let $d\mathbf{r}_3$ be a tangent vector along the u_3 curve, $|d\mathbf{r}_3| = ds_3$. Obviously $d\mathbf{r}_3 \cdot \mathbf{u}_3 = ds_3$, and

$$d\mathbf{r}_3 \cdot \mathbf{u}_3 = d\mathbf{r}_3 \cdot h_3 \nabla u_3$$

so that from (58)

$$ds_3 = h_3 \, du_3 \tag{83}$$

We see that h_3 is that quantity which must be multiplied into the differential coordinate du_3 so that arc length will result. For example, in polar coordinates $ds = r\,d\theta$ if we move on the θ-curve, so that $r = h_2$.

Similarly, $\mathbf{u}_1 = h_1 \nabla u_1$, $\mathbf{u}_2 = h_2 \nabla u_2$, so that

$$\begin{aligned} \mathbf{u}_1 &= \mathbf{u}_2 \times \mathbf{u}_3 = h_2h_3 \nabla u_2 \times \nabla u_3 \\ \mathbf{u}_2 &= \mathbf{u}_3 \times \mathbf{u}_1 = h_3h_1 \nabla u_3 \times \nabla u_1 \\ \mathbf{u}_3 &= \mathbf{u}_1 \times \mathbf{u}_2 = h_1h_2 \nabla u_1 \times \nabla u_2 \end{aligned} \tag{84}$$

and

$$\nabla u_1 \cdot \nabla u_2 \times \nabla u_3 = \frac{\mathbf{u}_1}{h_1} \cdot \frac{\mathbf{u}_2}{h_2} \times \frac{\mathbf{u}_3}{h_3} = (h_1h_2h_3)^{-1} \tag{85}$$

Note that the differential of volume is

$$dV = ds_1 \, ds_2 \, ds_3 = h_1h_2h_3 \, du_1 \, du_2 \, du_3$$

and making use of (85) as well as Prob. 9, Sec. 22,

$$dV = J\left(\frac{x, y, z}{u_1, u_2, u_3}\right) du_1 \, du_2 \, du_3 \tag{86}$$

Example 38. In cylindrical coordinates

$$ds^2 = dr^2 + r^2 \, d\theta^2 + dz^2$$

so that $h_1 = 1$, $h_2 = r$, $h_3 = 1$.

Example 39. If $f = f(u_1, u_2, u_3)$, then from Example 21,

$$\nabla f = \frac{\partial f}{\partial u_1} \nabla u_1 + \frac{\partial f}{\partial u_2} \nabla u_2 + \frac{\partial f}{\partial u_3} \nabla u_3$$

$$\nabla f = \frac{1}{h_1} \frac{\partial f}{\partial u_1} \mathbf{u}_1 + \frac{1}{h_2} \frac{\partial f}{\partial u_2} \mathbf{u}_2 + \frac{1}{h_3} \frac{\partial f}{\partial u_3} \mathbf{u}_3 \tag{87}$$

In cylindrical coordinates

$$\nabla f = \frac{\partial f}{\partial r} \mathbf{R} + \frac{1}{r} \frac{\partial f}{\partial \theta} \mathbf{P} + \frac{\partial f}{\partial z} \mathbf{k}$$

Our next attempt is to obtain an expression for the divergence of a vector when its components are known in an orthogonal curvilinear coordinate system. Now

$$\begin{aligned} \mathbf{f} &= f_1 \mathbf{u}_1 + f_2 \mathbf{u}_2 + f_3 \mathbf{u}_3 \\ &= f_1 h_2 h_3 \nabla u_2 \times \nabla u_3 + f_2 h_3 h_1 \nabla u_3 \times \nabla u_1 + f_3 h_1 h_2 \nabla u_1 \times \nabla u_2 \end{aligned}$$

from (84). Consequently

$$\begin{aligned} \nabla \cdot \mathbf{f} = \nabla(f_1 h_2 h_3) \cdot \nabla u_2 \times \nabla u_3 + f_1 h_2 h_3 \nabla \cdot (\nabla u_2 \times \nabla u_3) \\ + \nabla(f_2 h_3 h_1) \cdot \nabla u_3 \times \nabla u_1 + f_2 h_3 h_1 \nabla \cdot (\nabla u_3 \times \nabla u_1) \\ + \nabla(f_3 h_1 h_2) \cdot \nabla u_1 \times \nabla u_2 + f_3 h_1 h_2 \nabla \cdot (\nabla u_1 \times \nabla u_2) \end{aligned} \tag{88}$$

Now $\nabla(f_1 h_2 h_3) \cdot \nabla u_2 \times \nabla u_3 = \dfrac{\partial (f_1 h_2 h_3)}{\partial u_1} \nabla u_1 \cdot \nabla u_2 \times \nabla u_3$, and

$$\nabla \cdot (\nabla u_2 \times \nabla u_3) = 0$$

so that (88) reduces to

$$\nabla \cdot \mathbf{f} = \frac{1}{h_1 h_2 h_3} \left[\frac{\partial (h_2 h_3 f_1)}{\partial u_1} + \frac{\partial (h_3 h_1 f_2)}{\partial u_2} + \frac{\partial (h_1 h_2 f_3)}{\partial u_3} \right] \tag{89}$$

If we apply (89) to the vector ∇V as given by (87), we obtain

$$\nabla^2 V = \frac{1}{h_1 h_2 h_3} \left[\frac{\partial}{\partial u_1} \left(\frac{h_2 h_3}{h_1} \frac{\partial V}{\partial u_1} \right) + \frac{\partial}{\partial u_2} \left(\frac{h_3 h_1}{h_2} \frac{\partial V}{\partial u_2} \right) + \frac{\partial}{\partial u_3} \left(\frac{h_1 h_2}{h_3} \frac{\partial V}{\partial u_3} \right) \right] \tag{90}$$

X

This is the Laplacian in any orthogonal curvilinear coordinate system.

Example 40. In cylindrical coordinates

$$\nabla^2 V = \frac{1}{r}\left[\frac{\partial}{\partial r}\left(r\frac{\partial V}{\partial r}\right) + \frac{\partial}{\partial \theta}\left(\frac{1}{r}\frac{\partial V}{\partial \theta}\right) + \frac{\partial}{\partial z}\left(r\frac{\partial V}{\partial z}\right)\right] \tag{91}$$

Example 41. Solve $\nabla^2 V = 0$ assuming $V = V(r)$,

$$r = (x^2 + y^2)^{\frac{1}{2}}$$

From (91)

$$\frac{1}{r}\frac{d}{dr}\left(r\frac{dV}{dr}\right) = 0 \qquad \text{or} \qquad r\frac{dV}{dr} = c_1$$

and

$$V = c_1 \log r + c_2$$

Finally we obtain the curl of **f**.

$$\begin{aligned}\mathbf{f} &= f_1\mathbf{u}_1 + f_2\mathbf{u}_2 + f_3\mathbf{u}_3\\ &= f_1h_1\,\nabla u_1 + f_2h_2\,\nabla u_2 + f_3h_3\,\nabla u_3\end{aligned}$$

and

$$\nabla \times \mathbf{f} = \nabla(f_1h_1) \times \nabla u_1 + \nabla(f_2h_2) \times \nabla u_2 + \nabla(f_3h_3) \times \nabla u_3$$

since $\nabla \times (\nabla u_1) = \nabla \times (\nabla u_2) = \nabla \times (\nabla u_3) = 0$. Now

$$\nabla(f_1h_1) \times \nabla u_1 = \frac{\partial(f_1h_1)}{\partial u_1}\nabla u_1 \times \nabla u_1 + \frac{\partial(f_1h_1)}{\partial u_2}\nabla u_2 \times \nabla u_1 + \frac{\partial(f_1h_1)}{\partial u_3}\nabla u_3 \times \nabla u_1$$

Replacing $\nabla u_2 \times \nabla u_1$ by $-\dfrac{\mathbf{u}_3}{h_1h_2}$, etc., we obtain

$$\nabla \times \mathbf{f} = \frac{\mathbf{u}_1}{h_2h_3}\left[\frac{\partial(h_3f_3)}{\partial u_2} - \frac{\partial(h_2f_2)}{\partial u_3}\right] + \frac{\mathbf{u}_2}{h_3h_1}\left[\frac{\partial(h_1f_1)}{\partial u_3} - \frac{\partial(h_3f_3)}{\partial u_1}\right] + \frac{\mathbf{u}_3}{h_1h_2}\left[\frac{\partial(h_2f_2)}{\partial u_1} - \frac{\partial(h_1f_1)}{\partial u_2}\right] \tag{92}$$

Problems

1. For spherical coordinates, $ds^2 = dr^2 + r^2\,d\theta^2 + r^2 \sin^2\theta\,d\varphi^2$ where θ is the colatitude and φ the azimuthal angle. Show that

$$\nabla^2 V = \frac{1}{r^2 \sin\theta}\left[\frac{\partial}{\partial r}\left(r^2 \sin\theta \frac{\partial V}{\partial r}\right) + \frac{\partial}{\partial \theta}\left(\sin\theta \frac{\partial V}{\partial \theta}\right) + \frac{\partial}{\partial \varphi}\left(\frac{1}{\sin\theta}\frac{\partial V}{\partial \varphi}\right)\right]$$

2. Solve $\nabla^2 V = 0$ in spherical coordinates if $V = V(r)$.
3. Express $\nabla \cdot \mathbf{f}$ and $\nabla \times \mathbf{f}$ in cylindrical coordinates.
4. Express $\nabla \cdot \mathbf{f}$ and $\nabla \times \mathbf{f}$ in spherical coordinates by letting $\mathbf{a}, \mathbf{b}, \mathbf{c}$ be unit vectors in the r, θ, φ directions, respectively.
5. Write Eq. (92) in terms of a determinant.
6. Show that $\nabla \times [(r\,\nabla\theta)/\sin\theta] = \nabla\varphi$ where r, θ, φ are spherical coordinates.
7. If $\mathbf{a}, \mathbf{b}, \mathbf{c}$ are the vectors of Prob. 4, show that

$$\frac{\partial \mathbf{a}}{\partial r} = 0, \qquad \frac{\partial \mathbf{a}}{\partial \theta} = \mathbf{b}, \qquad \frac{\partial \mathbf{a}}{\partial \varphi} = \sin\theta\,\mathbf{c}$$

$$\frac{\partial \mathbf{b}}{\partial r} = 0, \qquad \frac{\partial \mathbf{b}}{\partial \theta} = -\mathbf{a}, \qquad \frac{\partial \mathbf{b}}{\partial \varphi} = \cos\theta\,\mathbf{c}$$

$$\frac{\partial \mathbf{c}}{\partial r} = 0, \qquad \frac{\partial \mathbf{c}}{\partial \theta} = 0, \qquad \frac{\partial \mathbf{c}}{\partial \varphi} = -\sin\theta\,\mathbf{a} - \cos\theta\,\mathbf{b}$$

8. If $x = r \sin\theta \cos\varphi$, $y = r \sin\theta \sin\varphi$, $z = r\cos\theta$, then the form $ds^2 = dx^2 + dy^2 + dz^2$ becomes

$$ds^2 = dr^2 + r^2\,d\theta^2 + r^2 \sin^2\theta\,d\varphi^2$$

Prove this. If, in general, $ds^2 = \sum_{\alpha=1}^{3} (dx^\alpha)^2$, and if

$$x^\alpha = x^\alpha(y^1, y^2, y^3)$$

$\alpha = 1, 2, 3$, show that

$$ds^2 = \sum_{\alpha,\beta,\gamma} \frac{\partial x^\alpha}{dy^\beta}\frac{\partial x^\alpha}{dy^\gamma}\,dy^\beta\,dy^\gamma$$

$$= \sum_{\beta,\gamma} g_{\beta\gamma}\,dy^\beta\,dy^\gamma$$

where

$$g_{\beta\gamma} = \sum_{\alpha=1}^{3} \frac{\partial x^\alpha}{\partial y^\beta} \frac{\partial x^\alpha}{\partial y^\gamma}$$

Check this result for the transformation to cylindrical coordinates:

$$\begin{aligned} x &= r \cos \theta \\ y &= r \sin \theta \\ z &= z \end{aligned}$$

and obtain $ds^2 = dr^2 + r^2\, d\theta^2 + dz^2$.

9. By making use of $\nabla^2\mathbf{V} = \nabla(\nabla \cdot \mathbf{V}) - \nabla \times (\nabla \times \mathbf{V})$, find $\nabla^2\mathbf{V}$ for $\mathbf{V} = v(r)\mathbf{e}_r$, $\mathbf{V}$ being purely radial (spherical coordinates). Find $\nabla^2\mathbf{V}$ for $\mathbf{V} = f(r)\mathbf{e}_r + \varphi(z)\mathbf{e}_z$ in cylindrical coordinates.

10. Find $\nabla^2\mathbf{V}$ if $\mathbf{V} = w(r)\mathbf{k} \times \mathbf{r}$.

11. Consider the equations

$$(\lambda + \mu)\nabla(\nabla \cdot \mathbf{s}) + \mu\, \nabla^2\mathbf{s} = \rho \frac{\partial^2 \mathbf{s}}{\partial t^2}$$

λ, μ, ρ constants. Assume $\mathbf{s} = e^{ipt}\mathbf{s}_1$, p constant, and show that

$$(\lambda + \mu)\nabla(\nabla \cdot \mathbf{s}_1) + (\mu + \rho p^2)\mathbf{s}_1 = 0$$

Next show that $[\nabla^2 + (\mu + \rho p^2)/(\lambda + \mu)](\nabla \cdot \mathbf{s}_1) = 0,$

$$\lambda + \mu \neq 0.$$

12. If $\mathbf{A} = \nabla \times (\psi\mathbf{r})$, $\nabla^2\psi = 0$, show that

$$\mathbf{A} \cdot \nabla \times \mathbf{A} = \frac{1}{\sin \theta}\left(\frac{\partial \psi}{\partial \varphi} \frac{\partial^2 \psi}{\partial \theta\, \partial r} - \frac{\partial \psi}{\partial \theta} \frac{\partial^2 \psi}{\partial \varphi\, \partial r}\right)$$

so that $\mathbf{A} \cdot \nabla \times \mathbf{A} = 0$ if, moreover, $\psi = R(r)\Theta(\theta)\Phi(\varphi)$.

13. Show that $\varphi_1 = Ae^x + Be^y + Ce^z$ satisfies $\nabla^2\varphi_1 = \varphi_1$, and show that if φ_2 satisfies $\nabla^2\varphi_2 = 0$, then $\varphi = \varphi_1 + \varphi_2$ also satisfies $\nabla^2\varphi = \varphi$. Find a solution of $\nabla^2\varphi = -\varphi$.

CHAPTER 3

DIFFERENTIAL GEOMETRY

24. Frenet-Serret Formulas. A three-dimensional curve in a Euclidean space can be represented by the locus of the end point of the position vector given by

$$\mathbf{r}(t) = x(t)\mathbf{i} + y(t)\mathbf{j} + z(t)\mathbf{k} \tag{93}$$

where t is a parameter ranging over a set of values $t_0 \leq t \leq t_1$. We assume that $x(t)$, $y(t)$, $z(t)$ have continuous derivatives of all orders and that they can be expanded in a Taylor series in the neighborhood of any point of the curve.

We have seen in Chap. 2, Sec. 16, that $\frac{d\mathbf{r}}{ds}$ is the unit tangent vector to the curve. Let $\mathbf{t} = \frac{d\mathbf{r}}{ds}$. Now $\mathbf{t}$ is a unit vector so that its derivative is perpendicular to $\mathbf{t}$. Moreover, this derivative, $\frac{d\mathbf{t}}{ds}$, tells us how fast the unit tangent vector is changing direction as we move along the curve. The principal normal to the curve is consequently defined by the equation

$$\frac{d\mathbf{t}}{ds} = \kappa\mathbf{n} \tag{94}$$

where κ is the magnitude of $\frac{d\mathbf{t}}{ds}$ and is called the curvature. The reciprocal of the curvature, $\rho = 1/\kappa$, is called the radius of curvature. It is important to note that (94) defines both κ and $\mathbf{n}$, κ being the length of $\frac{d\mathbf{t}}{ds}$ while $\mathbf{n}$ is the unit vector parallel to $\frac{d\mathbf{t}}{ds}$. At any point P of our curve we now have two vectors $\mathbf{t}$, $\mathbf{n}$ at right angles to each other (see Fig. 37). This enables us to set up a local coordinate system at P by defining a third vector at right angles to $\mathbf{t}$ and $\mathbf{n}$. We define as the binormal the vector

$$\mathbf{b} = \mathbf{t} \times \mathbf{n}$$

All vectors associated with the curve at the point P can be written as a linear combination of the three fundamental vectors $\mathbf{t}$, $\mathbf{n}$, $\mathbf{b}$, which form a trihedral at P.

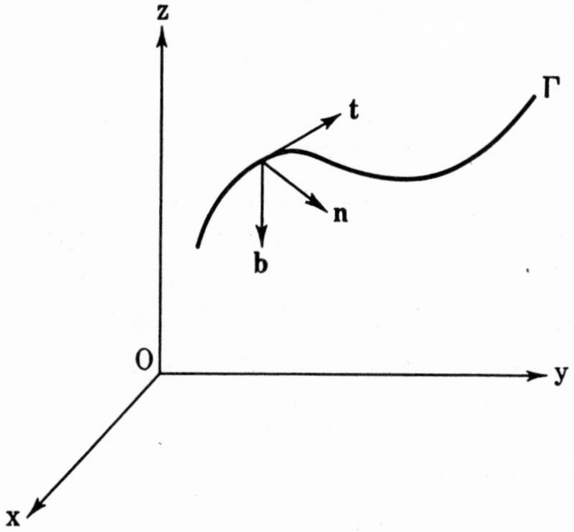

FIG. 37.

Let us now evaluate $\dfrac{d\mathbf{b}}{ds}$ and $\dfrac{d\mathbf{n}}{ds}\cdot$ Since $\mathbf{b}$ is a unit vector, its derivative is perpendicular to $\mathbf{b}$ and so lies in the plane of $\mathbf{t}$ and $\mathbf{n}$. Moreover, $\mathbf{b} \cdot \mathbf{t} = 0$ so that on differentiating we obtain

$$\frac{d\mathbf{b}}{ds} \cdot \mathbf{t} + \kappa \mathbf{b} \cdot \mathbf{n} = 0$$

or $\dfrac{d\mathbf{b}}{ds} \cdot \mathbf{t} = 0$. Hence $\dfrac{d\mathbf{b}}{ds}$ is also perpendicular to $\mathbf{t}$ so that $\dfrac{d\mathbf{b}}{ds}$ must be parallel to $\mathbf{n}$. Consequently, $\dfrac{d\mathbf{b}}{ds} = \tau \mathbf{n}$, where τ by definition is the magnitude of $\dfrac{d\mathbf{b}}{ds}\cdot$ τ is called the torsion of the curve.

Finally, to obtain $\frac{d\mathbf{n}}{ds}$, we note that $\mathbf{n} = \mathbf{b} \times \mathbf{t}$ so that

$$\frac{d\mathbf{n}}{ds} = \mathbf{b} \times \frac{d\mathbf{t}}{ds} + \frac{d\mathbf{b}}{ds} \times \mathbf{t} = \mathbf{b} \times \kappa\mathbf{n} + \tau\mathbf{n} \times \mathbf{t} = -\kappa\mathbf{t} - \tau\mathbf{b}$$

The famous Frenet-Serret formulas are

$$\begin{aligned} \frac{d\mathbf{t}}{ds} &= \kappa\mathbf{n} \\ \frac{d\mathbf{n}}{ds} &= -(\kappa\mathbf{t} + \tau\mathbf{b}) \\ \frac{d\mathbf{b}}{ds} &= \tau\mathbf{n} \end{aligned} \tag{95}$$

Successive derivatives are functions of $\mathbf{t}$, $\mathbf{n}$, $\mathbf{b}$ and the derivatives of κ and τ.

Example 42. The circular helix is given by

$$\mathbf{r} = a \cos t\, \mathbf{i} + a \sin t\, \mathbf{j} + bt\mathbf{k}$$

$$\mathbf{t} = \frac{d\mathbf{r}}{ds} = (-a \sin t\, \mathbf{i} + a \cos t\, \mathbf{j} + b\mathbf{k}) \frac{dt}{ds}$$

and

$$\begin{aligned} \mathbf{t} \cdot \mathbf{t} = 1 &= \left(\frac{dt}{ds}\right)^2 (a^2 \sin^2 t + a^2 \cos^2 t + b^2) \\ &= (a^2 + b^2) \left(\frac{dt}{ds}\right)^2 = 1 \end{aligned}$$

Hence

$$\mathbf{t} = (-a \sin t\, \mathbf{i} + a \cos t\, \mathbf{j} + b\mathbf{k})(a^2 + b^2)^{-\frac{1}{2}}$$

Now

$$\kappa\mathbf{n} = \frac{d\mathbf{t}}{ds} = (-a \cos t\, \mathbf{i} - a \sin t\, \mathbf{j})(a^2 + b^2)^{-1}$$

so that

$$\kappa = a(a^2 + b^2)^{-1}$$

Also

$$\begin{aligned} \mathbf{b} = \mathbf{t} \times \mathbf{n} &= \begin{vmatrix} \mathbf{i} & \mathbf{j} & \mathbf{k} \\ -a \sin t & a \cos t & b \\ -\cos t & -\sin t & 0 \end{vmatrix} (a^2 + b^2)^{-\frac{1}{2}} \\ &= (b \sin t\, \mathbf{i} - b \cos t\, \mathbf{j} + a\mathbf{k})(a^2 + b^2)^{-\frac{1}{2}} \end{aligned}$$

and

$$\frac{d\mathbf{b}}{ds} = \tau\mathbf{n} = (b \cos t\,\mathbf{i} + b \sin t\,\mathbf{j})(a^2 + b^2)^{-1}$$

so that

$$\tau = b(a^2 + b^2)^{-1}$$

Problems

1. Show that the radius of curvature of the twisted curve $x = \log \cos \theta$, $y = \log \sin \theta$, $z = \sqrt{2}\,\theta$ is $\rho = \sqrt{2} \csc 2\theta$.

2. Show that $\tau = 0$ is a necessary and sufficient condition that a curve be a plane curve.

3. Prove that $\tau = \dfrac{1}{\kappa^2}(\mathbf{r}'\mathbf{r}''\mathbf{r}''')$.

4. For the curve $x = a(3t - t^3)$, $y = 3at^2$, $z = a(3t + t^3)$, show that $\kappa = \tau = 1/3a(1 + t^2)^2$.

5. Prove that $\dfrac{d\mathbf{t}}{ds} \cdot \dfrac{d\mathbf{b}}{ds} = \kappa\tau$, $\dfrac{d\mathbf{n}}{ds} \cdot \dfrac{d\mathbf{b}}{ds} = 0$, $\dfrac{d\mathbf{t}}{ds} \cdot \dfrac{d\mathbf{n}}{ds} = 0$.

6. Prove that $\mathbf{r}''' = -\kappa^2\mathbf{t} + \kappa'\mathbf{n} - \tau\kappa\mathbf{b}$, where the primes mean differentiation with respect to arc length.

7. Prove that the shortest distance between the principal normals at consecutive points at a distance ds apart (s measured along the arc) is $ds\ \rho(\rho^2 + \tau^{-2})^{-\frac{1}{2}}$.

8. Find the curvature and torsion of the curve

$$x = a(u - \sin u), \qquad y = a(1 - \cos u), \qquad z = bu$$

9. For a plane curve given by $\mathbf{r} = x(t)\mathbf{i} + y(t)\mathbf{j}$, show that

$$\kappa = \frac{x'y'' - y'x''}{[(x')^2 + (y')^2]^{\frac{3}{2}}}$$

10. Prove that $(\mathbf{t}'\mathbf{t}''\mathbf{t}''') = \kappa^5 \dfrac{d}{ds}\left(\dfrac{\tau}{\kappa}\right)$.

11. Show that the line element $ds^2 = dx^2 + dy^2 + dz^2 - c^2\,dt^2$ remains invariant in form under the Lorentz transformation

$$x = \frac{\bar{x} - V\bar{t}}{[1 - (V^2/c^2)]^{\frac{1}{2}}}$$
$$y = \bar{y}$$
$$z = \bar{z}$$
$$t = \frac{\bar{t} - (V/c^2)\bar{x}}{[1 - (V^2/c^2)]^{\frac{1}{2}}}$$

V, c are constants. The transformation $ct = i\tau$, $i = \sqrt{-1}$, leads to the four-dimensional Euclidean line element

$$ds^2 = dx^2 + dy^2 + dz^2 + d\tau^2$$

12. If $x^\alpha = x^\alpha(s)$, $\alpha = 1, 2, \ldots, n$, represents a curve in an n-dimensional Euclidean space for which

$$ds^2 = (dx^1)^2 + (dx^2)^2 + \cdots + (dx^n)^2$$

define the unit tangent vector to this curve, this definition being a generalization of the definition of the tangent vector for the case $n = 3$. Show that the vector $\dfrac{d^2x^\alpha}{ds^2}$, $\alpha = 1, 2, \ldots, n$, is normal to the tangent vector, and define the unit principal normal $\mathbf{n}_1$ and curvature κ_1 by the equations

$$\frac{d^2x^\alpha}{ds^2} = \frac{dt^\alpha}{ds} = \kappa_1 n_1{}^\alpha, \quad \alpha = 1, 2, \ldots, n$$

Show that $\dfrac{dn_1{}^\alpha}{ds}$, $\alpha = 1, 2, \ldots, n$, is normal to n_1 and that $\displaystyle\sum_{\alpha=1}^{n} t^\alpha \frac{dn_1{}^\alpha}{ds} = -\kappa_1$. Define the second curvature κ_2 and unit normal $\mathbf{n}_2$ by the equations $\dfrac{dn_1{}^\alpha}{ds} = -\kappa_1 t^\alpha + \kappa_2 n_2{}^\alpha$, $\alpha = 1, 2, \ldots, n$, and show that $n_2{}^\alpha$ is normal to t^α and $n_1{}^\alpha$ if $\kappa_2 \neq 0$. Continue in this manner and obtain the generalization of the Frenet-Serret formulas.

25. Fundamental Planes. The plane containing the tangent and principal normal is called the osculating plane. Let $\mathbf{s}$ be a variable vector to any point in this plane and let $\mathbf{r}$ be the vector to the point P on the curve. $\mathbf{s} - \mathbf{r}$ lies in the plane and is consequently perpendicular to the binormal. The equation of the osculating plane is

$$(\mathbf{s} - \mathbf{r}) \cdot \mathbf{b} = 0 \tag{96}$$

The normal plane to the curve at P is defined as the plane through P perpendicular to the tangent vector. Its equation is easily seen to be

$$(\mathbf{s} - \mathbf{r}) \cdot \mathbf{t} = 0 \tag{97}$$

The third fundamental plane is the rectifying plane through P perpendicular to the normal **n.** Its equation is

$$(\mathbf{s} - \mathbf{r}) \cdot \mathbf{n} = 0 \tag{98}$$

Problems

1. Find the equations of the three fundamental planes for the curve

$$x = at, \qquad y = bt^2, \qquad z = ct^3$$

2. Show that the limiting position of the line of intersection of two adjacent normal planes is given by $(\mathbf{s} - \mathbf{r}) \cdot \mathbf{n} = \rho$ where $\mathbf{s}$ is the vector to any point on the line.

26. Intrinsic Equations of a Curve. The curvature and torsion of a curve depend on the point P of the curve and consequently on the arc parameter s. Let $\kappa = f(s)$, $\tau = F(s)$. These two equations are called the intrinsic equations of the curve. They owe their name to the fact that two curves with the same intrinsic equations are identical except possibly for orientation in space. Assume two curves with the same intrinsic equations. Let the trihedrals at a corresponding point P coincide; this can be done by a rigid motion.

Now

$$\begin{aligned} \frac{d}{ds}(\mathbf{t}_1 \cdot \mathbf{t}_2) &= \mathbf{t}_1 \cdot \kappa\mathbf{n}_2 + \kappa\mathbf{n}_1 \cdot \mathbf{t}_2 \\ \frac{d}{ds}(\mathbf{n}_1 \cdot \mathbf{n}_2) &= \mathbf{n}_1 \cdot (-\kappa\mathbf{t}_2 - \tau\mathbf{b}_2) + \mathbf{n}_2 \cdot (-\kappa\mathbf{t}_1 - \tau\mathbf{b}_1) \\ \frac{d}{ds}(\mathbf{b}_1 \cdot \mathbf{b}_2) &= \mathbf{b}_1 \cdot \tau\mathbf{n}_2 + \mathbf{b}_2 \cdot \tau\mathbf{n}_1 \end{aligned} \tag{99}$$

Adding, we obtain

$$\frac{d}{ds}(\mathbf{t}_1 \cdot \mathbf{t}_2 + \mathbf{n}_1 \cdot \mathbf{n}_2 + \mathbf{b}_1 \cdot \mathbf{b}_2) = 0$$

so that

$$\mathbf{t}_1 \cdot \mathbf{t}_2 + \mathbf{n}_1 \cdot \mathbf{n}_2 + \mathbf{b}_1 \cdot \mathbf{b}_2 = \text{constant} = 3 \qquad (100)$$

since at P

$$\mathbf{t}_1 = \mathbf{t}_2, \qquad \mathbf{n}_1 = \mathbf{n}_2, \qquad \mathbf{b}_1 = \mathbf{b}_2$$

Since (100) always maintains its maximum value, we must have $\mathbf{t}_1 \equiv \mathbf{t}_2$, $\mathbf{n}_1 \equiv \mathbf{n}_2$, $\mathbf{b}_1 \equiv \mathbf{b}_2$ so that $\frac{d\mathbf{r}_1}{ds} = \frac{d\mathbf{r}_2}{ds}$ or $\mathbf{r}_1 \equiv \mathbf{r}_2$ locally.

Hence the two curves are identical in a small neighborhood of P. Since we have assumed analyticity of the curves, they are identical everywhere.

Problems

1. Show that the intrinsic equations of $x = a(\theta - \sin\theta)$, $y = a(1 - \cos\theta)$, $z = 0$ are $\rho^2 + s^2 = 16a^2$, $\tau = 0$, where s is measured from the top of the arc of the cycloid.

2. Show that the intrinsic equation for the catenary

$$y = \frac{a}{2}\left(e^{x/a} + e^{-(x/a)}\right)$$

is $a\rho = s^2 + a^2$, where s is measured from the vertex of the catenary.

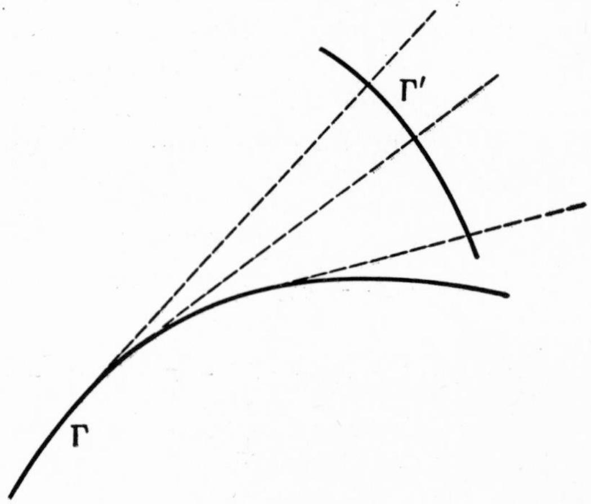

FIG. 38.

27. Involutes. Let us consider the space curve Γ. We construct the tangents to every point of Γ and define an involute as any curve which is normal to every tangent of Γ (see Fig. 38). From Fig. 39, it is evident that

$$\mathbf{r}_1 = \mathbf{r} + u\mathbf{t} \tag{101}$$

is the equation of the involute, u unknown. Differentiating (101), we obtain

$$\frac{d\mathbf{r}_1}{ds_1} = \mathbf{t}_1 = \left(\frac{d\mathbf{r}}{ds} + u\frac{d\mathbf{t}}{ds} + \frac{du}{ds}\mathbf{t}\right)\frac{ds}{ds_1} \tag{102}$$

where s is arc length along Γ and s_1 is arc length along Γ'. Using (95), (102) becomes

$$\mathbf{t}_1 = \left(\mathbf{t} + u\kappa\mathbf{n} + \frac{du}{ds}\mathbf{t}\right)\frac{ds}{ds_1} \tag{103}$$

Now $\mathbf{t} \cdot \mathbf{t}_1 = 0$ from the definition of the involute so that

$$1 + \frac{du}{ds} = 0 \quad \text{and} \quad u = c - s \tag{104}$$

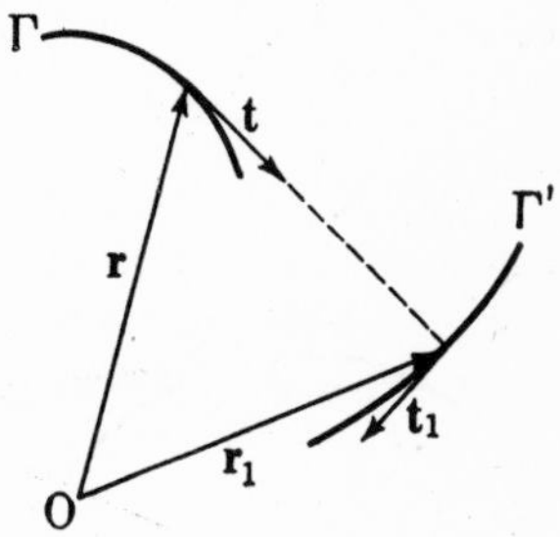

FIG. 39.

Therefore $\mathbf{r}_1 = \mathbf{r} + (c - s)\mathbf{t}$, and there exists an infinite family of involutes, one involute for each constant c. The distance between corresponding involutes remains a constant. An invo-

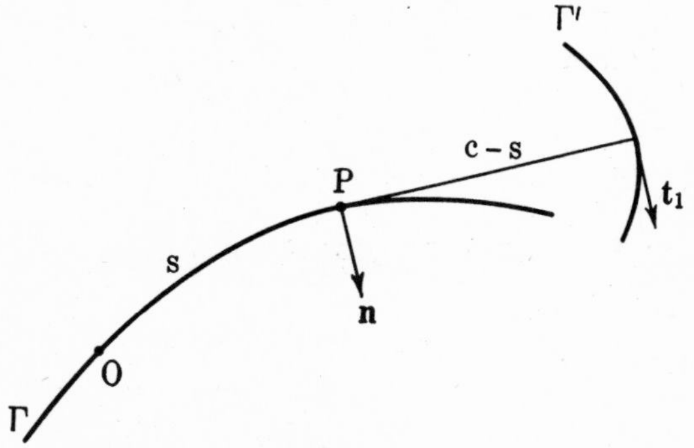

FIG. 40.

lute can be generated by unrolling a taut string of length c which has been wrapped along the curve. The end point of the string generates the involute (see Fig. 40). What are some properties of the involute?

$$\mathbf{r}_1 = \mathbf{r} + (c - s)\mathbf{t}$$

$$\mathbf{t}_1 = \frac{d\mathbf{r}_1}{ds_1} = \left[\frac{d\mathbf{r}}{ds} + (c - s)\frac{d\mathbf{t}}{ds} - \mathbf{t}\right]\frac{ds}{ds_1}$$

$$= (c - s)\,\kappa\frac{ds}{ds_1}\mathbf{n}$$

Hence the tangent to the involute is parallel to the corresponding normal of the curve. Since $\mathbf{t}_1$ and $\mathbf{n}$ are unit vectors, we must have $(c - s)\,\kappa\dfrac{ds}{ds_1} = 1$. The curvature of the involute is obtained from $\dfrac{d\mathbf{t}_1}{ds_1} = \kappa_1\mathbf{n}_1 = \dfrac{d\mathbf{n}}{ds}\dfrac{ds}{ds_1} = \dfrac{(-\kappa\mathbf{t} - \tau\mathbf{b})}{(c - s)\kappa}\cdot$ Hence

$$\kappa_1^2 = \frac{\kappa^2 + \tau^2}{\kappa^2(c - s)^2} \tag{105}$$

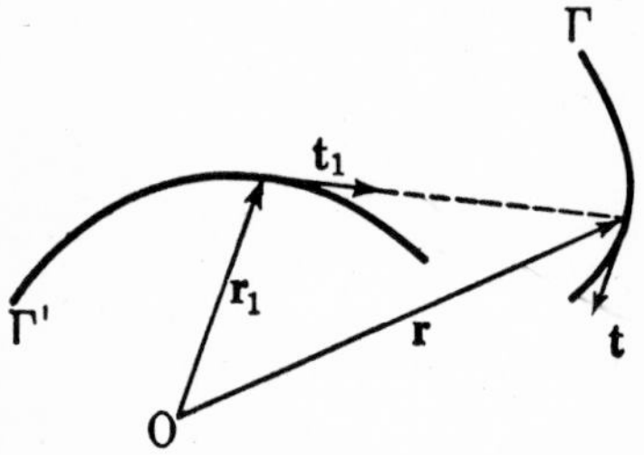

FIG. 41.

28. Evolutes. The curve Γ′ whose tangents are perpendicular to a given curve is called the evolute of the curve. The tangent to Γ′ must lie in the plane of $\mathbf{b}$ and $\mathbf{n}$ of Γ since it is perpendicular to $\mathbf{t}$. Consequently

$$\mathbf{r}_1 = \mathbf{r} + u\mathbf{n} + v\mathbf{b}$$

is the equation of the evolute. Differentiating, we obtain

$$\mathbf{t}_1 = \frac{d\mathbf{r}_1}{ds} = \left(\frac{d\mathbf{r}}{ds} + u\frac{d\mathbf{n}}{ds} + v\frac{d\mathbf{b}}{ds} + \frac{du}{ds}\mathbf{n} + \frac{dv}{ds}\mathbf{b}\right)\frac{ds}{ds_1}$$

$$= \left[\mathbf{t} + u(-\kappa\mathbf{t} - \tau\mathbf{b}) + v\tau\mathbf{n} + \frac{du}{ds}\mathbf{n} + \frac{dv}{ds}\mathbf{b}\right]\frac{ds}{ds_1}$$

Now $\mathbf{t}\cdot\mathbf{t}_1 = 0$, which implies $1 - u\kappa = 0$ or $u = \dfrac{1}{\kappa} = \rho$. Thus

$$\mathbf{t}_1 = \left[\left(-\tau u + \frac{dv}{ds}\right)\mathbf{b} + \left(v\tau + \frac{du}{ds}\right)\mathbf{n}\right]\frac{ds}{ds_1}$$

Also $\mathbf{t}_1$ is parallel to $\mathbf{r}_1 - \mathbf{r} = u\mathbf{n} + v\mathbf{b}$ (see Fig. 41). Therefore

$$\frac{(dv/ds) - u\tau}{u} = \frac{(du/ds) + v\tau}{v}$$

or

$$\tau = \frac{uv' - vu'}{u^2 + v^2} = \frac{d}{ds}\left(\tan^{-1}\frac{v}{u}\right)$$

Therefore

$$\varphi = \int_0^s \tau\, ds = \tan^{-1}\frac{v}{u} + c$$

and $v = \rho \tan(\varphi - c)$ since $u = \rho$. Therefore

$$\mathbf{r}_1 = \mathbf{r} + \rho\mathbf{n} + \rho\tan(\varphi - c)\mathbf{b} \tag{106}$$

and again we have a one-parameter family of evolutes to the curve Γ.

Problems

1. Show that the unit binormal to the involute is

$$\mathbf{b}_1 = \frac{\kappa\mathbf{b} - \tau\mathbf{t}}{(c - s)\kappa\kappa_1}$$

2. Show that the torsion of an involute has the value

$$\tau_1 = \left(\frac{d\kappa}{ds}\tau - \kappa\frac{d\tau}{ds}\right)[\kappa(\kappa^2 + \tau^2)(c - s)]^{-1}$$

3. Show that the principal normal to the evolute is parallel to the tangent of the curve Γ.

4. Show that the ratio of the torsion of the evolute to its curvature is $\tan(\varphi - c)$.

5. Show that if the principal normals of a curve are binormals (equal vectors not necessarily coincident) of another curve, then $c(\kappa^2 + \tau^2) = \kappa$ where c is a constant.

6. On the binormal of a curve of constant torsion τ, a point Q is taken at a constant distance c from the curve. Show that the binormal to the locus of Q is inclined to the binormal of the given curve at an angle

$$\tan^{-1}\frac{c\tau^2}{\kappa(c^2\tau^2 + 1)^{\frac{1}{2}}}$$

7. Consider two curves which have the same principal normals (equal vectors not necessarily coincident). Show that the tangents to the two curves are inclined at a constant angle.

29. Spherical Indicatrices

(*a*) When dealing with a family of unit vectors, it is often convenient to give them a common origin and then to consider the locus of their end points. This locus obviously lies on a unit sphere. Let us now consider the spherical indicatrix of the tangent vectors to a curve $\mathbf{r} = \mathbf{r}(s)$. The unit tangent vectors are $\mathbf{t}(s) = \dfrac{d\mathbf{r}}{ds}\cdot$ Let $\mathbf{r}_1 = \mathbf{t}$. Then

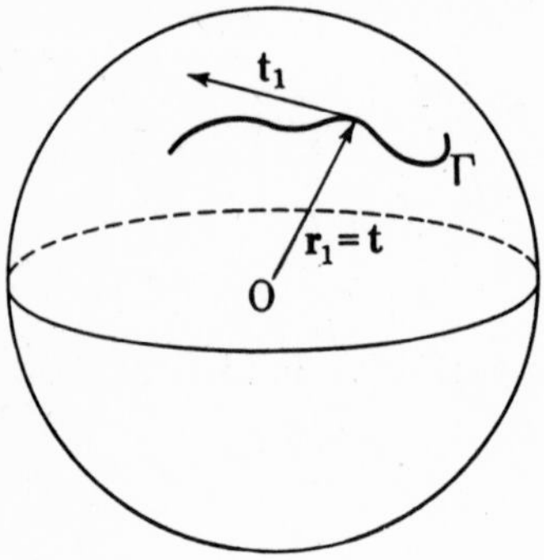

FIG. 42.

$$\mathbf{t}_1 = \frac{d\mathbf{r}_1}{ds_1} = \frac{d\mathbf{t}}{ds}\frac{ds}{ds_1} = \kappa\mathbf{n}\frac{ds}{ds_1}$$

Thus the tangent to the spherical indicatrix Γ is parallel to the normal of the curve at the corresponding point. Moreover, $1 = \kappa \dfrac{ds}{ds^1}$, $\mathbf{t}_1 = \mathbf{n}$. Let us now find the curvature κ_1 of the indicatrix. We obtain

$$\frac{d\mathbf{t}_1}{ds_1} = \kappa_1\mathbf{n}_1 = \frac{d\mathbf{n}}{ds}\frac{ds}{ds_1} = \frac{1}{\kappa}(-\kappa\mathbf{t} - \tau\mathbf{n})$$

and

$$\kappa_1{}^2 = \frac{\kappa^2 + \tau^2}{\kappa^2}$$

(*b*) The spherical indicatrix of the binormal, $\mathbf{r}_1 = \mathbf{b}$. Differentiating,

$$\mathbf{t}_1 = \frac{d\mathbf{r}_1}{ds_1} = \frac{d\mathbf{b}}{ds}\frac{ds}{ds_1} = \tau\mathbf{n}\frac{ds}{ds_1}$$

Therefore

$$\tau\frac{ds}{ds_1} = 1 \qquad \text{and} \qquad \mathbf{t}_1 = \mathbf{n}$$

Differentiating,

$$\frac{d\mathbf{t}_1}{ds_1} = \kappa_1\mathbf{n}_1 = \frac{d\mathbf{n}}{ds}\frac{ds}{ds_1} = \frac{1}{\tau}(-\kappa\mathbf{t} - \tau\mathbf{n})$$

and

$$\kappa_1{}^2 = \frac{\kappa^2 + \tau^2}{\tau^2}$$

Problems

1. Show that the torsion of the tangent indicatrix is

$$\tau_1 = \frac{\tau(d\kappa/ds) - \kappa(d\tau/ds)}{\kappa(\kappa^2 + \tau^2)}$$

2. Show that the torsion of the binormal indicatrix is

$$\tau_1 = \frac{\tau(d\kappa/ds) - \kappa(d\tau/ds)}{\tau(\kappa^2 + \tau^2)}$$

3. Find the curvature of the spherical indicatrix of the principal normal of a given curve.

30. Envelopes. Consider the one-parameter family of surfaces $F(x, y, z, c) = 0$. Two neighboring surfaces are

$$F(x, y, z, c) = 0$$

and $F(x, y, z, c + \Delta c) = 0$. These two surfaces will, in general, intersect in a curve. But these equations are equivalent to the equations $F(x, y, z, c) = 0$ and

$$\frac{F(x, y, z, c + \Delta c) - F(x, y, z, c)}{\Delta c} = 0$$

where $\Delta c \neq 0$. As $\Delta c \to 0$, the curve of intersection approaches a limiting position, called the characteristic curve, given by

$$\begin{aligned} F(x, y, z, c) &= 0 \\ \frac{\partial F(x, y, z, c)}{\partial c} &= 0 \end{aligned} \tag{107}$$

Each c determines a characteristic curve. The locus of all these curves [obtained by eliminating c from (107)] gives us a surface called the envelope of the one-parameter family. Now consider two neighboring characteristics

$$F(x, y, z, c) = 0 \qquad \frac{\partial F(x, y, z, c)}{\partial c} = 0$$

and (108)

$$F(x, y, z, c + \Delta c) = 0 \qquad \frac{\partial F(x, y, z, c + \Delta c)}{\partial c} = 0$$

which, in general, intersect at a point. The locus of these points is the envelope of the characteristics and is called the edge of regression. The edge of regression is given by the three simultaneous equations

$$\begin{aligned} F(x, y, z, c) &= 0 \\ \frac{\partial F(x, y, z, c)}{\partial c} &= 0 \\ \frac{\partial^2 F(x, y, z, c)}{\partial c^2} &= 0 \end{aligned} \tag{109}$$

Example 43. Let us consider the osculating plane at a point P. From (96) we have $[\mathbf{s} - \mathbf{r}(s)] \cdot \mathbf{b}(s) = 0$. If we let P vary, we obtain the one-parameter family of osculating planes given by

$$f(x, y, z, s) = [\mathbf{s} - \mathbf{r}(s)] \cdot \mathbf{b}(s) = 0$$

where s is the parameter and $\mathbf{s} = x\mathbf{i} + y\mathbf{j} + z\mathbf{k}$.

Now $\dfrac{\partial f}{\partial s} = -\dfrac{d\mathbf{r}}{ds} \cdot \mathbf{b} + (\mathbf{s} - \mathbf{r}) \cdot \dfrac{d\mathbf{b}}{ds} = (\mathbf{s} - \mathbf{r}) \cdot \tau\mathbf{n}$, and setting $\dfrac{\partial f}{\partial s} = 0$, we obtain $(\mathbf{s} - \mathbf{r}) \cdot \mathbf{n} = 0$. This locus is the rectifying plane. The intersection of $f = 0$ and $\dfrac{\partial f}{\partial s} = 0$ obviously yields the tangent lines which are the characteristics. Now

$$\frac{\partial^2 f}{\partial s^2} = -\mathbf{t} \cdot \mathbf{n} + (\mathbf{s} - \mathbf{r}) \cdot (-\kappa\mathbf{t} - \tau\mathbf{b}) + (\mathbf{s} - \mathbf{r}) \cdot \mathbf{n} \frac{d\tau}{ds}$$

It is easy to verify that $\mathbf{s} = \mathbf{r}$ satisfies $f = \dfrac{\partial f}{\partial s} = \dfrac{\partial^2 f}{\partial s^2} = 0$, so that the edge of regression is the original curve $\mathbf{r} = \mathbf{r}(s)$.

A developable surface, by definition, is the envelope of a one-parameter family of planes. The characteristics are straight lines, called generators. We have seen that the envelope of the osculating planes is the locus of the tangent line to the space curve Γ. In general, a developable surface is the tangent surface of a twisted curve. A contradiction to this is the case of a cylinder or cone.

31. Surfaces and Curvilinear Coordinates. Let us consider the equations

$$\begin{aligned} x &= x(u, v) \\ y &= y(u, v) \\ z &= z(u, v) \end{aligned} \tag{110}$$

where u and v are parameters ranging over a certain set of values. If we keep v fixed, the locus of (110) is a space curve. For each v, one such space curve exists, and if we let v vary, we shall obtain a locus of space curves which collectively form a surface. We shall consider those surfaces (110) for which x, y, z have continuous second-order derivatives. Equation (110) may be written

$$\mathbf{r}(u, v) = x(u, v)\mathbf{i} + y(u, v)\mathbf{j} + z(u, v)\mathbf{k} \tag{111}$$

where the end point of $\mathbf{r}$ generates the surface. The curves obtained by setting $v =$ constant are called the u curves, and similarly the v curves are obtained by setting $u =$ constant. The parameters u and v are called curvilinear coordinates, and the two curves are called the parametric curves.

32. Length of Arc on a Surface. If we move from the point $\mathbf{r}$ to the point $\mathbf{r} + d\mathbf{r}$ on the surface, the distance ds is given by

$$\begin{aligned} ds^2 = d\mathbf{r} \cdot d\mathbf{r} &= \left(\frac{\partial \mathbf{r}}{\partial u}\, du + \frac{\partial \mathbf{r}}{\partial v}\, dv\right)^2 \\ &= \left(\frac{\partial \mathbf{r}}{\partial u}\right)^2 du^2 + 2\,\frac{\partial \mathbf{r}}{\partial u} \cdot \frac{\partial \mathbf{r}}{\partial v}\, du\, dv + \left(\frac{\partial \mathbf{r}}{\partial v}\right)^2 dv^2 \end{aligned}$$

or

$$\underline{\overline{ds^2 = E\, du^2 + 2F\, du\, dv + G\, dv^2}} \tag{112}$$

where

$$E = \left(\frac{\partial \mathbf{r}}{\partial u}\right)^2, \qquad F = \frac{\partial \mathbf{r}}{\partial u} \cdot \frac{\partial \mathbf{r}}{\partial v}, \qquad G = \left(\frac{\partial \mathbf{r}}{\partial v}\right)^2$$

Equation (112) is called the first fundamental form for the surface $\mathbf{r} = \mathbf{r}(u, v)$. In particular, along the u curve, $dv = 0$, so that

$$(ds)_u = \sqrt{E}\, du$$

and similarly (113)

$$(ds)_v = \sqrt{G}\, dv$$

Now $\dfrac{\partial \mathbf{r}}{\partial u}$ and $\dfrac{\partial \mathbf{r}}{\partial v}$ are tangent vectors to the u and v curves, so that the parametric curves form an orthogonal system if and only if $\dfrac{\partial \mathbf{r}}{\partial u} \cdot \dfrac{\partial \mathbf{r}}{\partial v} = F = 0$.

Example 44. Consider the surface given by

$$\mathbf{r} = r \sin \theta \cos \varphi \, \mathbf{i} + r \sin \theta \sin \varphi \, \mathbf{j} + r \cos \theta \, \mathbf{k}, \quad r = \text{constant}$$

Differentiating,

$$\frac{\partial \mathbf{r}}{\partial \theta} = r \cos \theta \cos \varphi \, \mathbf{i} + r \cos \theta \sin \varphi \, \mathbf{j} - r \sin \theta \, \mathbf{k}$$

$$\frac{\partial \mathbf{r}}{\partial \varphi} = -r \sin \theta \sin \varphi \, \mathbf{i} + r \sin \theta \cos \varphi \, \mathbf{j}$$

and

$$E = \left(\frac{\partial \mathbf{r}}{\partial \theta}\right)^2 = r^2, \qquad F = \frac{\partial \mathbf{r}}{\partial \theta} \cdot \frac{\partial \mathbf{r}}{\partial \varphi} = 0, \qquad G = \left(\frac{\partial \mathbf{r}}{\partial \varphi}\right)^2 = r^2 \sin^2 \theta$$

so that $ds^2 = r^2 \, d\theta^2 + r^2 \sin^2 \theta \, d\varphi^2$ and the θ-curves are orthogonal to the φ-curves. Of course the surface is a sphere.

33. Surface Curves. By letting u and v be functions of a single variable t, we obtain

$$\mathbf{r} = \mathbf{r}[u(t), v(t)] \tag{114}$$

which represents a curve on the surface (111). Along this curve, $d\mathbf{r} = \left(\frac{\partial \mathbf{r}}{\partial u}\frac{du}{dt} + \frac{\partial \mathbf{r}}{\partial v}\frac{dv}{dt}\right) dt$. $d\mathbf{r}$ is completely determined when du and dv are specified, so that we will use the notation (du, dv) to specify a given direction on the surface. Now consider another curve such that $\delta \mathbf{r} = \frac{\partial \mathbf{r}}{\partial u} \delta u + \frac{\partial \mathbf{r}}{\partial v} \delta v$, where δu and δv are the differential changes of $u(t)$ and $v(t)$ for this new curve. Now

$$d\mathbf{r} \cdot \delta \mathbf{r} = E \, du \, \delta u + F(du \, \delta v + dv \, \delta u) + G \, dv \, \delta v \tag{115}$$

so that two curves are orthogonal if and only if

$$E \, du \, \delta u + F(du \, \delta v + dv \, \delta u) + G \, dv \, \delta v = 0$$

or

$$E + F\left(\frac{\delta v}{\delta u} + \frac{dv}{du}\right) + G \frac{dv}{du} \frac{\delta v}{\delta u} = 0 \tag{116}$$

If we have a system of curves on the surface given by the differential equation $P(u, v) \, \delta u + Q(u, v) \, \delta v = 0$, the differential equation for the orthogonal trajectories is given by

$$E + F\left(-\frac{P}{Q} + \frac{dv}{du}\right) - \frac{GP}{Q}\frac{dv}{du} = 0 \qquad (117)$$

since $\frac{\delta v}{\delta u} = -\frac{P}{Q}$.

Problems

1. Find the envelope and edge of regression of the one-parameter family of planes $x \sin c - y \cos c + z \tan \theta = c$, where c is the parameter and θ is a constant.

2. Show that any two v curves on the surface

$$\mathbf{r} = u \cos v\, \mathbf{i} + u \sin v\, \mathbf{j} + (v + \log \cos u)\mathbf{k}$$

cut equal segments from all the u curves.

3. Find the envelope and edge of regression of the family of ellipsoids $c^2\left(\frac{x^2}{a^2} + \frac{y^2}{b^2}\right) + \frac{z^2}{c^2} = 1$ where c is the parameter.

4. If θ is the angle between the two directions given by

$$P\,du^2 + Q\,du\,dv + R\,dv^2 = 0$$

show that $\tan \theta = H(Q^2 - 4PR)^{\frac{1}{2}}/(ER - FQ + GP)$, where $H = \left|\frac{\partial \mathbf{r}}{\partial u} \times \frac{\partial \mathbf{r}}{\partial v}\right|$.

5. Prove that the differential equations of the curves which bisect the angles between the parametric curves are

$$\sqrt{E}\,du - \sqrt{G}\,dv = 0$$

and $\sqrt{E}\,du + \sqrt{G}\,dv = 0$.

6. Given the curves $uv =$ constant on the surface $\mathbf{r} = u\mathbf{i} + v\mathbf{j}$, find the orthogonal trajectories.

7. Show that the area of a surface is given by

$$\iint (EG - F^2)^{\frac{1}{2}}\,du\,dv$$

34. Normal to a Surface. The vectors $\frac{\partial \mathbf{r}}{\partial u}$ and $\frac{\partial \mathbf{r}}{\partial v}$ are tangent to the surface $\mathbf{r}(u, v)$ along the u and v curves, respectively. Consequently, $\frac{\partial \mathbf{r}}{\partial u} \times \frac{\partial \mathbf{r}}{\partial v}$ is a vector normal to the surface. Note that $\frac{\partial \mathbf{r}}{\partial u}$ need not be a unit tangent vector to the u curve since the parameter u may not represent arc length. Since

$$(ds)_u = \sqrt{E}\, du$$

a necessary and sufficient condition for u to be arc length is that $E \equiv 1$. We define the unit normal to the surface as

$$\mathbf{n} = \frac{(\partial \mathbf{r}/\partial u) \times (\partial \mathbf{r}/\partial v)}{|(\partial \mathbf{r}/\partial u) \times (\partial \mathbf{r}/\partial v)|} \tag{118}$$

35. The Second Fundamental Form. Consider all the planes through a point P of the surface $\mathbf{r} = \mathbf{r}(u, v)$ which contain the normal $\mathbf{n}$. These planes intersect the surface in a family of curves, the normals to the curves being parallel to $\mathbf{n}$. We now compute the curvature of any one of these curves in the direction (du, dv). Let ds be length of arc along this curve. Now

$$\mathbf{t} = \frac{d\mathbf{r}}{ds} = \frac{\partial \mathbf{r}}{\partial u}\frac{du}{ds} + \frac{\partial \mathbf{r}}{\partial v}\frac{dv}{ds}$$

Therefore

$$\frac{d^2\mathbf{r}}{ds^2} = \frac{d\mathbf{t}}{ds} = \kappa_n \mathbf{n} = \frac{\partial^2 \mathbf{r}}{\partial u^2}\left(\frac{du}{ds}\right)^2 + 2\frac{\partial^2 \mathbf{r}}{\partial u\, \partial v}\frac{du}{ds}\frac{dv}{ds} + \frac{\partial^2 \mathbf{r}}{\partial v^2}\left(\frac{dv}{ds}\right)^2 + \frac{\partial \mathbf{r}}{\partial u}\frac{d^2u}{ds^2} + \frac{\partial \mathbf{r}}{\partial v}\frac{d^2v}{ds^2} \tag{119}$$

and

$$\kappa_n = \mathbf{n} \cdot (\kappa_n \mathbf{n}) = \left(\mathbf{n} \cdot \frac{\partial^2 \mathbf{r}}{\partial u^2}\right)\left(\frac{du}{ds}\right)^2 + 2\left(\mathbf{n} \cdot \frac{\partial^2 \mathbf{r}}{\partial u\, \partial v}\right)\frac{du}{ds}\frac{dv}{ds} + \left(\mathbf{n} \cdot \frac{\partial^2 \mathbf{r}}{\partial v^2}\right)\left(\frac{dv}{ds}\right)^2$$

since $\mathbf{n} \cdot \frac{\partial \mathbf{r}}{\partial u} = \mathbf{n} \cdot \frac{\partial \mathbf{r}}{\partial v} = 0$. Therefore

$$\kappa_n = \frac{e\, du^2 + 2f\, du\, dv + g\, dv^2}{ds^2}$$

$$\kappa_n = \frac{e\, du^2 + 2f\, du\, dv + g\, dv^2}{E\, du^2 + 2F\, du\, dv + G\, dv^2} \tag{120}$$

where we define

$$e = \mathbf{n} \cdot \frac{\partial^2 \mathbf{r}}{\partial u^2}, \qquad f = \mathbf{n} \cdot \frac{\partial^2 \mathbf{r}}{\partial u \, \partial v}, \qquad g = \mathbf{n} \cdot \frac{\partial^2 \mathbf{r}}{\partial v^2} \tag{121}$$

The quantity $e\,du^2 + 2f\,du\,dv + g\,dv^2$ is called the second fundamental form.

Now consider any curve Γ on the surface and let its normal be $\mathbf{n}_1$ at a point P, the direction of Γ being (du, dv) at P. Let Γ' be

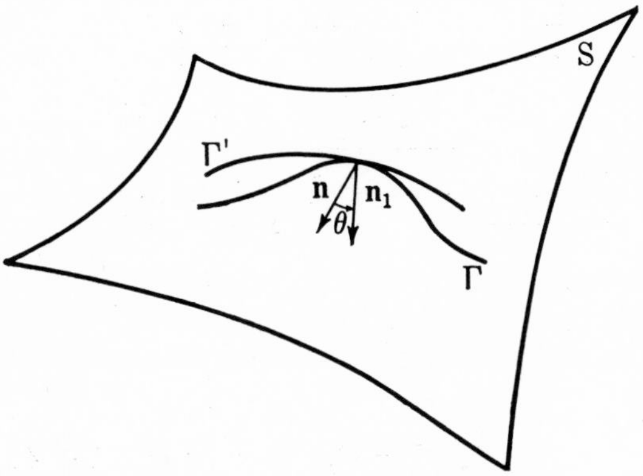

FIG. 43.

the normal curve in the same direction (du, dv) with normal $\mathbf{n}$ at P (Fig. 43). We have

$$\mathbf{n} \cdot \mathbf{n}_1 = \cos\theta = \mathbf{n} \cdot \frac{\mathbf{r}_1''}{\kappa} = \mathbf{n} \cdot \frac{\mathbf{r}''}{\kappa}$$

since $\mathbf{n} \cdot \mathbf{r}_1'' = \mathbf{r}'' \cdot \mathbf{n}$ for two curves with the same (du, dv) [see (119)]. Therefore

$$\cos\theta = \frac{\kappa_n}{\kappa}$$

so that

$$\boxed{\kappa = \kappa_n \sec\theta} \tag{122}$$

This is Meusnier's theorem.

36. Geometrical Significance of the Second Fundamental Form. We construct a tangent plane to the surface at the point $\mathbf{r}(u_0, v_0)$. What is the distance D of a neighboring point

$$\mathbf{r}(u_0 + \Delta u, v_0 + \Delta v)$$

on the surface, to the plane? It is $D = \Delta\mathbf{r} \cdot \mathbf{n}$. Now

$$\mathbf{r}(u_0 + \Delta u, v_0 + \Delta v) = \mathbf{r}(u_0, v_0) + \left(\frac{\partial \mathbf{r}}{\partial u}\,\Delta u + \frac{\partial \mathbf{r}}{\partial v}\,\Delta v\right) + \frac{1}{2!}\left(\frac{\partial^2 \mathbf{r}}{\partial u^2}\,\Delta u^2 + 2\,\frac{\partial^2 \mathbf{r}}{\partial u\,\partial v}\,\Delta u\,\Delta v + \frac{\partial^2 \mathbf{r}}{\partial v^2}\,\Delta v^2\right) + \cdots$$

from the calculus. Consequently

$$D = \Delta\mathbf{r} \cdot \mathbf{n} = \frac{1}{2!}\left(\mathbf{n} \cdot \frac{\partial^2 \mathbf{r}}{\partial u^2}\,\Delta u^2 + 2\mathbf{n} \cdot \frac{\partial^2 \mathbf{r}}{\partial u\,\partial v}\,\Delta u\,\Delta v + \mathbf{n} \cdot \frac{\partial^2 \mathbf{r}}{\partial v^2}\,\Delta v^2\right)$$

except for infinitesimals of higher order. Thus

$$2D = e\,du^2 + 2f\,du\,dv + g\,dv^2 \tag{123}$$

Problems

1. For the paraboloid of revolution

$$\mathbf{r} = u \cos v\,\mathbf{i} + u \sin v\,\mathbf{j} + u^2\mathbf{k}$$

show that $E = 1 + 4u^2$, $F = 0$, $G = u^2$, $e = 2(1 + 4u)^{-\frac{1}{2}}$, $f = 0$, $g = 2u^2(1 + 4u^2)^{-\frac{1}{2}}$, and find the normals to the surface and the normal curvature for the direction (du, dv).

2. What are the normal curvatures for directions along the parametric curves?

3. Find the second fundamental form for the sphere

$$\mathbf{r} = r \sin\theta \cos\varphi\,\mathbf{i} + r \sin\theta \sin\varphi\,\mathbf{j} + r \cos\theta\,\mathbf{k}$$

$r =$ constant.

4. Show that the curvature κ at any point P of the curve of intersection of two surfaces is given by

$$\kappa^2 \sin^2\theta = \kappa_1{}^2 + \kappa_2{}^2 - 2\kappa_1\kappa_2 \cos\theta$$

where κ_1, κ_2 are the normal curvatures of the surfaces in the direction of the curve at P, and θ is the angle between their normals.

5. Let us make a change of variable $u = u(\bar{u}, \bar{v})$, $v = v(\bar{u}, \bar{v})$. Show that E, F, G transform according to the law

$$\bar{E} = E\left(\frac{\partial u}{\partial \bar{u}}\right)^2 + 2F\,\frac{\partial u}{\partial \bar{u}}\,\frac{\partial v}{\partial \bar{u}} + G\left(\frac{\partial v}{\partial \bar{u}}\right)^2$$

$$\bar{F} = E\frac{\partial u}{\partial \bar{u}}\frac{\partial u}{\partial \bar{v}} + F\left(\frac{\partial u}{\partial \bar{u}}\frac{\partial v}{\partial \bar{v}} + \frac{\partial v}{\partial \bar{u}}\frac{\partial u}{\partial \bar{v}}\right) + G\frac{\partial v}{\partial \bar{u}}\frac{\partial v}{\partial \bar{v}}$$

$$\bar{G} = E\left(\frac{\partial u}{\partial \bar{v}}\right)^2 + 2F\frac{\partial u}{\partial \bar{v}}\frac{\partial v}{\partial \bar{v}} + G\left(\frac{\partial v}{\partial \bar{v}}\right)^2$$

and that $E\,du^2 + 2F\,du\,dv + G\,dv^2 = \bar{E}\,d\bar{u}^2 + 2\bar{F}\,d\bar{u}\,d\bar{v} + \bar{G}\,d\bar{v}^2$. Also show that

$$\bar{e} = \pm\left[e\left(\frac{\partial u}{\partial \bar{u}}\right)^2 + 2f\frac{\partial u}{\partial \bar{u}}\frac{\partial v}{\partial \bar{u}} + g\left(\frac{\partial v}{\partial \bar{u}}\right)^2\right]$$

$$\bar{f} = \pm\left[e\frac{\partial u}{\partial \bar{u}}\frac{\partial u}{\partial \bar{v}} + f\left(\frac{\partial u}{\partial \bar{u}}\frac{\partial v}{\partial \bar{v}} + \frac{\partial v}{\partial \bar{u}}\frac{\partial u}{\partial \bar{v}}\right) + g\frac{\partial v}{\partial \bar{u}}\frac{\partial v}{\partial \bar{v}}\right]$$

$$\bar{g} = \pm\left[e\left(\frac{\partial u}{\partial \bar{v}}\right)^2 + 2f\frac{\partial u}{\partial \bar{v}}\frac{\partial v}{\partial \bar{v}} + g\left(\frac{\partial v}{\partial \bar{v}}\right)^2\right]$$

37. Principal Directions. From (120) we have

$$(\kappa_n E - e)\,du^2 + 2(\kappa_n F - f)\,du\,dv + (\kappa_n G - g)\,dv^2 = 0 \quad (124)$$

or

$$A\,du^2 + 2B\,du\,dv + C\,dv^2 = 0$$

This quadratic equation has two directions (du, dv), $(\delta u, \delta v)$, which give the same value for κ_n. These two directions will coincide if the quadratic equation (124) has a double root. This is true if and only if

$$B^2 - AC = (\kappa_n F - f)^2 - (\kappa_n E - e)(\kappa_n G - g) = 0$$

or

$$\kappa_n^2(F^2 - EG) + \kappa_n(eG + gE - 2fF) + (f^2 - eg) = 0 \quad (125)$$

Moreover, we have $\dfrac{du}{dv} = -\dfrac{B}{A}$ and $\dfrac{dv}{du} = -\dfrac{B}{C}$ if $B^2 - AC = 0$, so that

$$\begin{aligned}(\kappa_n E - e)\,du + (\kappa_n F - f)\,dv &= 0\\ (\kappa_n F - f)\,du + (\kappa_n G - g)\,dv &= 0\end{aligned} \quad (126)$$

The solutions of (124) give the two directions for a given κ_n. When κ_n is eliminated between (124) and (125), the two directions coincide and satisfy

$$(Ef - Fe)\,du^2 + (Eg - Ge)\,du\,dv + (Fg - Gf)\,dv^2 = 0 \quad (127)$$

The two directions, solutions of (127), are called principal directions and are the only ones with a unique normal curvature, that is, no other direction can have the same curvature. The normal curvatures in these two directions are called the principal curvatures at the point. The average of the two principal curvatures is

$$\boxed{H = \frac{eG + gE - 2fF}{2(EG - F^2)}} \tag{128}$$

which is obtained by taking one-half of the sum of the roots of (125). The Gaussian curvature K is defined as the product of the curvatures, that is,

$$\boxed{K = \frac{f^2 - eg}{F^2 - EG}} \tag{129}$$

A line of curvature is a curve whose tangent at any point has a direction coinciding with a principal direction at that point. The lines of curvature are obtained by solving the differential equation (126). The curvature of a line of curvature is not a principal curvature since the line of curvature need not be a normal curve.

Example 45. Let us consider the right helicoid

$$\mathbf{r} = u \cos \varphi \, \mathbf{i} + u \sin \varphi \, \mathbf{j} + c\varphi\mathbf{k}$$

We have

$$\frac{\partial \mathbf{r}}{\partial u} = \cos \varphi \, \mathbf{i} + \sin \varphi \, \mathbf{j}, \qquad \frac{\partial \mathbf{r}}{\partial \varphi} = -u \sin \varphi \, \mathbf{i} + u \cos \varphi \, \mathbf{j} + c\mathbf{k}$$

$$\frac{\partial^2 \mathbf{r}}{\partial u^2} = 0, \qquad \frac{\partial^2 \mathbf{r}}{\partial u \, \partial \varphi} = -\sin \varphi \, \mathbf{i} + \cos \varphi \, \mathbf{j},$$

$$\frac{\partial^2 \mathbf{r}}{\partial \varphi^2} = -u \cos \varphi \, \mathbf{i} - u \sin \varphi \, \mathbf{j}$$

Hence

$$E = \left(\frac{\partial \mathbf{r}}{\partial u}\right)^2 = 1, \qquad F = \frac{\partial \mathbf{r}}{\partial u} \cdot \frac{\partial \mathbf{r}}{\partial \varphi} = 0, \qquad G = \left(\frac{\partial \mathbf{r}}{\partial \varphi}\right)^2 = u^2 + c^2$$

Also

$$\mathbf{n} = \frac{(\partial \mathbf{r}/\partial u) \times (\partial \mathbf{r}/\partial \varphi)}{|(\partial \mathbf{r}/\partial u) \times (\partial \mathbf{r}/\partial \varphi)|} = (c \sin \varphi\, \mathbf{i} - c \cos \varphi\, \mathbf{j} + u\mathbf{k})(c^2 + u^2)^{-\frac{1}{2}}$$

$$e = \mathbf{n} \cdot \frac{\partial^2 \mathbf{r}}{\partial u^2} = 0, \qquad f = \mathbf{n} \cdot \frac{\partial^2 \mathbf{r}}{\partial u\, \partial \varphi} = -c(c^2 + u^2)^{-\frac{1}{2}}$$

$$g = \mathbf{n} \cdot \frac{\partial^2 \mathbf{r}}{\partial \varphi^2} = 0$$

Equation (125) yields

$$-(u^2 + c^2)\kappa_n + c^2(c^2 + u^2)^{-1} = 0$$

whence

$$\kappa_n = \pm \frac{c}{u^2 + c^2}$$

The average curvature is $H = \frac{1}{2}\left(\frac{c}{u^2 + c^2} - \frac{c}{u^2 + c^2}\right) = 0$, and the Gaussian curvature is $K = \frac{-c^2}{(u^2 + c^2)^2}$. The differential equation (126) for the lines of curvature becomes

$$-c(c^2 + u^2)^{-\frac{1}{2}}\, du^2 + c(c^2 + u^2)^{\frac{1}{2}}\, d\varphi^2 = 0$$

so that

$$d\varphi = \pm \frac{du}{(c^2 + u^2)^{\frac{1}{2}}} \qquad \text{and} \qquad \varphi = \pm \log\left(u + \sqrt{u^2 + c^2}\right) + \alpha$$

and the lines of curvature are given by

$$\mathbf{r} = u \cos\left[\pm \log\left(u + \sqrt{u^2 + c^2}\right) + \alpha\right]\mathbf{i} + u \sin \varphi\, \mathbf{j} + c\varphi\mathbf{k}.$$

Referring to (126) for the two principal directions, we have

$$\begin{aligned} \frac{dv}{du} + \frac{\delta v}{\delta u} &= -\frac{Eg - Ge}{Fg - Gf} \\ \frac{dv}{du}\frac{\delta v}{\delta u} &= \frac{Ef - Fe}{Fg - Gf} \end{aligned} \tag{130}$$

Substituting (130) into (116) we obtain

$$E - F\left(\frac{Eg - Ge}{Fg - Gf}\right) + G\left(\frac{Ef - Fe}{Fg - Gf}\right) \equiv 0$$

so that the principal directions are orthogonal.

Now let us choose the principal curves as the parametric lines. Thus u = constant, v = constant are to represent the principal curves. These two curves satisfy the equation $du\,dv = 0$, so that from (127) we must have

$$\begin{aligned} Ef - Fe &= 0 \\ Fg - Gf &= 0 \\ Eg - Ge &\neq 0 \end{aligned}$$

From these equations we conclude that

$$f(Eg - Ge) = gfE - feG = Feg - eFg = 0$$

and $F(Eg - Ge) = 0$, so that $f = F = 0$. We have shown that a necessary and sufficient condition that the lines of curvature be parametric curves is that

$$f = F = 0 \tag{131}$$

Problems

1. Find the lines of curvature on the surface

$$x = a(u + v), \qquad y = b(u - v), \qquad z = uv$$

2. Show that the principal radii of curvature of the right conoid $x = u \cos v$, $y = u \sin v$, $z = f(v)$ are given by the roots of

$$f'^2\kappa^2 - uf''(u^2 + f'^2)^{\frac{1}{2}}\kappa - (u^2 + f'^2)^2 = 0$$

3. The surface generated by the binormals of the curve $\mathbf{r} = \mathbf{r}(s)$ is given by $\mathbf{R} = \mathbf{r} + u\mathbf{b}$. Show that the Gauss curvature is $K = -\tau^2/(1 + \tau^2u^2)^2$. Also show that the differential equation of the lines of curvature is

$$-\tau^2\,du^2 - \left(\kappa + \kappa\tau^2u^2 + \frac{d\tau}{ds}u\right)du\,ds + (1 + \tau^2u^2)\tau\,ds^2 = 0$$

38. Conjugate Directions. Let P and Q be neighboring points on a surface. The tangent planes at P and Q will intersect in a straight line l. Now let Q approach P along some fixed direction. The line l will approach a limiting position l'. The directions PQ and l' are called conjugate directions.

We now compute the analytical expression for two directions to be conjugate. Let $\mathbf{n}$ be the normal at P and $\mathbf{n} + d\mathbf{n}$ the normal at Q, where $d\mathbf{r} = \overrightarrow{PQ} = \frac{\partial \mathbf{r}}{\partial u} du + \frac{\partial \mathbf{r}}{\partial v} dv$. Let the direction of l' be given by $\delta\mathbf{r} = \frac{\partial \mathbf{r}}{\partial u} \delta u + \frac{\partial \mathbf{r}}{\partial v} \delta v$. Since $\delta\mathbf{r}$ lies in both planes, we must have $\delta\mathbf{r} \cdot \mathbf{n} = 0$ and $\delta\mathbf{r} \cdot (\mathbf{n} + d\mathbf{n}) = 0$. These two equations imply $\delta\mathbf{r} \cdot d\mathbf{n} = 0$, or

$$\left(\frac{\partial \mathbf{r}}{\partial u} \delta u + \frac{\partial \mathbf{r}}{\partial v} \delta v\right) \cdot \left(\frac{\partial \mathbf{n}}{\partial u} du + \frac{\partial \mathbf{n}}{\partial v} dv\right) = 0 \tag{132}$$

Expanding, we obtain

$$\left(\frac{\partial \mathbf{r}}{\partial u} \cdot \frac{\partial \mathbf{n}}{\partial v}\right) du\, \delta u + \left[\left(\frac{\partial \mathbf{r}}{\partial v} \cdot \frac{\partial \mathbf{n}}{\partial u}\right) \delta v\, du + \left(\frac{\partial \mathbf{r}}{\partial u} \cdot \frac{\partial \mathbf{n}}{\partial v}\right) \delta u\, dv\right] + \left(\frac{\partial \mathbf{r}}{\partial v} \cdot \frac{\partial \mathbf{n}}{\partial v}\right) \delta v\, dv = 0 \tag{133}$$

Now $\mathbf{n} \cdot \frac{\partial \mathbf{r}}{\partial u} = 0$, so that by differentiating we see that

$$\frac{\partial \mathbf{n}}{\partial u} \cdot \frac{\partial \mathbf{r}}{\partial u} + \mathbf{n} \cdot \frac{\partial^2 \mathbf{r}}{\partial u^2} = 0$$

which implies

$$\frac{\partial \mathbf{n}}{\partial u} \cdot \frac{\partial \mathbf{r}}{\partial u} = -\mathbf{n} \cdot \frac{\partial^2 \mathbf{r}}{\partial u^2} = -e$$

Similarly

$$\frac{\partial \mathbf{n}}{\partial v} \cdot \frac{\partial \mathbf{r}}{\partial u} = \frac{\partial \mathbf{n}}{\partial u} \cdot \frac{\partial \mathbf{r}}{\partial v} = -f$$

$$\frac{\partial \mathbf{n}}{\partial v} \cdot \frac{\partial \mathbf{r}}{\partial v} = -g$$

so that (133) becomes

$$e\, du\, \delta u + f(du\, \delta v + dv\, \delta u) + g\, \delta v\, dv = 0 \tag{134}$$

If the direction (du, dv) is given, there is only one corresponding conjugate direction $(\delta u, \delta v)$, obtained by solving (134).

Now consider the lines of curvature taken as parametric curves. Their directions are $(du, 0)$, $(0, \delta v)$. Equation (134) is satisfied by these directions since $f = 0$ for lines of curvature, so that the lines of curvature are conjugate directions.

39. Asymptotic Lines. The directions which are self-conjugate are called asymptotic directions. Those curves whose tangents are asymptotic directions are called asymptotic lines. If a direction is self-conjugate, $\dfrac{dv}{du} = \dfrac{\delta v}{\delta u}$, so that (134) becomes

$$e\,du^2 + 2f\,du\,dv + g\,dv^2 = 0 \tag{135}$$

We see that the asymptotic directions are those for which the second fundamental form vanishes. Moreover, the normal curvature κ_n vanishes for this direction.

If $e = g = 0$, $f \neq 0$, the solution of (135) is $u =$ constant, $v =$ constant, so that the parametric curves are asymptotic lines if and only if $e = g = 0$, $f \neq 0$.

Example 46. Let us find the lines of curvature and asymptotic lines of the surface of revolution $z = x^2 + y^2$. Let $x = u \cos v$, $y = u \sin v$, $z = u^2$, and

$$\mathbf{r} = u \cos v\,\mathbf{i} + u \sin v\,\mathbf{j} + u^2\mathbf{k}$$

We obtain

$$\frac{\partial \mathbf{r}}{\partial u} = \cos v\,\mathbf{i} + \sin v\,\mathbf{j} + 2u\mathbf{k}, \qquad \frac{\partial \mathbf{r}}{\partial v} = -u \sin v\,\mathbf{i} + u \cos v\,\mathbf{j}$$

$$\frac{\partial^2 \mathbf{r}}{\partial u^2} = 2\mathbf{k}, \qquad \frac{\partial^2 \mathbf{r}}{\partial u\,\partial v} = -\sin v\,\mathbf{i} + \cos v\,\mathbf{j}$$

$$\frac{\partial^2 \mathbf{r}}{\partial v^2} = -u \cos v\,\mathbf{i} - u \sin v\,\mathbf{j}$$

$$\mathbf{n} = \frac{(\partial \mathbf{r}/\partial u) \times (\partial \mathbf{r}/\partial v)}{|(\partial \mathbf{r}/\partial u) \times (\partial \mathbf{r}/\partial v)|}$$
$$= (-2u^2 \cos v\,\mathbf{i} - 2u^2 \sin v\,\mathbf{j} + u\mathbf{k})u^{-1}(1 + 4u^2)^{-\frac{1}{2}}$$

Therefore

$$e = \mathbf{n} \cdot \frac{\partial^2 \mathbf{r}}{\partial u^2} = 2(1 + 4u)^{-\frac{1}{2}}, \qquad f = \mathbf{n} \cdot \frac{\partial^2 \mathbf{r}}{\partial u\,\partial v} = 0$$

$$g = \mathbf{n} \cdot \frac{\partial^2 \mathbf{r}}{\partial v^2} = 2u^2(1 + 4u^2)^{-\frac{1}{2}}$$

Also $F = \frac{\partial \mathbf{r}}{\partial u} \cdot \frac{\partial \mathbf{r}}{\partial v} = 0$, so that $f = F = 0$, and from (131) the parametric curves are the lines of curvature. The asymptotic lines are given by $du^2 + u^2\,dv^2 = 0$. These are imaginary, so that the surface possesses no asymptotic lines.

Problems

1. Show that the asymptotic lines of the hyperboloid

$$\mathbf{r} = a \cos \theta \sec \psi\, \mathbf{i} + b \sin \theta \sec \psi\, \mathbf{j} + c \tan \psi\, \mathbf{k}$$

are given by $\theta \pm \psi =$ constant.

2. The parametric equations of the helicoid are

$$x = u \cos v, \qquad y = u \sin v, \qquad z = cv$$

Show that the asymptotic lines are the parametric curves, and that the lines of curvature are $u + \sqrt{u^2 + c^2} = Ae^{\pm v}$. Show that the principal radii of curvature are $\pm(u^2 + c^2)c^{-1}$.

3. Prove that, at any point of a surface, the sum of the normal curvature in conjugate directions is constant.

4. Find the asymptotic lines on the surface $z = y \sin x$.

40. Geodesics. The distance between two points on a surface (we are allowed to move only on the surface) is given by

$$s = \int_{t_0}^{t_1} \left[E \left(\frac{du}{dt}\right)^2 + 2F \frac{du}{dt}\frac{dv}{dt} + G \left(\frac{dv}{dt}\right)^2 \right]^{\frac{1}{2}} dt \qquad (136)$$

Among the many curves on the surface that join the two fixed points will be those that make (136) an extremal. Such curves are called geodesics. We wish now to determine the geodesics. To do this, we require the use of the calculus of variations, and so we say a few words about this important method.

Let us first consider the integral

$$\int_{P(x_2, y_2)}^{Q(x_1, y_1)} (1 + y'^2)^{\frac{1}{2}}\, dx \qquad (137)$$

We might ask what must be the function $y = y(x)$ joining the two points P and Q which will make (137) a minimum. The reader might be tempted to say, $y' = 0$ or $y =$ constant, since the integrand is then a minimum. But we find that $y =$ constant will not, in general, pass through the two fixed points. Hence the solution to this problem is not trivial. We now formulate a more general problem: to find $y = y(x)$ such that

$$\int_a^b f(x, y, y')\, dx \tag{138}$$

is an extremal. The function $f(x, y, y')$ is given. It is $y(x)$ and so also $y'(x)$ that are unknown. Let $y = y(x)$ be that function which makes (138) an extremal. Now let $\bar{Y}(x, \alpha) = y(x) + \alpha\varphi(x)$, where α is arbitrary and independent of x and $\varphi(x)$ is any function with continuous first derivative having the property that $\varphi(a) = \varphi(b) = 0$ (see Fig. 44). Under our assumption,

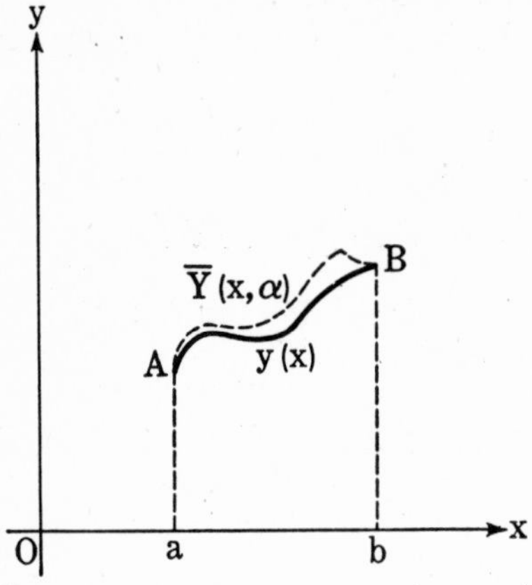

FIG. 44.

$$J(\alpha) = \int_a^b f(x, \bar{Y}, \bar{Y}')\, dx \tag{139}$$

is an extremal for $\alpha = 0$. Consequently $\left.\dfrac{dJ}{d\alpha}\right|_{\alpha=0} = 0$ or

$$\left.\frac{dJ}{d\alpha}\right|_{\alpha=0} = \int_a^b \left(\frac{\partial f}{\partial y}\varphi + \frac{\partial f}{\partial y'}\varphi'\right) dx = 0 \tag{140}$$

since

$$\frac{\partial f}{\partial \alpha} = \frac{\partial f}{\partial \bar{Y}}\frac{\partial \bar{Y}}{\partial \alpha} + \frac{\partial f}{\partial \bar{Y}'}\frac{\partial \bar{Y}'}{\partial \alpha} = \frac{\partial f}{\partial \bar{Y}}\varphi + \frac{\partial f}{\partial \bar{Y}'}\varphi'$$

and for $\alpha = 0$,

$$\frac{\partial f}{\partial \bar{Y}} = \frac{\partial f}{\partial y}, \qquad \frac{\partial f}{\partial \bar{Y}'} = \frac{\partial f}{\partial y'}$$

We now integrate the right-hand term of (140) by parts and obtain

$$\int_a^b \frac{\partial f}{\partial y}\varphi\, dx + \left[\frac{\partial f}{\partial y'}\varphi\right]_a^b - \int_a^b \frac{d}{dx}\left(\frac{\partial f}{\partial y'}\right)\varphi\, dx = 0 \tag{141}$$

Now $\varphi(a) = \varphi(b) = 0$ by construction of $\varphi(x)$, so that

$$\int_a^b \left[\frac{\partial f}{\partial y} - \frac{d}{dx}\left(\frac{\partial f}{\partial y'}\right)\right] \varphi \, dx = 0 \tag{142}$$

Now let us assume that $\frac{\partial f}{\partial y} - \frac{d}{dx}\left(\frac{\partial f}{\partial y'}\right)$ is continuous. If $\frac{\partial f}{\partial y} - \frac{d}{dx}\left(\frac{\partial f}{\partial y'}\right)$ is not identically zero on the interval (a, b), it will be positive or negative at some point. If it is positive at $x = c$, it will be positive in a neighborhood of $x = c$ from continuity (see Secs. 42 and 43). We can construct φ to be positive on this interval and zero elsewhere. Then

$$\int_a^b \left[\frac{\partial f}{\partial y} - \frac{d}{dx}\left(\frac{\partial f}{\partial y'}\right)\right] \varphi \, dx > 0$$

so that we have a contradiction to (142). Consequently, the function of $y(x)$ must satisfy the Euler-Lagrange differential equation

$$\boxed{\frac{d}{dx}\left(\frac{\partial f}{\partial y'}\right) - \frac{\partial f}{\partial y} = 0} \tag{143}$$

If $f = f(y, y')$, we can immediately arrive at an integral of (143). Let us consider

$$\begin{aligned}\frac{d}{dx}\left(f - y'\frac{\partial f}{\partial y'}\right) &= \frac{\partial f}{\partial y}y' + \frac{\partial f}{\partial y'}y'' - y''\frac{\partial f}{\partial y'} - y'\frac{d}{dx}\left(\frac{\partial f}{\partial y'}\right) \\ &= y'\left[\frac{\partial f}{\partial y} - \frac{d}{dx}\left(\frac{\partial f}{\partial y'}\right)\right] = 0 \qquad \text{from (143)}\end{aligned}$$

Hence

$$\boxed{f - y'\frac{\partial f}{\partial y'} = \text{constant}} \tag{144}$$

is an integral of (143) if $f = f(y, y')$.

Example 47. To extremalize (137), we have $f = (1 + y'^2)^{\frac{1}{2}}$, which is independent of x. From (144),

$$f - y'\frac{\partial f}{\partial y'} = \text{constant} = \frac{1}{\alpha}$$

so that

$$(1 + y'^2)^{\frac{1}{2}} - y' \frac{y'}{(1 + y'^2)^{\frac{1}{2}}} = \frac{1}{\alpha}$$

and

$$y' = \pm \sqrt{\alpha^2 - 1}$$

and finally

$$y = \pm \sqrt{\alpha^2 - 1}\, x + \beta \tag{145}$$

The constants α and β are determined by noting that the straight line (145) passes through the two given fixed points.

Example 48. If $f = f\left(x^1, x^2, \ldots, x^n, \frac{dx^1}{dt}, \frac{dx^2}{dt}, \ldots, \frac{dx^n}{dt}, t\right)$, then $\int_{t_0}^{t_1} f\,dt$ is an extremal when the $x^\alpha(t)$ satisfy

$$\boxed{\frac{d}{dt}\left(\frac{\partial f}{\partial \dot{x}^\alpha}\right) - \frac{\partial f}{\partial x^\alpha} = 0} \tag{146}$$

for $\alpha = 1, 2, \ldots, n$ with $\dot{x}^\alpha = \frac{dx^\alpha}{dt}$. The superscripts are not powers but labels that enable us to distinguish between the various variables. The formulas (146) are a consequence of the fact that $\int_{t_0}^{t_1} f\,dt$ must be an extremal when $x^i(t)$ is allowed to vary while we keep all other x^j fixed, $j = 1, 2, \ldots, i - 1, i + 1, \ldots, n$.

Let us now try to find the differential equations that $u(t)$ and $v(t)$ must satisfy to make (136) an extremal. We write

$$s = \int_{t_0}^{t_1} (E\dot{u}^2 + 2F\dot{u}\dot{v} + G\dot{v}^2)^{\frac{1}{2}}\, dt$$

and apply (146), where $x^1 = u$ and $x^2 = v$. We thus obtain

$$\frac{d}{dt}\left(\frac{\partial f}{\partial \dot{u}}\right) - \frac{\partial f}{\partial u} = 0 \tag{147}$$

$$\frac{d}{dt}\left(\frac{\partial f}{\partial \dot{v}}\right) - \frac{\partial f}{\partial v} = 0 \tag{148}$$

where

$$f = (E\dot{u}^2 + 2F\dot{u}\dot{v} + G\dot{v}^2)^{\frac{1}{2}} = \frac{ds}{dt}, \qquad E = E(u, v), \text{ etc.}$$

Now

$$\frac{\partial f}{\partial \dot{u}} = \frac{E\dot{u} + F\dot{v}}{f}, \qquad \frac{\partial f}{\partial u} = \frac{\dot{u}^2 \dfrac{\partial E}{\partial u} + 2\dot{u}\dot{v} \dfrac{\partial F}{\partial u} + \dot{v}^2 \dfrac{\partial G}{\partial u}}{2f}$$

so that (147) becomes

$$\frac{d}{dt}\left(\frac{E\dot{u} + F\dot{v}}{ds/dt}\right) = \frac{\dot{u}^2 + 2\dot{u}\dot{v}(\partial F/\partial u) + \dot{v}^2(\partial G/\partial u)}{2\, ds/dt} \tag{149}$$

and if we choose for the parameter t the arc length s, then $t = s$ and $\dfrac{ds}{dt} = 1$, so that (149) reduces to

$$\left.\begin{aligned} &\frac{d}{ds}(E\dot{u} + F\dot{v}) = \frac{1}{2}\left(\dot{u}^2 \frac{\partial E}{\partial u} + 2\dot{u}\dot{v} \frac{\partial F}{\partial u} + \dot{v}^2 \frac{\partial G}{\partial u}\right) \\ &\text{while similarly (148) yields} \\ &\frac{d}{ds}(F\dot{u} + G\dot{v}) = \frac{1}{2}\left(\dot{u}^2 \frac{\partial E}{\partial v} + 2\dot{u}\dot{v} \frac{\partial F}{\partial v} + \dot{v}^2 \frac{\partial G}{\partial v}\right) \end{aligned}\right\} \tag{150}$$

In Chap. 8 we shall derive by tensor methods a slightly different system of differential equations.

Example 49. Consider the sphere given by

$$\mathbf{r} = a \sin\theta \cos\varphi\, \mathbf{i} + a \sin\theta \sin\varphi\, \mathbf{j} + a \cos\theta\, \mathbf{k}$$

where $ds^2 = a^2\, d\theta^2 + a^2 \sin^2\theta\, d\varphi^2$ so that $E = a$, $F = 0$,

$$G = a^2 \sin^2\theta,$$

and

$$\frac{\partial E}{\partial \theta} = \frac{\partial E}{\partial \varphi} = \frac{\partial F}{\partial \theta} = \frac{\partial F}{\partial \varphi} = \frac{\partial G}{\partial \varphi} = 0, \qquad \frac{\partial G}{\partial \theta} = a^2 \sin 2\theta$$

Hence (150) reduces to

$$\begin{aligned} \frac{d}{ds}\left(a \frac{d\theta}{ds}\right) &= \frac{1}{2}\left(\frac{d\varphi}{ds}\right)^2 \sin 2\theta \\ \frac{d}{ds}\left(a \sin^2\theta \frac{d\varphi}{ds}\right) &= 0 \end{aligned} \tag{151}$$

Integrating (151) we have $\sin^2\theta \dfrac{d\varphi}{ds} = \text{constant}$. We can choose our coordinate system so that the coordinates of the fixed points are a, 0, 0 and a, θ, 0. Hence $\sin^2 \theta \frac{d\varphi}{ds} = 0$, and $\frac{d\varphi}{ds} = 0$, so that $\varphi \equiv 0$. Hence the geodesic is the arc of the great circle joining the two fixed points.

Example 50. Let us find $y(x)$ which extremalizes

$$\int y(1 + y'^2)^{\frac{1}{2}}\, dx$$

Since $f = y(1 + y'^2)^{\frac{1}{2}} = f(y, y')$, we can apply (144) to obtain a first integral. We obtain $y(1 + y'^2)^{\frac{1}{2}} - y'^2 y(1 + y'^2)^{-\frac{1}{2}} = \alpha^{-1}$, and simplifying this expression yields $y' = \pm(\alpha^2 y^2 - 1)^{\frac{1}{2}}$. A further integration yields $\alpha y = \cosh(\beta \pm \alpha x)$. These are the curves (catenoids) which have minimum surfaces of revolution.

Problems

1. Find the geodesics on the ellipsoid of revolution

$$\frac{x^2 + z^2}{a^2} + \frac{y^2}{b^2} = 1$$

Hint: Let $x = u \cos v$, $z = u \sin v$.

2. Show that the differential equation of the geodesics for the right helicoid $x = u \cos v$, $y = u \sin v$, $z = cv$ is

$$\frac{du}{dv} = \pm \frac{1}{h} [(u^2 + c^2)(u^2 + c^2 - h^2)^{\frac{1}{2}}], \quad h = \text{constant}$$

3. Prove that the geodesics on a right circular cylinder are helices.

4. Show that the perpendicular from the vertex of a right circular cone to the tangents of a given geodesic is of constant length.

5. Find $y(x)$ which extremalizes $\int_a^b [(1 + y'^2)/y]^{\frac{1}{2}}\, dx$.

CHAPTER 4

INTEGRATION

41. Point-set Theory. In geometry and analysis the student has frequently made use of the concept of a point and of the notion of a set or collection of elements (objects, points, numbers, etc.). We shall not define these concepts, but shall take their notion as intuitive. We may be interested in the points subject to the condition $x^2 + y^2 < r^2$. These will be the points interior to the circle of radius r with center at the origin. We might also be interested in the rational points of the one-dimensional continuum. For convenience, we shall consider only points of the real-number axis in what follows. Any set of real numbers will be called a linear set. The integers form such a set, as do the rationals and irrationals. All the definitions and theorems proved for linear sets can easily be extended to any finite-dimensional space.

Closed Interval. The set of points $\{x\}$ satisfying $a \leqq x \leqq b$ will be called a closed interval. If we omit the end points, that is, consider those x that satisfy $a < x < b$, we say that the interval is *open* (open at both ends). For example, $0 \leqq x \leqq 1$ is a closed interval, while $0 < x < 1$ is an open interval.

Bounded Set. A linear set of points will be said to be bounded if there exists an open interval containing the set. It must be emphasized that the ends of the interval are to be finite numbers, which thus excludes $-\infty, +\infty$.

An alternative definition would be the following: A set of numbers S is bounded if there exists a finite number N such that $-N < x < N$ for all x in S.

The set of numbers whose squares are less than 3 is certainly bounded, for if $x^2 < 3$ then obviously $-2 < x < 2$. However, the set of numbers whose cubes are less than 3 is unbounded, for $x^3 < 3$ is at least satisfied by all the negative numbers. This set is, however, bounded above. By this we mean that there exists a finite number N such that $x < N$ if $x^3 < 3$. Certainly $N = 2$

does the trick. Specifically, a set of elements S is bounded above if a finite number N exists such that $x < N$ for all x in S. Let the student frame a definition for sets bounded below.

We shall, in the main, be concerned with sets that contain an infinite number of distinct points. The rational numbers in the interval $0 < x < 1$ form such a collection. Let the reader prove that between any two rationals there exists another rational.

Limit Point. A point P will be called a limit point of a set S if every open interval containing P contains an infinite number of distinct elements of S. For example, let S be the set of numbers $(1/2, 1/3, 1/4, \ldots, 1/n, \ldots)$. It is easy to verify that any open interval containing the origin, 0, contains an infinite number of S. In this case the limit point 0 does not belong to S. It is at once apparent that a set S containing only a finite number of points cannot have a limit point.

Neighborhood. A neighborhood of a point is any open interval containing that point.

Interior Point. A point P is said to be an interior point of a set S if it belongs to S and if a neighborhood N of P exists, every element of N belonging to S. If S is the set of points $0 \leqq x \leqq 1$, then 0 and 1 are not interior points of S since every neighborhood of 0 or 1 contains points that are not in S. However, all other points of S are interior points.

Boundary Point. A point P is a boundary point of a set S if every neighborhood of P contains points in S and points not in S. If S is the set $0 \leqq x \leqq 1$, then 0 and 1 are the only boundary points. A boundary point need not belong to the set. 1 is a boundary point of the set S for which $x > 1$, but 1 is not in S since $1 \not> 1$.

Complement. The complement of a set S is the set of points not in S. The complement $C(S)$ has a relative meaning, for it depends on the set T in which S is embedded. If S, for example, is the set of numbers $-1 \leqq x \leqq 1$, then the complement of S relative to the real axis is the set of points $|x| > 1$. But the complement of $-1 \leqq x \leqq 1$ relative to the set $-1 \leqq x \leqq 1$ is the null set (no elements). The complement of the set of rationals relative to the reals is the set of irrationals, and conversely.

Open Set. A set of points S is said to be an open set (not to be confused with open interval) if every point of S is an interior point of S. For example, the set S consisting of points which satisfy either $0 < x < 1$ or $6 < x < 8$ is open.

Closed Set. A set containing all its limit points is called a closed set. For example, the set $(0, 1/2, 1/3, 1/4, \ldots, 1/n, \ldots)$ is closed, since its only limit point is 0, which it contains.

Problems

1. What are the limit points of the set $0 \leqq x < 1$? Is the set closed? Open? What are the boundary points?

2. Repeat Prob. 1 with the point $x = \frac{1}{2}$ removed.

3. Show that the set of all boundary points (the boundary) of a set S is closed.

4. Prove that the set of all limit points of a set S is closed.

5. Prove that the complement of an open set is closed, and conversely.

6. Why is every finite set closed?

7. Prove that the set of points common to two closed sets is closed. The set of points belonging to both S_1 and S_2 is called the intersection of S_1 and S_2, written $S_1 \cap S_2$.

8. Prove that the set of points which belong to either S_1 or S_2 is open if S_1 and S_2 are open. This set is called the union of S_1 and S_2, written $S_1 \cup S_2$.

9. An infinite union of closed sets is not necessarily closed. Give an example which verifies this.

10. An infinite intersection of open sets is not necessarily open. Give an example which verifies this.

Supremum. A number s is said to be the supremum of a set of points S if

1. x in S implies $x \leqq s$
2. $t < s$ implies an x in S such that $x > t$

Example 51. Let S be the set of rationals less than 1. Then 1 is the supremum of S, for (1) obviously holds from the definition of S, and if $t < 1$, it is possible to find a rational $r < 1$ such that $t < r$, so that (2) holds. We give a proof of this statement in a later paragraph.

Example 52. Let S be the set of rationals whose squares are less than 3, that is, $S|x^2 < 3$. Certainly we expect the $\sqrt{3}$ to be the supremum of this set. However, we cannot prove this without postulating the existence of irrationals. We overcome this by postulating

Every nonempty set of points has a supremum (152)

Hence the rationals whose squares are less than 3 have a supremum. It is obvious that we should define this supremum as the square root of 3.

The supremum of a set may be $+\infty$ as in the case of the set of all integers, or it may be $-\infty$ as in the case of the null set.

The *infemum* of a set S is the number s such that

1. x in S implies $s \leqq x$
2. $t > s$ implies an x in S such that $x < t$

Example 53. Let $a > 0$ and consider the sequence a, $2a$, $3a$, . . . , na, ... If this set is bounded, there exists a finite supremum s. Hence an integer r exists such that $ra > s - (a/2)$ so that $(r + 1)a > s + (a/2) > s$, a contradiction, since $na \leqq s$ for all n. Hence the sequence $\{na\}$ is unbounded. This is the Archimedean ordering postulate.

Example 54. To prove that a rational exists between any two numbers a, b. Assumed: $a > b > 0$, so that $a - b > 0$. From example 53, an integer q exists such that $q(a - b) > 1$, or $qa > qb + 1$. Also an integer p exists such that $p \cdot 1 = p > qb$. Choose the smallest p. Thus $p > qb \geqq p - 1$. Hence

$$qa > qb + 1 \geqq p > qb,$$

and $a > p/q > b$. Q.E.D.

With the aid of (152) we are in a position to prove the well-known *Weierstrass-Bolzano theorem.*

"*Every infinite bounded set of points S has a limit point.*" The proof proceeds as follows: We construct a new set T. Into T we place all points which are less than an infinite number of S. T is not empty since S is bounded below. From (152), T has a supremum; call it s. We now show that s is a limit point of S. Consider any neighborhood N of s. The points in N which are less than s are less than an infinite number of points of S, whereas those points in N which are greater than s are less than a finite number of points of S. Hence N contains an infinite number of S, so that the theorem is proved. We have at the same time proved that s is the greatest limit point of S. A limit point may or may not belong to the set.

Problems

1. The set $1/2, 1/3, \ldots, 1/n, \ldots, 1, 2, 3, \ldots$ is unbounded. Does it have any limit points? Does this violate the Weierstrass-Bolzano theorem?

2. Prove that every bounded monotonic (either decreasing or increasing) sequence has a unique limit.

3. Prove that $\lim r^n = 0$ if $|r| < 1$. Hint: The sequence $r, r^2, \ldots, r^n, \ldots$ is bounded and monotonic decreasing for $r > 0$, and $r^{n+1} = rr^n$.

4. Show that if P is a limit point of a set S, we can pick out a subsequence of S which converges to P.

5. Show that (152) implies that every set has an infemum.

6. Show that removing a finite number of elements from a set cannot affect the limit points.

7. Prove the Weierstrass-Bolzano theorem for a bounded set of points lying in a two-dimensional plane.

8. Let the sequence of numbers $s_1, s_2, \ldots, s_n, \ldots$ satisfy the following criterion: for any $\epsilon > 0$ there exists an integer N such that $|s_{n+p} - s_n| < \epsilon$ for $n \geqq N, p \geqq 0$. Prove that a unique limit point exists for the sequence.

9. A set of numbers is said to be countable if they can be written as a sequence, that is, if the set can be put into one-to-one correspondence with the positive integers. Show that a countable collection of countable sets is countable. Prove that the rationals are countable.

10. Show that the set S consisting of x satisfying $0 \leqq x \leqq 1$ is uncountable by assuming that the set S is countable, the numbers x being written in decimal form.

Theorem of Nested Sets. Consider an infinite sequence of nonempty closed and bounded sets $S_1, S_2, \ldots, S_n, \ldots$ such that S_n contains S_{n+1}, that is, $S_1 \supset S_2 \supset S_3 \supset \cdots$. There exists a point P which belongs to every S_i, $i = 1, 2, 3, \ldots$. The proof is easy. Let P_1 be any point of S_1, P_2 any point of S_2, etc. Now consider the sequence of points $P_1, P_2, \ldots, P_n, \ldots$. This infinite set belongs to S_1 and has a limit point P which belongs to S_1 since S_1 is closed. But P is also a limit of $P_n, P_{n+1}, \ldots$, so that P belongs to S_n. Hence P is in all S_n, $n = 1, 2, \ldots$.

Diameter of a Set. The diameter of a set S is the supremum of all distances between points of the set. For example, if S is the set of numbers x which satisfy $\frac{1}{2} < x \leqq 1$, $3 \leqq x < 7$, the diameter of S is $7 - \frac{1}{2} = 6\frac{1}{2}$. There are pairs of points in S whose distance apart can be made as close to $6\frac{1}{2}$ as we please.

Problems

1. If a set is closed and bounded, the diameter is actually attained by the set. Prove this.

2. If, in the theorem of nested sets, the diameters of the S_n approach zero, then P is unique. Prove this.

The Heine-Borel Theorem. Let S be any closed and bounded set, and let T be any collection of open intervals having the property that if x is any element of S, then there exists an open interval T_x of the collection T such that x is contained in T_x. The theorem states that there exists a subcollection T' of T which is finite in number and such that every element x of S is contained in one of the finite collection of open intervals that comprise T'.

Before proceeding to the proof we point out that (1) both the set S and the collection T are given beforehand, since it is no great feat to pick out a single open interval which completely covers a bounded set S; (2) S must be closed, for consider the set $S(1, 1/2, 1/3, \ldots, 1/n, \ldots)$ and let T consist of the following set of open intervals:

$$T_1|x \text{ such that } |x - 1| < \frac{1}{2^2}$$

$$T_2|x \text{ such that } |x - \tfrac{1}{2}| < \frac{1}{3^2}$$

$$\vdots$$

$$T_n|x \text{ such that } \left|x - \frac{1}{n}\right| < \frac{1}{(n+1)^2}$$

$$\vdots$$

It is very easy to see that we cannot reduce the covering of S by eliminating any of the given T_n, for there is no overlapping of these open intervals. Each T_n is required to cover the point $1/n$ of S contained in it.

Proof of the theorem: Let S be contained in the interval $-N \leqq x \leqq N$. This is possible since S is bounded. Now divide this closed interval into two equal intervals (1) $-N \leqq x \leqq 0$ and (2) $0 \leqq x \leqq N$. Any element x of S will belong to either (1) or (2). Now if the theorem is false, it will not be possible to cover the points of S in both (1) and (2) by a finite number of the given collection T, so that the points of S in either (1) or (2) require an infinite covering. Assume that the elements of S in (1) still require an infinite covering. We subdivide this interval into two equal parts and repeat the above argument. In this way we construct a sequence of sets $S_1 \supset S_2 \supset S_3 \supset \cdots$, such that each S_i is closed and bounded and such that the diameters of the $S_i \to 0$. From the theorem of nested sets there exists a unique point P which is contained in each S_i. Since P is in S, one of the open intervals of T, say T_p, will cover P. This T_p has a finite nonzero diameter so that eventually one of the S_i will be contained in T_p, since the diameters of the $S_i \to 0$. But by assumption all the elements of this S_i require an infinite number of the $\{T\}$ to cover them. This is a direct contradiction to the fact that a single T_p covers them. Hence our original assumption is wrong, and the theorem is proved.

42. Uniform Continuity. A real, single-valued function $f(x)$ is said to be continuous at a point $x = c$ if, given any positive number $\epsilon > 0$, there exists a positive number $\delta > 0$ such that $|f(x) - f(c)| < \epsilon$ whenever $|x - c| < \delta$. The δ may well depend on the ϵ and the point $x = c$. The function $f(x)$ is said to be continuous over a set of points S if it is continuous at every point of S.

We now prove that if $f(x)$ is continuous on a closed and bounded interval, it is uniformly continuous. We define uniform continuity as follows: If, for any $\epsilon > 0$ there exists a $\delta > 0$ such that $|f(x_1) - f(x_2)| < \epsilon$ whenever $|x_1 - x_2| < \delta$, then $f(x)$ is said to be uniformly continuous. It is important to notice that δ is independent of any x on the interval. We make use of the Heine-Borel theorem to prove uniform continuity. Choose an $\epsilon/2 > 0$. Then at every point c of our closed and bounded set there exists a $\delta(c, \epsilon/2)$ such that $|f(x) - f(c)| < \epsilon/2$ for $c - \delta < x < c + \delta$. Hence every point of S is covered by a 2δ-neighborhood, and so also by a δ-neighborhood. By the Heine-Borel theorem we can pick out a finite number of these neighborhoods which will cover S. Let δ_1 be the diameter of the smallest of this finite collection of δ-neighborhoods. Now consider any two points x_1 and x_2 of S whose distance apart is less than δ_1. Let x_0 be the center of the δ-neighborhood which covers x_1. From continuity $|f(x_1) - f(x_0)| < \epsilon/2$. But also x_2 differs from x_0 by less than 2δ, so that $|f(x_2) - f(x_0)| < \epsilon/2$. Consequently $|f(x_1) - f(x_2)| < \epsilon$. Q.E.D.

43. Some Properties of Continuous Functions

(*a*) Assume $f(x)$ continuous on the closed and bounded interval $a \leqq x \leqq b$. As a consequence of uniform continuity, we can prove that $f(x)$ is bounded. Choose any $\epsilon > 0$ and consider the corresponding $\delta > 0$ such that $|f(x_1) - f(x_2)| < \epsilon$ whenever $|x_1 - x_2| < \delta$. Now subdivide the interval (a, b) into a finite number of δ-intervals, say n of them. It is easily seen that $\max |f(x)| \leqq |f(a)| + n\epsilon$.

(*b*) If $f(a) < 0$ and $f(b) > 0$, there exists a c such that $f(c) = 0$, $a < c < b$. Let $\{x\}$ be the set of all points on (a, b) for which $f(x) < 0$. The set is bounded and nonvacuous since a belongs to $\{x\}$, for $f(a) < 0$. The set $\{x\}$ will have a supremum; call it c. Assume $f(c) > 0$. Choose $\epsilon = f(c)/2$. From continuity, a $\delta > 0$ exists such that $|f(x) - f(c)| < f(c)/2$ if $|x - c| < \delta$. Hence $f(x) \geqq \frac{1}{2}f(c)$ for all x in some neighborhood of $x = c$. Hence c is not the supremum of $\{x\}$. Similarly $f(c) < 0$ is impossible, so that $f(c) = 0$.

(*c*) We prove that $f(x)$ attains its maximum. In (*a*) we showed that $f(x)$ was bounded. Let s be the supremum of $f(x)$, $a \leqq x \leqq b$. As a consequence, $f(x) \leqq s$ for all x on (a, b). Now consider the set $s - \epsilon, s - \epsilon/2, \ldots, s - \epsilon/n, \ldots,$ where $\epsilon > 0$. Since s is a supremum, there exist $x_1, x_2, \ldots, x_n, \ldots$ such that $f(x_1) > s - \epsilon$, $f(x_2) > s - \epsilon/2, \ldots,$

$$f(x_n) > s - \frac{\epsilon}{n}, \ldots .$$

The set $\{x_i\}$ will have a limit point c. Let $\{x_n'\}$ be a subsequence of $\{x_n\}$ which converges to c. Then $\lim\limits_{x_n' \to c} f(x_n') \geqq s$, since $\epsilon/n \to 0$ as $n \to \infty$. But from continuity

$$\lim_{x_n' \to c} f(x_n') = f(c).$$

Hence $f(c) = s$. Q.E.D.

Problems

1. In the proof of (c) we exclude $f(c) > s$. Why?

2. If $f(x)$ is continuous on (a, b), show that the set of values $\{f(x)\}$ is closed.

3. If $f(x)$ has a derivative at every point of (a, b), show that $f(x)$ is continuous on (a, b).

4. If $f(x)$ has a derivative at every point of (a, b), show that a c exists such that $f'(c) = 0$, $a < c < b$, when $f(a) = f(b) = 0$.

5. If $f(x)$ has a derivative at every point of (a, b), show that a c exists such that $f(b) - f(a) = (b - a)f'(c)$, $a \leqq c \leqq b$.

6. Show that if two continuous functions $f(x)$, $g(x)$ exist such that $f(x) = g(x)$ for the rationals on (a, b), then $f(x) \equiv g(x)$ on (a, b).

7. Given the function $f(x) = 0$ when x is irrational, $f(x) = 1/q$ when x is rational and equal to p/q (p, q integers and relatively prime), prove that $f(x)$ is continuous at the irrational points of $(0, 1)$ and that $f(x)$ is discontinuous at the rational points of this interval.

44. Cauchy Criterion for Sequences. Let $x_1, x_2, \ldots, x_n, \ldots$ be a sequence of real numbers. We say that L is the limit of this sequence, or that the sequence converges to L, if, given any $\epsilon > 0$, there exists an integer N depending on ϵ such that $|L - x_n| < \epsilon$ whenever $n \geqq N(\epsilon)$. However, in most cases we do not know L, so that we need the Cauchy convergence criterion. This states that a necessary and sufficient condition that a sequence converge to a limit L is that given any $\epsilon > 0$, there exists an integer N such that $|x_n - x_m| < \epsilon$ for $n \geqq N$, $m \geqq N$. That the condition is necessary is obvious, for

$$|L - x_n| < \epsilon/2,$$

$|L - x_m| < \epsilon/2$ for $m, n \geqq N$ implies $|x_n - x_m| < \epsilon$ for $m, n \geqq N$. The proof of the converse is not as trivial. Choose any $\epsilon/2 > 0$. Then we assume an N exists such that $|x_n - x_m| < \epsilon/2$ for m, $n \geqq N$. Hence $|x_n| < |x_N| + (\epsilon/2)$, $n \geqq N$, so that the sequence is bounded. We ignore $x_1, x_2, \ldots, x_{N-1}$ since a finite number of elements cannot affect a limit point. From the Weierstrass-Bolzano theorem this infinite bounded set has at least one limit point L. Hence, given an $\epsilon/2$, there exists an x_n, with $n > N$, such that $|L - x_n| < \epsilon/2$. But we also have

$|x_m - x_n| < \epsilon/2$ for all $m, n \geqq N$. Hence $|L - x_m| < \epsilon$ for all $m \geqq N$. Q.E.D.

Problems

1. Show how the convergence of a series can be transformed into a problem involving the convergence of a sequence.
2. Show that the Cauchy criterion implies that the nth term of a convergent series must approach zero as $n \to \infty$.
3. Show that the sequence $1, 1/2, 1/3, \ldots, 1/n, \ldots$ converges by applying the Cauchy test.

45. Regular Arcs in the Plane. Consider the set of points in the two-dimensional plane such that the set can be represented in some coordinate system by $x = f(t)$, $y = \varphi(t)$, $\alpha \leqq t \leqq \beta$, where $f(t)$ and $\varphi(t)$ are continuous and have continuous first derivatives. Such curves are called regular arcs. A regular curve is a set of points consisting of a finite number of regular arcs joined one after the other (see Fig. 45).

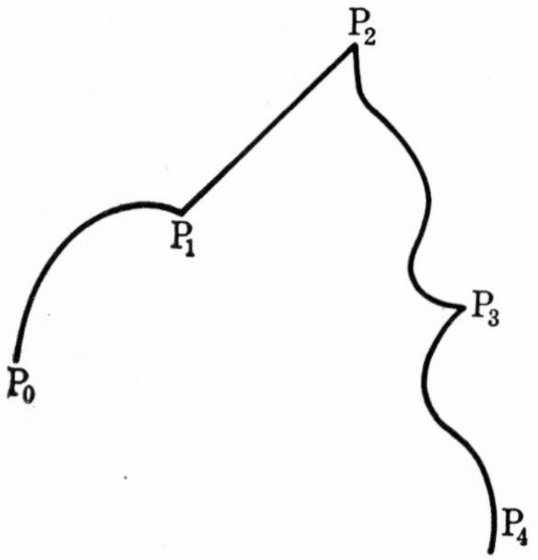

FIG. 45.

P_0P_1, P_1P_2, P_2P_3, P_3P_4 are the regular arcs joined at P_1, P_2, P_3. Notice that there are at most a finite number of discontinuities of the first derivatives. In Fig. 45 the derivatives are discontinuous at P_1, P_2, P_3.

46. Jordan Curves. The locus $\begin{cases} x = f(t) \\ y = \varphi(t), \end{cases}$ $\alpha \leqq t \leqq \beta$, will be called a Jordan curve provided that $f(t)$ and $\varphi(t)$ are continuous and that two distinct points on the curve correspond to two distinct values of the parameter t (no multiple points).

A closed Jordan curve is a continuous curve having $f(\alpha) = f(\beta)$, $\varphi(\alpha) = \varphi(\beta)$ but otherwise no multiple points.

From this we see that a Jordan curve is always "oriented," that is, it is always clear which part of the curve lies between two points on the arc, and which points precede a given point. We shall be interested in those curves which are rectifiable, or, in other words, we shall attempt to assign a definite length to a given Jordan arc or curve.

47. Functions of Bounded Variation. Let $f(x)$ be defined on the interval $a \leqq x \leqq b$. Subdivide this interval into a finite number of parts, say $a = x_0, x_1, \ldots x_i, \ldots, x_{n-1}, x_n = b$. Now consider the sum

$$|f(x_1) - f(x_0)| + |f(x_2) - f(x_1)| + \cdots + |f(x_n) - f(x_n - 1)| = \sum_{i=1}^{n} |f(x_i) - f(x_{i-1})|$$

If the sums of this type for all possible finite subdivisions are bounded, that is, if

$$\sum_{i=1}^{n} |f(x_i) - f(x_{i-1})| < A < \infty \tag{153}$$

we say that $f(x)$ is of bounded variation on (a, b).

A finite, monotonic nondecreasing function is always of bounded variation since

$$\sum_{i=1}^{n} |f(x_i) - f(x_{i-1})| = \sum_{i=1}^{n} [f(x_i) - f(x_{i-1})] = f(b) - f(a) = A$$

An example of a continuous function that is not of bounded variation is the following:

$$f(x) = x \sin \frac{\pi}{2x}, \qquad 0 < x \leqq 1$$

$$f(0) = 0$$

Let us subdivide $0 \leqq x \leqq 1$ into the intervals

$$\frac{1}{(n+1)} \leqq x \leqq \frac{1}{n},$$

$n = 1, 2, \ldots, N$. Now $f(1/n) = (1/n) \sin (\pi n/2)$ so that

$$\sum_{n=1}^{N} \left| f\left(\frac{1}{n}\right) - f\left(\frac{1}{n+1}\right) \right| = 1 + \frac{2}{3} + \frac{2}{5} + \cdots + \frac{2}{N}$$

We cannot bound this sum for all finite N since the series diverges as $N \to \infty$. N was chosen as an odd integer.

48. Arc Length. Consider the curve given by $x = f(t)$, $y = \varphi(t)$, and assume no multiple points. Divide the parameter t in any manner into n parts, say

$$\alpha = t_0 < t_1 < t_2 < \cdots < t_n = \beta$$

and consider

$$S_n = \sum_{i=1}^{n} \{[x(t_i) - x(t_{i-1})]^2 + [y(t_i) - y(t_{i-1})]^2\}^{\frac{1}{2}}$$

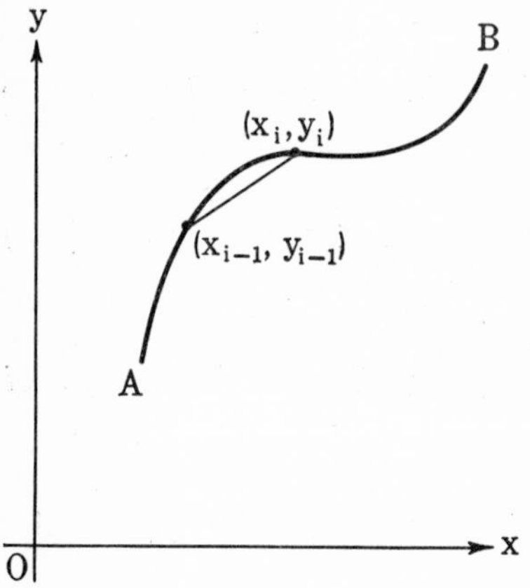

FIG. 46.

This is the length of the straight-line segments joining the points $x(t_i)$, $y(t_i)$, $i = 0, 1, 2, \ldots, n$ (see Fig. 46). If the set of all such lengths, obtained by all finite methods of subdivision, is bounded, we say that the curve is rectifiable and define the length of the curve as the supremum of these lengths.

An important theorem is the following: A necessary and sufficient condition that the curve described by

$$x = f(t), \qquad y = \varphi(t)$$

be rectifiable is that $f(t)$ and $\varphi(t)$ be of bounded variation. Let

$$A = \sum_{i=1}^{n} [|f(t_i) - f(t_{i-1})|^2 + |\varphi(t_i) - \varphi(t_{i-1})|^2]^{\frac{1}{2}}$$

$$B = \sum_{i=1}^{n} \{|f(t_i) - f(t_{i-1})| + |\varphi(t_i) - \varphi(t_{i-1})|\}$$

so that

$$A \leqq B \leqq \sqrt{2}\, A$$

Consequently if the curve is rectifiable, $f(t)$ and $\varphi(t)$ are of bounded variation, and conversely, if $f(t)$ and $\varphi(t)$ are of bounded variation, A is bounded, and hence the curve is rectifiable.

If $f'(t)$ and $\varphi'(t)$ are continuous, then from the law of the mean, $|f(t_i) - f(t_{i-1})| = |f'(\bar{t}_i)(t_i - t_{i-1})| < A|t_i - t_{i-1}|$, where A is the supremum of $|f'(t)|$, $\alpha \leqq t \leqq \beta$ and $t_{i-1} \leqq \bar{t}_i \leqq t_i$.

$$\sum_{i=1}^{n} |f(t_i) - f(t_{i-1})| \leqq A \sum_{i=1}^{n} |(t_i - t_{i-1}| = A(\beta - \alpha)$$

Similarly, $\varphi(t)$ is of bounded variation, so that the curve is rectifiable. Under these conditions it can easily be shown that the arc length is given by

$$s = \int_\alpha^\beta \left[\left(\frac{df}{dt}\right)^2 + \left(\frac{d\varphi}{dt}\right)^2\right]^{\frac{1}{2}} dt \tag{154}$$

49. The Riemann Integral. We now develop the theory of the Riemann integral in connection with line integrals. We need a curve over which an integration can be performed and a function to be integrated over this curve. Let $\Gamma \begin{matrix} x = \varphi(t), \\ y = \psi(t) \end{matrix}$, $\alpha \leqq t \leqq \beta$, be a rectifiable arc. Let $f(x, y)$ be a function continuous at all points of the curve Γ. Subdivide the parameter t into n parts, $\alpha = t_0 < t_1 < t_2 < \cdots < t_n = \beta$. Let the coordinates of P_i be $[\varphi(t_i), \psi(t_i)]$, and let Δs_i be the length of arc joining P_{i-1} to P_i. Since $f(x, y)$ is continuous, it will take on both its minimum and maximum for each segment $P_{i-1}P_i$. Multiply each arc length by the maximum value of $f(x, y)$ on this arc, say f_M, and form the sum

$$J = \sum_{i=1}^{n} f_M(x_i, y_i)\, \Delta s_i$$

Let $\bar{J}$ be the infemum of all such sums. Similarly, let $\bar{K}$ be the supremum of all sums when the minimum value of $f(x, y)$, say f_m, is used. If $\bar{J} = \bar{K}$, we say that $f(x, y)$ is Riemann-integrable over the curve Γ, and write

$$\bar{J} = \bar{K} = \int_\Gamma f(x, y)\, ds = \int_\alpha^\beta f[\varphi(t), \psi(t)](\varphi'^2 + \psi'^2)^{\frac{1}{2}}\, dt$$

whenever φ' and ψ' are continuous.

That $J = \bar{K}$ for a continuous function defined over a rectifiable curve is not difficult to prove. For,

$$J - K = \sum_{i=1}^{n} (f_M - f_m)\, \Delta s_i$$

and from the uniform continuity of $f(x, y)$ we can subdivide the curve Γ into arcs such that the difference between the maximum and minimum values of $f(x, y)$ on any arc is less than any given $\epsilon > 0$ so that

$$|J - K| \leqq \epsilon \Sigma |\Delta s_i| \leqq \epsilon L \tag{155}$$

We leave it as an exercise for the reader to prove that we can make J as close to $\bar{J}$ as we please and similarly that the difference

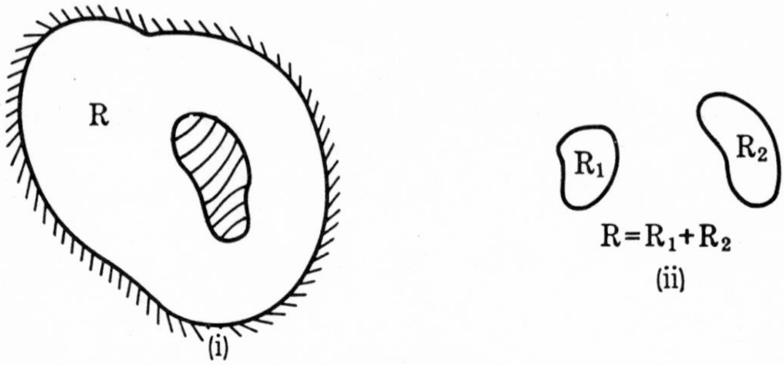

FIG. 47.

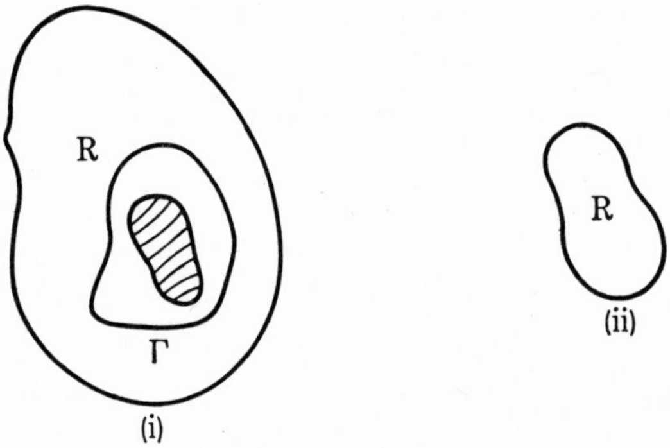

FIG. 48.

between K and $\bar{K}$ can be made arbitrarily small, for a sufficiently large number of subdivisions. Since the ϵ of (155) is arbitrary, we can make the difference between $\bar{J}$ and $\bar{K}$ as small as we please so that $\bar{J} = \bar{K}$ since they are fixed numbers.

50. Connected and Simply Connected Regions. A region R is said to be connected if, given any two points of the region, we can join them by an arc, every point of the arc belonging to the region R. In Fig. 47, (i) is connected; (ii) is not connected. In (i), R is the nonshaded region.

If every closed curve of a connected region R can be continuously shrunk to a point of R, we say that the region is simply connected. In Fig. 48, (i) is connected, but not simply connected; (ii) is simply connected. In (i) the curve Γ cannot be continuously shrunk to a point. An analytic expression for simple connectedness can be set up, but we shall omit this.

51. The Line Integral. Let

$$\mathbf{f} = X(x, y, z)\mathbf{i} + Y(x, y, z)\mathbf{j} + Z(x, y, z)\mathbf{k}$$

and consider the line integral $\int \left(\mathbf{f} \cdot \frac{d\mathbf{r}}{ds}\right) ds$ along a rectifiable space curve Γ given by $\mathbf{r} = \mathbf{r}(s)$. Since

$$\frac{d\mathbf{r}}{ds} ds = d\mathbf{r} = dx\,\mathbf{i} + dy\,\mathbf{j} + dz\,\mathbf{k}$$

we have

$$\int_\Gamma \left(\mathbf{f} \cdot \frac{d\mathbf{r}}{ds}\right) ds = \int_\Gamma X\,dx + Y\,dy + Z\,dz \tag{156}$$

We use (156) as a means of evaluating the line integral. If the space curve is given by $x = x(t)$, $y = y(t)$, $z = z(t)$, then (156) reduces to

$$\int_\Gamma \mathbf{f} \cdot d\mathbf{r} = \int_{t_0}^{t_1} \left[X(t)\frac{dx}{dt} + Y(t)\frac{dy}{dt} + Z(t)\frac{dz}{dt}\right] dt \tag{157}$$

In general, the line integral will depend on the path joining the two end points of integration. If $\int \mathbf{f} \cdot d\mathbf{r}$ does not depend on the curve Γ joining the end points of integration, we say that $\mathbf{f}$ is a conservative vector field. If $\mathbf{f}$ is a force field, we define $\int_A^B \mathbf{f} \cdot d\mathbf{r}$ as the work done by $\mathbf{f}$ as the unit particle moves from A to B. We shall now work out a few examples for the reader.

Example 55. Let $\mathbf{f} = x^2\mathbf{i} + y^3\mathbf{j}$, and let the path of integration be the parabola $y = x^2$, the integration being performed from (0, 0) to (1, 1). We exhibit three methods of solution.

(*a*) Let $x = t$, so that $y = t^2$. Thus

$$\mathbf{f} = t^2\mathbf{i} + t^6\mathbf{j}, \qquad \mathbf{r} = t\mathbf{i} + t^2\mathbf{j}, \qquad d\mathbf{r} = (\mathbf{i} + 2t\mathbf{j})\,dt$$

and

$$\int_{t=0}^{t=1} \mathbf{f} \cdot d\mathbf{r} = \int_0^1 (t^2 + 2t^7)\, dt = \tfrac{7}{12}$$

(*b*) Since $y = x^2$ everywhere along Γ, $\mathbf{f} = x^2\mathbf{i} + x^6\mathbf{j}$ along Γ, and $d\mathbf{r} = dx\,\mathbf{i} + dy\,\mathbf{j} = (\mathbf{i} + 2x\mathbf{j})\, dx$, so that

$$\int_\Gamma \mathbf{f} \cdot d\mathbf{r} = \int_0^1 (x^2 + 2x^7)\, dx = \tfrac{7}{12}$$

(*c*)

$$\int_{(0,0)}^{(1,1)} \mathbf{f} \cdot d\mathbf{r} = \int_{(0,0)}^{(1,1)} x^2\, dx + y^3\, dy = \left.\frac{x^3}{3}\right|_0^1 + \left.\frac{y^4}{4}\right|_0^1 = \frac{7}{12}$$

(*c*) shows that the integral from (0, 0) to (1, 1) is independent of the path since $x^2\, dx + y^3\, dy$ is a perfect differential, that is, $x^2\, dx + y^3\, dy = d[(x^3/3) + (y^4/4)]$.

Example 56. $\mathbf{f} = y\mathbf{i} - x\mathbf{j}$, and let the path of integration be $y = x^2$ from (0, 0) to (1, 1). Then

$$\int_{(0,0)}^{(1,1)} \mathbf{f} \cdot d\mathbf{r} = \int_{(0,0)}^{(1,1)} y\, dx - x\, dy = \int_0^1 x^2\, dx - x(2x\, dx) = -\tfrac{1}{3}$$

Next we compute the integral by moving along the x axis from $x = 0$ to $x = 1$ and then along the line $x = 1$ from $y = 0$ to $y = 1$. We have

$$\int_{(0,0)}^{(1,1)} \mathbf{f} \cdot d\mathbf{r} = \int_{(0,0)}^{(1,0)} \mathbf{f} \cdot d\mathbf{r} + \int_{(1,0)}^{(1,1)} \mathbf{f} \cdot d\mathbf{r}$$

Along the first part of the path, $\mathbf{f} = -x\mathbf{j}$ and $d\mathbf{r} = dx\,\mathbf{i}$, since $y = dy = 0$. Along the second part of the path, $\mathbf{f} = y\mathbf{i} - \mathbf{j}$, $d\mathbf{r} = dy\,\mathbf{j}$ since $x = 1$ and $dx = 0$. Thus

$$\begin{aligned}\int_{(0,0)}^{(1,1)} \mathbf{f} \cdot d\mathbf{r} &= \int_{x=0}^{x=1} -x\mathbf{j} \cdot dx\,\mathbf{i} + \int_{y=0}^{y=1} (y\mathbf{i} - \mathbf{j}) \cdot dy\,\mathbf{j} \\ &= \int_0^1 -dy = -1\end{aligned}$$

The line integral does depend on the path for the vector field $\mathbf{f} = y\mathbf{i} - x\mathbf{j}$. We say that $\mathbf{f}$ is a nonconservative field.

52. Line Integral (*Continued*). Let us assume that $\mathbf{f}$ can be written as the gradient of a scalar $\varphi(x, y, z)$, that is, $\mathbf{f} = \nabla\varphi$. Then the line integral $\int_A^B \mathbf{f} \cdot d\mathbf{r}$ is independent of the path of integration from A to B since

$$\int_A^B \mathbf{f}\cdot d\mathbf{r} = \int_A^B \nabla\varphi\cdot d\mathbf{r} = \int_A^B d\varphi = \varphi(B) - \varphi(A) \qquad (158)$$

Our final result in (158) depends only on the value of $\varphi(x, y, z)$ when evaluated at the points $A(x_0, y_0, z_0)$, $B(x_1, y_1, z_1)$ and in no way depends on the path of integration. If our path is closed, then $A = B$ and $\varphi(B) = \varphi(A)$, so that the line integral around any closed path vanishes if $\mathbf{f} = \nabla\varphi$. However, the region for which $\mathbf{f} = \nabla\varphi$ must be simply connected. Let us consider the following example.

Example 57. Let

$$\mathbf{f} = \frac{-y\mathbf{i}}{x^2 + y^2} + \frac{x\mathbf{j}}{x^2 + y^2}$$

Then $\mathbf{f} = \nabla\varphi$ where $\varphi = \tan^{-1}(y/x)$, and if we integrate $\mathbf{f}$ over the unit circle with center at the origin, we have

$$\oint \mathbf{f}\cdot d\mathbf{r} = \oint d\left(\tan^{-1}\frac{y}{x}\right) = \int_0^{2\pi} d\theta = 2\pi$$

so that our line integral does not vanish. The region for which $\mathbf{f} = \nabla\varphi$ is not simply connected since φ is not defined at the origin.

We now prove that if $\int \mathbf{f}\cdot d\mathbf{r}$ is independent of the path, then $\mathbf{f}$ is the gradient of a scalar φ. Let

$$\varphi(x, y, z) = \int_{P_0(x_0, y_0, z_0)}^{P(x, y, z)} \mathbf{f}\cdot d\mathbf{r} = \int_{P_0}^{P} \left(\mathbf{f}\cdot\frac{d\mathbf{r}}{ds}\right) ds \qquad (159)$$

Now

$$\frac{\varphi(x + \Delta x, y, z) - \varphi(x, y, z)}{\Delta x} = \frac{1}{\Delta x}\int_{P(x, y, z)}^{Q(x+\Delta x, y, z)} \left(\mathbf{f}\cdot\frac{d\mathbf{r}}{ds}\right) ds$$

and since the line integral is independent of the path joining P to Q, we choose the straight line from P to Q as our path of integration, that is, $d\mathbf{r} = dx\,\mathbf{i}$. Thus

$$\lim_{\Delta x\to 0} \frac{\varphi(x + \Delta x, y, z) - \varphi(x, y, z)}{\Delta x} = \frac{\partial\varphi}{\partial x}$$

$$= \lim_{\Delta x\to 0} \frac{\int_x^{x+\Delta x} X(x, y, z)\, dx}{\Delta x} = X(x, y, z)$$

from the calculus. We have assumed that **f** is continuous. Now similarly

$$\frac{\partial \varphi}{\partial y} = Y(x, y, z), \qquad \frac{\partial \varphi}{\partial z} = Z(x, y, z)$$

so that

$$\mathbf{f} = \frac{\partial \varphi}{\partial x}\mathbf{i} + \frac{\partial \varphi}{\partial y}\mathbf{j} + \frac{\partial \varphi}{\partial z}\mathbf{k} = \nabla\varphi \tag{160}$$

If **f** has continuous derivatives, we can easily conclude whether **f** is the gradient of a scalar or not. Assume $\mathbf{f} = \nabla\varphi$, or

$$(1)\ X = \frac{\partial \varphi}{\partial x}, \qquad (2)\ Y = \frac{\partial \varphi}{\partial y}, \qquad (3)\ Z = \frac{\partial \varphi}{\partial z}$$

Differentiating (1) with respect to y and (2) with respect to x, we see that

Similarly

$$\left.\begin{aligned} \frac{\partial X}{\partial y} &= \frac{\partial Y}{\partial x} \\ \frac{\partial Y}{\partial z} &= \frac{\partial Z}{\partial y} \\ \frac{\partial Z}{\partial x} &= \frac{\partial X}{\partial z} \end{aligned}\right\} \tag{161}$$

This is the condition that $\nabla \times \mathbf{f} = 0$. Conversely, assume $\nabla \times \mathbf{f} = 0$. Let

$$\varphi(x, y, z) = \int_{x_0}^{x} X(x, y, z)\, dx + \int_{y_0}^{y} Y(x_0, y, z)\, dy + \int_{z_0}^{z} Z(x_0, y_0, z)\, dz \tag{162}$$

Now

$$\begin{aligned} \frac{\partial \varphi}{\partial x} &= X(x, y, z) \\ \frac{\partial \varphi}{\partial y} &= \int_{x_0}^{x} \frac{\partial X}{\partial y}\, dx + Y(x_0, y, z) \\ &= \int_{x_0}^{x} \frac{\partial Y}{\partial x}\, dx + Y(x_0, y, z) \\ &= Y(x, y, z) - Y(x_0, y, z) + Y(x_0, y, z) \\ &= Y(x, y, z) \end{aligned}$$

Similarly $\dfrac{\partial \varphi}{\partial z} = Z(x, y, z)$. Consequently

$$\mathbf{f} = \frac{\partial \varphi}{\partial x}\mathbf{i} + \frac{\partial \varphi}{\partial y}\mathbf{j} + \frac{\partial \varphi}{\partial z}\mathbf{k} = \nabla \varphi$$

We have proved that a necessary and sufficient condition for $\mathbf{f}$ to be the gradient of a scalar is that

$$\boxed{\nabla \times \mathbf{f} = 0} \tag{163}$$

If $\mathbf{f} = \nabla \varphi$ or $\nabla \times \mathbf{f} = 0$, $\mathbf{f}$ is said to be an irrotational vector.

Example 58. Let $\mathbf{f} = 2xye^z\mathbf{i} + x^2e^z\mathbf{j} + x^2ye^z\mathbf{k}$. Then

$$\nabla \times \mathbf{f} = \begin{vmatrix} \mathbf{i} & \mathbf{j} & \mathbf{k} \\ \dfrac{\partial}{\partial x} & \dfrac{\partial}{\partial y} & \dfrac{\partial}{\partial z} \\ 2xye^z & x^2e^z & x^2ye^z \end{vmatrix} \equiv 0$$

and

$$\begin{aligned} \varphi &= \int_0^x 2xye^z\,dx + \int_0^y 0^2e^z\,dy + \int_0^z 0^2 \cdot 0 \cdot e^z\,dz \\ &= x^2ye^z \end{aligned}$$

so that $\mathbf{f} = \nabla(x^2ye^z + \text{constant})$.

Problems

1. Given $\mathbf{f} = xy\mathbf{i} - x\mathbf{j}$, evaluate $\int \mathbf{f} \cdot d\mathbf{r}$ along the curve $y = x^3$ from the origin to the point $P(1, 1)$.

2. Show that if the line integral around every closed path is zero, that is, if $\oint \mathbf{f} \cdot d\mathbf{r} = 0$, then $\mathbf{f} = \nabla \varphi$.

3. Show that $\oint \mathbf{r} \cdot d\mathbf{r} = 0$.

4. Show that the inverse-square force field $\mathbf{f} = -\mathbf{r}/r^3$ is conservative. The origin is excepted.

5. $\mathbf{f} = (y + \sin z)\mathbf{i} + x\mathbf{j} + x \cos z\,\mathbf{k}$. Show that $\mathbf{f}$ is conservative and find φ so that $\mathbf{f} = \nabla \varphi$.

6. Evaluate $\int x\,dy - y\,dx$ around the unit circle with center at the origin.

7. If A is a constant vector, show that $\oint \mathbf{A} \cdot d\mathbf{r} = 0$,

$$\oint \mathbf{A} \times d\mathbf{r} = 0$$

53. Stokes's Theorem. We begin by considering a surface of the type encountered in Chap. 3. Now consider a closed rectifiable curve Γ that lies on the surface. As we move along the curve Γ, keeping our head in the same direction as the normal $\dfrac{\partial \mathbf{r}}{\partial u} \times \dfrac{\partial \mathbf{r}}{\partial v}$, we keep track of the area to our left. It is this surface that we shall keep in touch with, and Γ will be the boundary of this surface. We neglect the rest of the surface $\mathbf{r}(u, v)$. We now consider a mesh of networks on the surface formed by a collection of parametric curves. Of course, the boundary Γ will not, in general, consist of arcs of these parametric curves (see Fig. 49). Consider the mesh $ABCD$. Let the surface coordinates of A be (u, v), so that $A(u, v)$, $B(u + du, v)$,

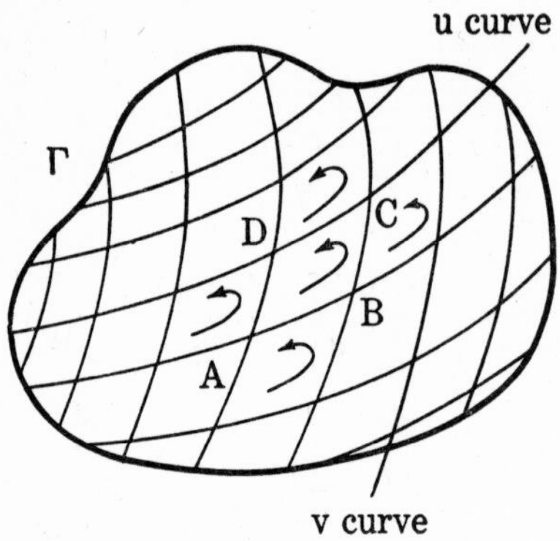

FIG. 49.

$$C(u + du, v + dv)$$

$D(u, v + dv)$ are the coordinates of A, B, C, D. We also assume that the parametric curves are rectifiable. Now consider

$$\oint_{ABCD} \mathbf{f} \cdot d\mathbf{r}$$

The value of $\mathbf{f}$ at A is $\mathbf{f}(u, v)$; at B it is $\mathbf{f}(u + du, v)$; at D it is $\mathbf{f}(u, v + dv)$.

Now

$$\begin{aligned} \mathbf{f}(u + du, v) &= \mathbf{f}(u, v) + d\mathbf{f}_u \\ &= \mathbf{f}(u, v) + (d\mathbf{r}_u \cdot \nabla)\mathbf{f} \\ &= \mathbf{f}(u, v) + \left(\frac{\partial \mathbf{r}}{\partial u} du \cdot \nabla\right)\mathbf{f} \end{aligned}$$

except for infinitesimals of higher order. Similarly

$$\mathbf{f}(u, v + dv) = \mathbf{f}(u, v) + \left(\frac{\partial \mathbf{r}}{\partial v} dv \cdot \nabla\right)\mathbf{f}$$

Hence, but for infinitesimals of higher order,

$$\oint_{ABCD} \mathbf{f} \cdot d\mathbf{r} = \mathbf{f} \cdot \frac{\partial \mathbf{r}}{\partial u} du + \left[\mathbf{f} + \left(\frac{\partial \mathbf{r}}{\partial u} du \cdot \nabla\right)\mathbf{f}\right] \cdot \frac{\partial \mathbf{r}}{\partial v} dv - \mathbf{f} \cdot \frac{\partial \mathbf{r}}{\partial v} dv$$
$$- \left[\mathbf{f} + \left(\frac{\partial \mathbf{r}}{\partial v} dv \cdot \nabla\right)\mathbf{f}\right] \cdot \frac{\partial \mathbf{r}}{\partial u} du$$
$$= \left\{\left[\left(\frac{\partial \mathbf{r}}{\partial u} \cdot \nabla\right)\mathbf{f}\right] \cdot \frac{\partial \mathbf{r}}{\partial v} - \left[\left(\frac{\partial \mathbf{r}}{\partial v} \cdot \nabla\right)\mathbf{f}\right] \cdot \frac{\partial \mathbf{r}}{\partial u}\right\} du\, dv \tag{164}$$

Now

$$(\nabla \times \mathbf{f}) \cdot \left(\frac{\partial \mathbf{r}}{\partial u} \times \frac{\partial \mathbf{r}}{\partial v}\right) = \left[(\nabla \times \mathbf{f}) \times \frac{\partial \mathbf{r}}{\partial u}\right] \cdot \frac{\partial \mathbf{r}}{\partial v}$$
$$= \left[\left(\frac{\partial \mathbf{r}}{\partial u} \cdot \nabla\right)\mathbf{f}\right] \cdot \frac{\partial \mathbf{r}}{\partial v} - \left[\left(\frac{\partial \mathbf{r}}{\partial v} \cdot \nabla\right)\mathbf{f}\right] \cdot \frac{\partial \mathbf{r}}{\partial u}$$

Hence

$$\oint_{ABCD} \mathbf{f} \cdot d\mathbf{r} = (\nabla \times \mathbf{f}) \cdot \frac{\partial \mathbf{r}}{\partial u} \times \frac{\partial \mathbf{r}}{\partial v} du\, dv \tag{165}$$

Now $\frac{\partial \mathbf{r}}{\partial u} \times \frac{\partial \mathbf{r}}{\partial v} du\, dv$ = area of sector $ABCD$ in magnitude, and its direction is along the normal. We define

$$d\boldsymbol{\sigma} = \frac{\partial \mathbf{r}}{\partial u} \times \frac{\partial \mathbf{r}}{\partial v} du\, dv \tag{166}$$

so that

$$\int_{ABCD} \mathbf{f} \cdot d\mathbf{r} = (\nabla \times \mathbf{f}) \cdot d\boldsymbol{\sigma}$$

We now sum over the entire network. Interior line integrals will cancel out in pairs leaving only $\oint_\Gamma \mathbf{f} \cdot d\mathbf{r}$. Also

$$\sum_{\text{over surface } S} (\nabla \times \mathbf{f}) \cdot d\boldsymbol{\sigma} \to \iint_S (\nabla \times \mathbf{f}) \cdot d\boldsymbol{\sigma}$$

as the areas approach zero in size. We thus have Stokes's theorem:

$$\oint_\Gamma \mathbf{f} \cdot d\mathbf{r} = \iint_S (\nabla \times \mathbf{f}) \cdot d\boldsymbol{\sigma} \tag{167}$$

Comments

1. The reader may well be aware that (165) does not hold for a mesh that has Γ as part of its boundary. This is true, but fortunately we need not worry about the inequality. The line integrals cancel out no matter what subdivisions we use, and for a fine network the contributions of those areas next to Γ contribute little to $\iint (\nabla \times \mathbf{f}) \cdot d\boldsymbol{\sigma}$. The limiting process takes care of this apparent negligence.

2. We have proved Stokes's theorem for a surface of the type $\mathbf{r}(u, v)$ discussed in Chap. 3. The theorem is easily seen to be true if we have a finite number of these surfaces connected continuously (edges).

3. Stokes's theorem is also true for conical points, where no $d\boldsymbol{\sigma}$ can be defined. We just neglect to integrate over a small area covering this point. Since the area can be made arbitrarily small, it cannot affect the integral.

4. The reader is referred to the text of Kellogg, "Foundations of Potential Theory," for a much more rigorous proof of Stokes's theorem.

54. Examples of Stokes's Theorem

Example 59. Let Γ be a closed Jordan curve in the x-y plane. Let $\mathbf{f} = -y\mathbf{i} + x\mathbf{j}$. Applying Stokes's theorem, we have

$$\oint \mathbf{f} \cdot d\mathbf{r} = \oint x\,dy - y\,dx = \iint_S (\nabla \times \mathbf{f}) \cdot d\boldsymbol{\sigma}$$

$$= \iint_S \begin{vmatrix} \mathbf{i} & \mathbf{j} & \mathbf{k} \\ \dfrac{\partial}{\partial x} & \dfrac{\partial}{\partial y} & \dfrac{\partial}{\partial z} \\ -y & x & 0 \end{vmatrix} \cdot \mathbf{k}\,dy\,dx$$

$$= 2 \iint_S dy\,dx = 2A$$

or

$$\text{Area } A = \tfrac{1}{2} \oint x\,dy - y\,dx \tag{168}$$

For the ellipse $x = a \cos t$, $y = b \sin t$, $dx = -a \sin t\,dt$,

$$dy = b \cos t\,dt$$

and

$$A = \tfrac{1}{2} \int_0^{2\pi} ab(\cos^2 t + \sin^2 t)\,dt = \pi ab$$

Example 60. If $\mathbf{f}$ has continuous derivatives, then a necessary and sufficient condition that $\oint \mathbf{f} \cdot d\mathbf{r} = 0$ around every closed path is that $\nabla \times \mathbf{f} = 0$.

If $\nabla \times \mathbf{f} = 0$, then $\oint \mathbf{f} \cdot d\mathbf{r} = \iint_S (\nabla \times \mathbf{f}) \cdot d\boldsymbol{\sigma} = 0$. Conversely, assume $\oint \mathbf{f} \cdot d\mathbf{r} = 0$ for every closed path. If $\nabla \times \mathbf{f} \not\equiv 0$, then $\nabla \times \mathbf{f} \neq 0$ at some point P. From continuity, $\nabla \times \mathbf{f} \neq 0$ in some region about P. Choose a small plane surface S in this region, the normal to the plane being parallel to $\nabla \times \mathbf{f}$. Then

$$\oint \mathbf{f} \cdot d\mathbf{r} = \iint_S \nabla \times \mathbf{f} \cdot d\boldsymbol{\sigma} > 0, \quad \text{a contradiction}$$

Example 61. We see that an irrotational field is characterized by any one of the three conditions:

$$\begin{aligned} &\text{(i)} && \mathbf{f} = \nabla\varphi \\ &\text{(ii)} && \nabla \times \mathbf{f} = 0 \\ &\text{(iii)} && \oint \mathbf{f} \cdot d\mathbf{r} = 0 \quad \text{for every closed path} \end{aligned} \tag{169}$$

Any of these conditions implies the other two.

Example 62. Assume $\mathbf{f}$ not irrotational. Perhaps a scalar $\mu(x, y, z)$ exists such that $\mu\mathbf{f}$ is irrotational, that is, $\nabla \times (\mu\mathbf{f}) = 0$. If this is so,

$$\mu \nabla \times \mathbf{f} + \nabla\mu \times \mathbf{f} = 0 \tag{170}$$

and dotting (170) with $\mathbf{f}$, we have, since $\mathbf{f} \cdot \nabla\mu \times \mathbf{f} = 0$, the equation

$$\mathbf{f} \cdot (\nabla \times \mathbf{f}) = 0 \tag{171}$$

If $\mathbf{f} = X\mathbf{i} + Y\mathbf{j} + Z\mathbf{k}$, (171) may be written

$$\begin{vmatrix} X & Y & Z \\ \dfrac{\partial}{\partial x} & \dfrac{\partial}{\partial y} & \dfrac{\partial}{\partial z} \\ X & Y & Z \end{vmatrix} = 0 \tag{172}$$

In texts on differential equations it is shown that (172) is also sufficient for $\mu(x, y, z)$ to exist. We call $\mu(x, y, z)$ an integrating factor.

Example 63. Let $\mathbf{f} = f(x, y, z)\mathbf{a}$, where $\mathbf{a}$ is any constant vector. Applying Stokes's theorem, we have

$$\oint \mathbf{f} \cdot d\mathbf{r} = \iint_S \nabla \times (f\mathbf{a}) \cdot d\boldsymbol{\sigma}$$
$$= \iint_S (\nabla f \times \mathbf{a}) \cdot d\boldsymbol{\sigma}$$

or

$$\mathbf{a} \cdot \oint \mathbf{f}\, d\mathbf{r} = \mathbf{a} \cdot \iint_S d\boldsymbol{\sigma} \times \nabla f$$

or

$$\mathbf{a} \cdot \left(\oint f\, d\mathbf{r} - \iint_S d\boldsymbol{\sigma} \times \nabla f \right) = 0$$

Since **a** has arbitrary direction, we have

$$\oint f\, d\mathbf{r} = \iint_S d\boldsymbol{\sigma} \times \nabla f \tag{173}$$

Example 64. Let $\mathbf{f} = \mathbf{a} \times \mathbf{g}$, where **a** is any constant vector. Applying (167), we have

$$\mathbf{a} \cdot \oint \mathbf{g} \times d\mathbf{r} = \iint_S \nabla \times (\mathbf{a} \times \mathbf{g}) \cdot d\boldsymbol{\sigma}$$
$$= \iint_S [\mathbf{a}(\nabla \cdot \mathbf{g}) - (\mathbf{a} \cdot \nabla)\mathbf{g}] \cdot d\boldsymbol{\sigma}$$
$$= \mathbf{a} \cdot \iint_S (\nabla \cdot \mathbf{g}) d\boldsymbol{\sigma} - \mathbf{a} \cdot \iint_S \nabla(\mathbf{g} \cdot d\boldsymbol{\sigma})$$

remembering that ∇ does not operate on $d\boldsymbol{\sigma}$ (summed). Therefore

$$-\mathbf{a} \cdot \oint \mathbf{g} \times d\mathbf{r} = \mathbf{a} \cdot \iint_S (d\boldsymbol{\sigma} \times \nabla) \times \mathbf{g}$$

and

$$\oint d\mathbf{r} \times \mathbf{g} = \iint_S (d\boldsymbol{\sigma} \times \nabla) \times \mathbf{g} \tag{174}$$

We notice that in all cases

$$\oint d\mathbf{r} * \mathbf{f} = \iint_S (d\boldsymbol{\sigma} \times \nabla) * \mathbf{f} \tag{175}$$

The star (*) can denote dot or cross or ordinary multiplication. In the latter case, **f** becomes a scalar f.

Problems

1. Prove that $\oint d\mathbf{r} = 0$ from (175).

2. Show that $\left|\oint d\mathbf{r} \times \mathbf{r}\right|$ taken around a curve in the x-y plane is twice the area enclosed by the curve.

3. If $\mathbf{f} = \cos y\,\mathbf{i} + x(1 + \sin y)\mathbf{j}$, find the value of $\oint \mathbf{f} \cdot d\mathbf{r}$ around a circle of radius r in the x-y plane.

4. Prove that $\oint \mathbf{r} \cdot d\mathbf{r} = 0$.

5. Prove that $\iint_S d\boldsymbol{\sigma} \times \mathbf{r} = \frac{1}{2} \oint r^2\, d\mathbf{r}$.

6. Prove that $\oint u\, \nabla v \cdot d\mathbf{r} = -\oint v\, \nabla u \cdot d\mathbf{r}$.

7. Prove that $\oint u\, \nabla v \cdot d\mathbf{r} = \iint_S \nabla u \times \nabla v \cdot d\boldsymbol{\sigma}$.

8. If a vector is normal to a surface at each point, show that its curl either is zero or is tangent to the surface at each point.

9. If a vector is zero at each point of a surface, show that its curl either is zero or is tangent to the surface.

10. Show that $\oint \mathbf{a} \times \mathbf{r} \cdot d\mathbf{r} = 2\mathbf{a} \cdot \iint_S d\boldsymbol{\sigma}$, if $\mathbf{a}$ is constant.

11. If $\oint \mathbf{E} \cdot d\mathbf{r} = -\dfrac{1}{c}\dfrac{\partial}{\partial t} \iint_S \mathbf{B} \cdot d\boldsymbol{\sigma}$ for all closed curves, show that $\nabla \times \mathbf{E} = -\dfrac{1}{c}\dfrac{\partial \mathbf{B}}{\partial t}$

12. By Stokes's theorem prove that $\nabla \times (\nabla\varphi) = 0$.

13. Show that $\oint d\mathbf{r}/r = \iint_S (\mathbf{r}/r^3) \times d\boldsymbol{\sigma}$ where $r = |\mathbf{r}|$.

14. Find the vector $\mathbf{f}$ such that $xy = \int_{(0,0,0)}^{(x,y,z)} \mathbf{f} \cdot d\mathbf{r}$.

15. If $\mathbf{f} = \mathbf{r}/r^3$, show that $\nabla \times \mathbf{f} = 0$ and find the potential φ such that $\mathbf{f} = \nabla\varphi$.

16. Show that the vector $\mathbf{f} = (-y\mathbf{i} + x\mathbf{j})/(x^2 + y^2)$ is irrotational and that $\oint_\Gamma \mathbf{f} \cdot d\mathbf{r} = 2\pi$, where Γ is a circle containing the origin. Does this contradict Stokes's theorem? Explain.

17. Show that $\iint_S \nabla \times \mathbf{f} \cdot d\boldsymbol{\sigma} = 0$, where S is a closed surface.

18. Let C_1 and C_2 be two closed curves bounding the surfaces S_1 and S_2. Show that

$$\int_{C_2}\int_{C_1} r_{12}^2\, d\mathbf{r}_1 \cdot d\mathbf{r}_2 = -4 \iint\limits_{S_1} d\mathbf{d}_1 \cdot \iint\limits_{S_2} d\mathbf{d}_2$$

$$\int_{C_2}\int_{C_1} r_{12}^2\, d\mathbf{r}_1 \times d\mathbf{r}_2 = -2 \iint\limits_{S_1} d\mathbf{d}_1 \times \iint\limits_{S_2} d\mathbf{d}_2$$

where r_{12} is the distance between points on the two curves.

55. The Divergence Theorem (Gauss). Let us consider a region V over which $\mathbf{f}$ and $\nabla \cdot \mathbf{f}$ are continuous. We shall assume that V is bounded by a finite number of surfaces such that at each surface there is a well-defined continuous normal. We shall also assume that $\mathbf{f}$ can be integrated over the total surface bounding V. Now, no matter what physical significance $\mathbf{f}$ has, if any, we can always imagine $\mathbf{f}$ to be the product of the density and velocity of some fluid. We have seen previously (Sec. 20) that the net loss of fluid per unit volume per unit time is given by $\nabla \cdot \mathbf{f}$. Consequently, the total loss per unit time is given by

$$\iiint\limits_V (\nabla \cdot \mathbf{f})\, d\tau \tag{176}$$

Now since $\mathbf{f}$ and $\nabla \cdot \mathbf{f}$ are continuous, there cannot be any point in the region V at which fluid is being manufactured or destroyed; that is, no sources or sinks appear. Consequently, the total loss of fluid must be due to the flow of fluid through the boundary S of the region V. We might station a great many observers on the boundary S, let each observer measure the outward flow, and then sum up each observer's recorded data. At a point on the surface with normal vector area $d\mathbf{d}$, the component of the velocity perpendicular to the surface is $\mathbf{V} \cdot \mathbf{N}$, where $\mathbf{N}$ is the unit outward normal vector. It is at once apparent that $\rho\mathbf{V} \cdot d\mathbf{d} \equiv \mathbf{f} \cdot d\mathbf{d}$ represents the outward flow of mass per unit time. Hence the total loss of mass per unit time is given by

$$\iint\limits_S \mathbf{f} \cdot d\mathbf{d} \tag{177}$$

Equating (176) and (177), we have the divergence theorem:

$$\iint\limits_S \mathbf{f} \cdot d\mathbf{d} = \iiint\limits_V (\nabla \cdot \mathbf{f})\, d\tau \tag{178}$$

For a more detailed and rigorous proof, see Kellogg, "Foundations of Potential Theory." We now derive Gauss's theorem by a different method. Let **f** be a differentiable vector inside a connected region R with rectifiable surface S. Surround any point P of R by a small element of volume $d\tau$ having a surface area ΔS. Form the surface integral $\iint_{\Delta S} \mathbf{f} \cdot d\boldsymbol{\sigma}$ and consider the limit,

$$\lim_{\Delta\tau \to 0} \frac{\iint_{\Delta S} \mathbf{f} \cdot d\boldsymbol{\sigma}}{\Delta\tau}$$

If this limit exists independent of the approach of $\Delta\tau$ to zero, we define

$$\operatorname{div} \mathbf{f} = \lim_{\Delta\tau \to 0} \frac{\iint_{\Delta S} \mathbf{f} \cdot d\boldsymbol{\sigma}}{\Delta\tau} \tag{179}$$

We can write

$$\Delta\tau \operatorname{div} \mathbf{f} = \iint_{\Delta S} \mathbf{f} \cdot d\boldsymbol{\sigma} + \epsilon\, \Delta\tau \tag{180}$$

where $\epsilon \to 0$ as $\Delta\tau \to 0$. If we now subdivide our region into many elementary volumes, we obtain formulas of the type (180) for all of these regions. Summing up (180) for all volumes and then passing to the limit, we have

$$\iiint_R \operatorname{div} \mathbf{f}\, d\tau = \iint_S \mathbf{f} \cdot d\boldsymbol{\sigma} \tag{181}$$

In the derivation of (181) use has been made of the fact that for each internal $d\boldsymbol{\sigma}$ there is a $-d\boldsymbol{\sigma}$, so that all interior surface integrals cancel in pairs, leaving only the boundary surface S as a contributing factor. The sum of the $\epsilon_i\, \Delta\tau_i$ vanishes in the limit, for $\left|\Sigma \epsilon_i\, \Delta\tau_i\right| \leqq \Sigma \left|\epsilon\right|_{\max} \Delta\tau \leqq \left|\epsilon\right|_{\max} V$, and if div **f** is continuous, $\left|\epsilon\right|_{\max} \to 0$ as $\Delta\tau \to 0$.

By choosing rectangular parallelepipeds and using the method of Sec. 20, we can show that

$$\operatorname{div} \mathbf{f} = \lim_{\Delta\tau \to 0} \frac{\iint_{\Delta S} \mathbf{f} \cdot d\boldsymbol{\sigma}}{\Delta\tau} \equiv \frac{\partial u}{\partial x} + \frac{\partial v}{\partial y} + \frac{\partial w}{\partial z} \tag{182}$$

for $\mathbf{f} = u\mathbf{i} + v\mathbf{j} + w\mathbf{k}$, which corresponds to our original definition of the divergence.

Example 65. Consider a sphere with center at O and radius a. Take $\mathbf{f} = \mathbf{r} = x\mathbf{i} + y\mathbf{j} + z\mathbf{k}$,

$$d\mathbf{d} = \left(\frac{\mathbf{r}}{a}\right) dS = \left(\frac{1}{a}\right) (x\mathbf{i} + y\mathbf{j} + z\mathbf{k})\, dS$$

Now $\nabla \cdot \mathbf{f} = 3$, $\mathbf{f} \cdot d\mathbf{d} = (1/a)(x^2 + y^2 + z^2)\, dS = a\, dS$ on the sphere. Applying (178), $\iiint 3\, d\tau = \iint a\, dS$, or $3V = aS$, where S is the surface area of the sphere. If V is known to be $\frac{4}{3}\pi a^3$, then $S = 4\pi a^2$.

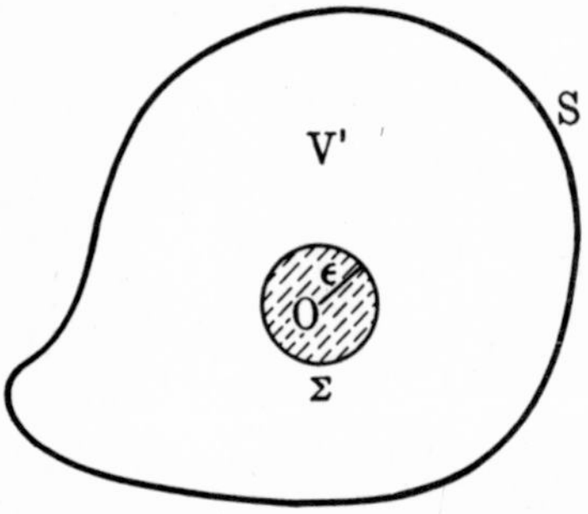

FIG. 50.

Example 66. Let $\mathbf{f} = q\mathbf{r}/r^3$ and let V be a region surrounding the origin and let S be its surface. We cannot apply the divergence theorem to this region since $\mathbf{f}$ is discontinuous at $r = 0$. We overcome this difficulty by surrounding the origin by a small sphere Σ of radius ϵ (see Fig. 50). The divergence theorem can be applied to the connected region V'. The region V' has two boundaries, S and Σ. Applying (178),

$$\iiint_{V'} \nabla \cdot \left(\frac{q\mathbf{r}}{r^3}\right) d\tau = \iint_S \frac{q\mathbf{r} \cdot d\mathbf{d}}{r^3} + \iint_\Sigma \frac{q\mathbf{r} \cdot d\mathbf{d}}{r^3} \tag{183}$$

In Example 25 we saw that $\nabla \cdot (\mathbf{r}/r^3) = 0$. This implies that (183) reduces to

$$\iint_S \frac{q\mathbf{r}}{r^3} \cdot d\mathbf{d} = - \iint_\Sigma \frac{q\mathbf{r} \cdot d\mathbf{d}}{r^3} \tag{184}$$

Now for the sphere Σ, $d\mathbf{d} = -\mathbf{r}\, dS/\epsilon$, since the outward normal to the region V' is directed toward the origin and is parallel to the radius vector. Hence

$$\iint_S \frac{q\mathbf{r}}{r^3} \cdot d\mathbf{d} = \iint_S \mathbf{f} \cdot d\mathbf{d} = \iint \frac{q}{\epsilon^2}\, dS = 4\pi q$$

since $\iint dS = 4\pi\epsilon^2$. The integral $\iint_S \mathbf{f} \cdot d\boldsymbol{\sigma}$ is called the *flux* of the vector field $\mathbf{f}$ over the surface S. We have, for an inverse-square force $\mathbf{f} = q\mathbf{r}/r^3$,

$$\iint_S \mathbf{f} \cdot d\boldsymbol{\sigma} = 4\pi q \tag{185}$$

Example 67. A vector field $\mathbf{f}$ whose flux over every closed surface vanishes is called a solenoidal vector field. From (178) it is easy to verify that $\nabla \cdot \mathbf{f} = 0$ for such fields. Hence a solenoidal vector is characterized by either $\oint\!\!\oint \mathbf{f} \cdot d\boldsymbol{\sigma} = 0$ or $\nabla \cdot \mathbf{f} = 0$.

Now assume $\mathbf{f} = \nabla \times \mathbf{g}$, which implies $\nabla \cdot \mathbf{f} = \nabla \cdot (\nabla \times \mathbf{g}) = 0$. Is the converse true? If $\nabla \cdot \mathbf{f} = 0$, can we write $\mathbf{f}$ as the curl of some vector $\mathbf{g}$? The answer is "Yes"! We call $\mathbf{g}$ the vector potential of $\mathbf{f}$. This theorem is of importance in electricity theory, as we shall see later. Notice that $\mathbf{g}$ is not uniquely determined since $\nabla \times (\mathbf{g} + \nabla\varphi) = \nabla \times \mathbf{g}$. We now show the existence of $\mathbf{g}$. Let $\mathbf{f} = X\mathbf{i} + Y\mathbf{j} + Z\mathbf{k}$ and assume

$$\mathbf{g} = \alpha\mathbf{i} + \beta\mathbf{j} + \gamma\mathbf{k}$$

We wish to find α, β, γ such that $\mathbf{f} = \nabla \times \mathbf{g}$, and hence

$$\begin{aligned} X &= \frac{\partial\gamma}{\partial y} - \frac{\partial\beta}{\partial z} \\ Y &= \frac{\partial\alpha}{\partial z} - \frac{\partial\gamma}{\partial x} \\ Z &= \frac{\partial\beta}{\partial x} - \frac{\partial\alpha}{\partial y} \end{aligned} \tag{186}$$

Now assume $\alpha = 0$. Then $X = \dfrac{\partial\gamma}{\partial y} - \dfrac{\partial\beta}{\partial z}$, $Y = -\dfrac{\partial\gamma}{\partial x}$, $Z = \dfrac{\partial\beta}{\partial x}$. Consequently if there is a solution with $\alpha = 0$, of necessity

$$\begin{aligned} \beta &= \int_{x_0}^{x} Z\,dx + \sigma(y, z) \\ \gamma &= -\int_{x_0}^{x} Y\,dx + \tau(y, z) \end{aligned}$$

Now $\nabla \cdot \mathbf{f} = 0$ or $\dfrac{\partial X}{\partial x} = -\left(\dfrac{\partial Y}{\partial y} + \dfrac{\partial Z}{\partial z}\right)$ by assumption.

Hence $$\frac{\partial \gamma}{\partial y} - \frac{\partial \beta}{\partial z} = -\int_{x_0}^{x} \left(\frac{\partial Y}{\partial y} + \frac{\partial Z}{\partial z} \right) dx + \frac{\partial \tau}{\partial y} - \frac{\partial \sigma}{\partial z}$$

$$= \int_{x_0}^{x} \frac{\partial X}{\partial x}\, dx + \frac{\partial \tau}{\partial y} - \frac{\partial \sigma}{\partial z}$$

$$= X(x, y, z) - X(x_0, y, z) + \frac{\partial \tau}{\partial y} - \frac{\partial \sigma}{\partial z}$$

so that (186) is satisfied by $\alpha = 0$, $\beta = \int_{x_0}^{x} Z\, dx + \sigma(y, z)$, $\gamma = -\int_{x_0}^{x} Y\, dx$ by choosing $\tau = 0$, $\sigma(y, z) = -\int_{z_0}^{z} X(x_0, y, z)\, dZ$. Hence $\mathbf{f} = \nabla \times \mathbf{g}$ where $\mathbf{g} = \left[\int_{x_0}^{x} Z\, dx + \sigma(y, z) \right] \mathbf{j} - \int_{x_0}^{x} Y\, dx\, \mathbf{k}$. In general

$$\mathbf{g} = \left[\int_{x_0}^{x} Z\, dx + \sigma(y, z) \right] \mathbf{j} - \int_{x_0}^{x} Y\, dx\, \mathbf{k} + \nabla \varphi \tag{187}$$

For example, if $\mathbf{f} = \nabla(1/r)$, then $\nabla \cdot \mathbf{f} = 0$. Now

$$X = -\frac{x}{r^3}, \qquad Y = -\frac{y}{r^3}, \qquad Z = -\frac{z}{r^3}$$

where $r^2 = x^2 + y^2 + z^2$. Applying (187)

$$\mathbf{g} = \int_0^x \frac{-z\, dx}{(x^2 + y^2 + z^2)^{\frac{3}{2}}}\, \mathbf{j} + \int_0^x \frac{y\, dx}{(x^2 + y^2 + z^2)^{\frac{3}{2}}}\, \mathbf{k} + \nabla \varphi,$$

φ arbitrary

and

$$\mathbf{g} = \frac{x}{(x^2 + y^2 + z^2)^{\frac{1}{2}} (y^2 + z^2)} (-z\mathbf{j} + y\mathbf{k}) + \nabla \varphi$$

Example 68. *Green's theorem.* We apply (178) to $\mathbf{f}_1 = u \nabla v$ and $\mathbf{f}_2 = v \nabla u$ and obtain

$$\iiint\limits_R \nabla \cdot (u \nabla v)\, d\tau = \iiint\limits_R (u \nabla^2 v + \nabla u \cdot \nabla v)\, d\tau = \iint\limits_S u \nabla v \cdot d\boldsymbol{\sigma} \tag{188}$$

$$\iiint\limits_R \nabla \cdot (v \nabla u)\, d\tau = \iiint\limits_R (v \nabla^2 u + \nabla v \cdot \nabla u)\, d\tau = \iint\limits_S v \nabla u \cdot d\boldsymbol{\sigma}$$

Subtracting, we obtain

$$\iiint\limits_R (u \nabla^2 v - v \nabla^2 u)\, d\tau = \iint\limits_S (u \nabla v - v \nabla u) \cdot d\boldsymbol{\sigma} \tag{189}$$

Example 69. *Uniqueness theorem.* Assume two functions which satisfy Laplace's equation everywhere inside a region and which take on the same values over the boundary surface S. The functions are identical. Let $\nabla^2\varphi = \nabla^2\psi = 0$ inside R and $\varphi = \psi$ on S. Now from (188) we have

$$\iiint_R \theta\,\nabla^2\theta\,d\tau + \iiint_R \nabla\theta\cdot\nabla\theta\,d\tau = \iint_S \theta\,\nabla\theta\cdot d\boldsymbol{\sigma}$$

Define $\theta \equiv \varphi - \psi$ so that $\nabla^2\theta = 0$ over R and $\theta = 0$ on S. Hence $\displaystyle\iiint_R (\nabla\theta)^2\,d\tau = 0$, which implies $\nabla\theta = 0$ inside R. Hence $\theta \equiv \text{constant} = \varphi - \psi$, and since $\varphi = \psi$ on S, we must have $\varphi \equiv \psi$. We have assumed the existence of $\nabla\varphi$, $\nabla\psi$ on S.

Example 70. *Another uniqueness theorem.* Let $\mathbf{f}$ be a vector whose curl and divergence are known in a simply connected region R, and whose normal components are given on the surface S which bounds R. We now prove that $\mathbf{f}$ is unique. Let $\mathbf{f}_1$ be another vector such that $\nabla\cdot\mathbf{f} \equiv \nabla\cdot\mathbf{f}_1$, $\nabla\times\mathbf{f} \equiv \nabla\times\mathbf{f}_1$, and $\mathbf{f}\cdot d\boldsymbol{\sigma} = \mathbf{f}_1\cdot d\boldsymbol{\sigma}$ on S. We now construct the vector $\mathbf{g} = \mathbf{f} - \mathbf{f}_1$. We immediately have that $\nabla\cdot\mathbf{g} = \nabla\times\mathbf{g} = \mathbf{g}\cdot d\boldsymbol{\sigma} = 0$. In Sec. 52 we saw that if $\nabla\times\mathbf{g} = 0$, then $\mathbf{g}$ is the gradient of a scalar φ, $\mathbf{g} = \nabla\varphi$. Consequently, $\nabla\cdot\mathbf{g} = \nabla^2\varphi = 0$. Applying (188) with $u = v = \varphi$, we have

$$\iiint_R [\varphi\,\nabla^2\varphi + (\nabla\varphi)^2]\,d\tau = \iint_S \varphi\,\nabla\varphi\cdot d\boldsymbol{\sigma} = \int \varphi\mathbf{g}\cdot d\boldsymbol{\sigma} = 0$$

and hence $\displaystyle\iiint_R (\nabla\varphi)^2\,d\tau = 0$ so that $\nabla\varphi \equiv 0$ inside R, and $\mathbf{g} = \nabla\varphi \equiv 0$, so that $\mathbf{f} = \mathbf{f}_1$ inside R.

We can also prove that $\mathbf{f}$ is uniquely determined if its divergence and curl are known throughout all of space, provided that $\mathbf{f}$ tends to zero like $1/r^2$ as $r \to \infty$. We duplicate the above proof and need $\displaystyle\lim \iint_{S\to\infty} \varphi\,\nabla\varphi\cdot d\boldsymbol{\sigma} = 0$. If φ tends to zero like $1/r$, then $\nabla\varphi$ tends to zero like $1/r^2$, and $\displaystyle\iint_S \varphi\,\nabla\varphi\cdot d\boldsymbol{\sigma}$ tends to zero like $1/r$ as $r \to \infty$.

Example 71. Let $\mathbf{f} = f(x, y, z)\mathbf{a}$, where $\mathbf{a}$ is constant. Applying (178), we obtain

$$\mathbf{a} \cdot \iint_S f\, d\boldsymbol{\sigma} = \iiint_V \nabla \cdot (f\mathbf{a})\, d\tau = \mathbf{a} \cdot \iiint_V \nabla f\, d\tau$$

Hence

$$\iint_S f\, d\boldsymbol{\sigma} = \iiint_V \nabla f\, d\tau \tag{190}$$

We leave it to the reader to prove that

$$\boxed{\iint_S d\boldsymbol{\sigma} * \mathbf{f} = \iiint_V (\nabla * \mathbf{f})\, d\tau} \tag{191}$$

Problems

1. Prove that $\iint_S d\boldsymbol{\sigma} \times \mathbf{f} = \iiint_V (\dot{\nabla} \times \mathbf{f})\, d\tau$.

2. Prove that $\iint_S d\boldsymbol{\sigma} = 0$ over a closed surface S.

3. If $\mathbf{f} = ax\mathbf{i} + by\mathbf{j} + cz\mathbf{k}$, a, b, c constants, show that $\iint_S \mathbf{f} \cdot d\boldsymbol{\sigma} = \frac{4}{3}\pi(a + b + c)$, where S is the surface of a unit sphere.

4. By defining $\operatorname{grad} f = \lim_{\Delta\tau \to 0} \dfrac{\iint_{\Delta S} f\, d\boldsymbol{\sigma}}{\Delta\tau}$, show that

$$\operatorname{grad} f = \frac{\partial f}{\partial x}\mathbf{i} + \frac{\partial f}{\partial y}\mathbf{j} + \frac{\partial f}{\partial z}\mathbf{k}$$

5. By defining $\operatorname{div} \mathbf{f} = \lim_{\Delta\tau \to 0} \dfrac{\iint_{\Delta S} \mathbf{f} \cdot d\boldsymbol{\sigma}}{\Delta\tau}$, show that

$$\operatorname{div} \mathbf{f} = \frac{1}{r^2 \sin\theta}\left[\frac{\partial}{\partial r}(r^2 \sin\theta\, f_r) + \frac{\partial}{\partial\theta}(r \sin\theta\, f_\theta) + \frac{\partial}{\partial\varphi}(r f_\varphi)\right]$$

for spherical coordinates.

6. Show that

$$\iiint_V \mathbf{f} \cdot \nabla\varphi\, d\tau = \iint_S \varphi\mathbf{f} \cdot d\boldsymbol{\sigma} - \iiint_V \varphi\, \nabla \cdot \mathbf{f}\, d\tau.$$

7. If $\mathbf{w} = \frac{1}{2}\nabla \times \mathbf{v}$, $\mathbf{v} = \nabla \times \mathbf{u}$, show that

$$\frac{1}{2}\iiint_R \mathbf{v}^2\, d\tau = \frac{1}{2}\iint_S \mathbf{u} \times \mathbf{v} \cdot d\boldsymbol{\sigma} + \iiint_R \mathbf{u} \cdot \mathbf{w}\, d\tau.$$

8. Show that

$$\iiint_R w \nabla u \cdot \nabla v\, d\tau = \iint_S uw \nabla v \cdot d\boldsymbol{\sigma} - \iiint_R u \nabla \cdot (w \nabla v)\, d\tau$$

9. If $\mathbf{v} = \nabla\varphi$ and $\nabla \cdot \mathbf{v} = 0$, show that for a closed surface

$$\iiint_R \mathbf{v}^2\, d\tau = \iint_S \varphi\mathbf{v} \cdot d\boldsymbol{\sigma}.$$

10. Show that $\iint_S |\mathbf{r}|^2\mathbf{r} \cdot d\boldsymbol{\sigma} = 5\iiint_R \mathbf{r}^2\, d\tau.$

11. If $\mathbf{f} = x\mathbf{i} - y\mathbf{j} + (z^2 - 1)\mathbf{k}$, find the value of $\iint_S \mathbf{f} \cdot d\boldsymbol{\sigma}$ over the closed surface bounded by the planes $z = 0$, $z = 1$ and the cylinder $x^2 + y^2 = 1$.

12. If $\mathbf{f}$ is directed along the normal at each point of the boundary of a region V, show that $\iiint_V (\nabla \times \mathbf{f})\, d\tau = 0.$

13. Show that $\iint_S \mathbf{r} \times d\boldsymbol{\sigma} = 0$ over a closed surface.

14. Given $\mathbf{f} = (xye^z + \log z - \sin x)\mathbf{k}$, find the value of $\iint_S \nabla \times \mathbf{f} \cdot d\boldsymbol{\sigma}$ over the part of the sphere $x^2 + y^2 + z^2 = 1$ above the x-y plane.

15. Show that $(x\mathbf{i} + y\mathbf{j})/(x^2 + y^2)$ is solenoidal.

16. If $\mathbf{f}_1$ and $\mathbf{f}_2$ are irrotational, show that $\mathbf{f}_1 \times \mathbf{f}_2$ is solenoidal.

17. Find a vector $\mathbf{A}$ such that

$$\mathbf{f} \equiv yz\mathbf{i} - zx\mathbf{j} + (x^2 + y^2)\mathbf{k} = \nabla \times \mathbf{A}.$$

18. If r, θ, z are cylindrical coordinates, show that $\nabla\theta$ and $\nabla \log r$ are solenoidal vectors. Find the vector potentials.

19. Let S_1 and S_2 be the surface boundaries of two regions V_1 and V_2. Let r be the distance between two elementary

volumes $d\tau_1$ and $d\tau_2$ of V_1 and V_2. Show that

$$\int_{S_2} d\boldsymbol{\delta}_2 \cdot \int_{S_1} r^m \, d\boldsymbol{\delta}_1 = -m(m+1) \int_{V_2} d\tau_2 \int_{V_1} r^{m-2} \, d\tau_1$$

and that

$$\int_{S_2} d\boldsymbol{\delta}_2 \cdot \int_{S_1} \log r \, d\boldsymbol{\delta}_1 = -\int_{V_2} d\tau_2 \int_{V_1} \frac{d\tau_1}{r}$$

20. If $\nabla \cdot \mathbf{f} = \varphi$, $\nabla \times \mathbf{f} = \psi_1 \mathbf{i} + \psi_2 \mathbf{j} + \psi_3 \mathbf{k}$, $\mathbf{f} = X\mathbf{i} + Y\mathbf{j} + Z\mathbf{k}$, show that $\nabla^2 X = \dfrac{\partial \varphi}{\partial x} - \dfrac{\partial \psi_3}{\partial y} - \dfrac{\partial \psi_2}{\partial z}$, and find similar expressions for $\nabla^2 Y$, $\nabla^2 Z$. Find a vector $\mathbf{f}$ such that $\nabla \cdot \mathbf{f} = 2x + y - 1$, $\nabla \times \mathbf{f} = z\mathbf{i}$.

56. Conjugate Functions. Let us consider the two-dimensional vector field $\mathbf{w} = u(x, y)\mathbf{i} + v(x, y)\mathbf{j}$ and an orthogonal vector field $\mathbf{w}_1 = v(x, y)\mathbf{i} - u(x, y)\mathbf{j}$. Obviously $\mathbf{w} \cdot \mathbf{w}_1 = 0$. What are the conditions on $u(x, y)$, $v(x, y)$ which will make $\mathbf{w}$ and $\mathbf{w}_1$ irrotational? From Stokes's theorem

$$\begin{aligned} \oint \mathbf{w} \cdot d\mathbf{r} &= \iint_S \nabla \times \mathbf{w} \cdot d\boldsymbol{\delta} = \iint_S \left(\frac{\partial v}{\partial x} - \frac{\partial u}{\partial y} \right) dy \, dx \\ \oint \mathbf{w}_1 \cdot d\mathbf{r} &= \iint_S \nabla \times \mathbf{w}_1 \cdot d\boldsymbol{\delta} = \iint_S \left(-\frac{\partial u}{\partial x} - \frac{\partial v}{\partial y} \right) dy \, dx \end{aligned} \tag{192}$$

A necessary and sufficient condition that both $\mathbf{w}$ and $\mathbf{w}_1$ be irrotational is that $\dfrac{\partial v}{\partial x} - \dfrac{\partial u}{\partial y} = 0$ and $-\dfrac{\partial u}{\partial x} - \dfrac{\partial v}{\partial y} = 0$ (see Sec. 52). This yields

$$\begin{aligned} \frac{\partial v}{\partial x} &= \frac{\partial u}{\partial y} \\ \frac{\partial v}{\partial y} &= -\frac{\partial u}{\partial x} \end{aligned} \tag{193}$$

The reader who is familiar with complex-variable theory will immediately recognize (193) as the Cauchy-Riemann equations, which must be satisfied for the analyticity of the complex function $w = v(x, y) + iu(x, y)$, $i = \sqrt{-1}$.

On differentiating (193), we obtain

$$\left.\begin{aligned} \nabla^2 u &= \frac{\partial^2 u}{\partial x^2} + \frac{\partial^2 u}{\partial y^2} = 0 \\ \nabla^2 v &= \frac{\partial^2 v}{\partial x^2} + \frac{\partial^2 v}{\partial y^2} = 0 \\ \text{and} \qquad & \\ \frac{\partial u}{\partial x}\frac{\partial v}{\partial x} &+ \frac{\partial u}{\partial y}\frac{\partial v}{\partial y} = 0 \end{aligned}\right\} \tag{194}$$

If functions $u(x, y)$, $v(x, y)$ satisfy Eqs. (193) we say that they are harmonic conjugates. The importance of such func-

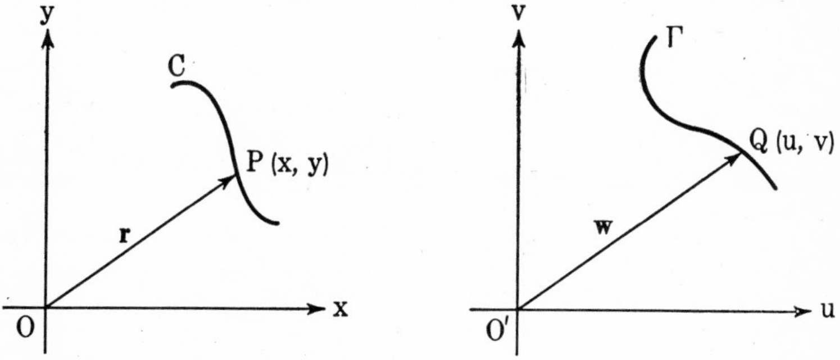

FIG. 51.

tions is due to the fact that they satisfy the two-dimensional Laplace's equation given by (194). If u satisfies $\nabla^2 u = 0$, we say that $u(x, y)$ is harmonic.

Let us now consider two rectangular cartesian coordinate systems, the x-y plane and the u-v plane (Fig. 51). Let

$$\mathbf{r} = x\mathbf{i} + y\mathbf{j}.$$

Now to every point $P(x, y)$ there corresponds a point $Q(u, v)$ given by the transformation $u = u(x, y)$, $v = v(x, y)$. Hence the vector $\mathbf{w} = u(x, y)\mathbf{i} + v(x, y)\mathbf{j}$ corresponds to the vector

$$\mathbf{r} = x\mathbf{i} + y\mathbf{j}.$$

If now $P(x, y)$ traverses a curve C in the x-y plane, $Q(u, v)$ will trace out a corresponding curve Γ in the u-v plane.

The curve $u(x, y) = \text{constant}$ in the x-y plane transforms into the straight line $u = \text{constant}$ in the u-v plane. Similarly,

$v(x, y)$ = constant transforms into the straight line v = constant. The two straight lines are orthogonal. Do the curves

$$u(x, y) = \text{constant}$$

$v(x, y)$ = constant intersect orthogonally? The answer is "Yes"! The normal to the curve $u(x, y)$ = constant is the vector

$$\nabla u = \frac{\partial u}{\partial x}\mathbf{i} + \frac{\partial u}{\partial y}\mathbf{j}$$

and the normal to the curve $v(x, y)$ = constant is $\nabla v = \dfrac{\partial v}{\partial x}\mathbf{i} + \dfrac{\partial v}{\partial y}\mathbf{j}$

so that $\nabla u \cdot \nabla v = \dfrac{\partial u}{\partial x}\dfrac{\partial v}{\partial x} + \dfrac{\partial u}{\partial y}\dfrac{\partial v}{\partial y} \equiv 0$ from (194).

Example 72. Consider the vector field $\mathbf{w} = 2xy\mathbf{i} + (x^2 - y^2)\mathbf{j}$. Here $u = 2xy$, $v = x^2 - y^2$, and

$$\frac{\partial v}{\partial x} = \frac{\partial u}{\partial y} = 2x$$

$$\frac{\partial v}{\partial y} = -\frac{\partial u}{\partial x} = -2y$$

so that u and v are conjugate harmonics. The curves

$$u = 2xy = \text{constant}$$

and $v(x, y) = x^2 - y^2$ = constant are orthogonal hyperbolas which transform into the straight lines u = constant, v = constant, in the u-v plane (Fig. 52).

Example 73. Consider

$$\mathbf{w} = \left(\tan^{-1}\frac{y}{x}\right)\mathbf{i} + \tfrac{1}{2}\log(x^2 + y^2)\mathbf{j}$$

Here $u(x, y) = \tan^{-1}(y/x)$, $v = \frac{1}{2}\log(x^2 + y^2)$, and

$$\frac{\partial u}{\partial x} = -\frac{\partial v}{\partial y} = -\frac{y}{x^2 + y^2}$$

$$\frac{\partial u}{\partial y} = \frac{\partial v}{\partial x} = \frac{x}{x^2 + y^2}$$

so that u and v are conjugate harmonics.

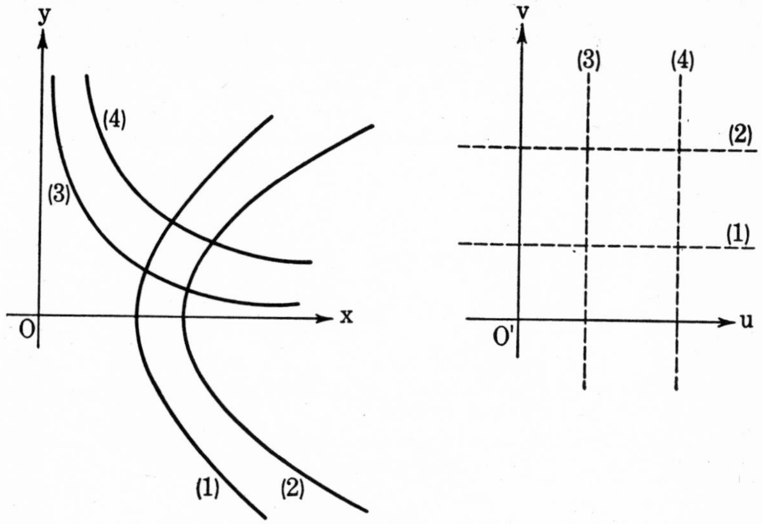

FIG. 52.

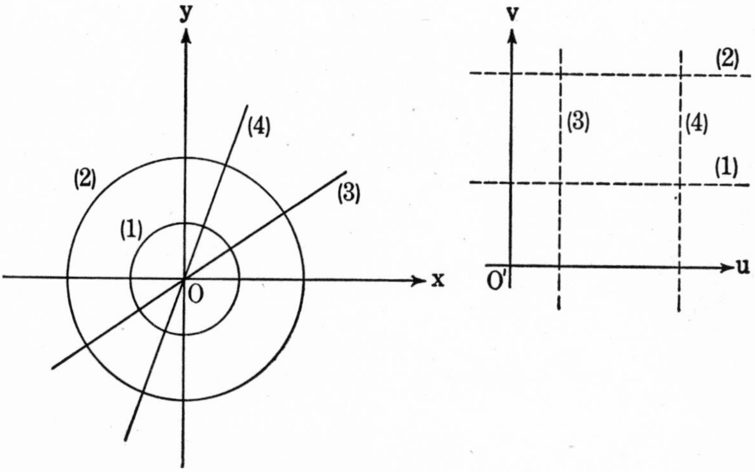

FIG. 53.

The circles $x^2 + y^2 =$ constant transform into the straight lines $v = \frac{1}{2} \log c$, while the straight lines $y = mx$ transform into the straight lines $u = \tan^{-1} m$ (Fig. 53).

Example 74. If $u(x, y)$ is given as harmonic, we can find its conjugate $v(x, y)$. If $v(x, y)$ does exist satisfying (193), then

$$dv = \frac{\partial v}{\partial x}\,dx + \frac{\partial v}{\partial y}\,dy = \frac{\partial u}{\partial y}\,dx - \frac{\partial u}{\partial x}\,dy$$

Now consider the vector field $\mathbf{f} = \frac{\partial u}{\partial y}\,\mathbf{i} - \frac{\partial u}{\partial x}\,\mathbf{j}$. We have that $\nabla \times \mathbf{f} = -\left(\frac{\partial^2 u}{\partial x^2} + \frac{\partial^2 u}{\partial y^2}\right)\mathbf{k} = 0$ by our assumption about $u(x, y)$. Hence $\mathbf{f}$ is irrotational and so is the gradient of the scalar v, $\mathbf{f} = \nabla v$, and

$$v = \int_a^x \frac{\partial u}{\partial y}\,dx - \int_b^y \frac{\partial u}{\partial x}\,dy + c \tag{195}$$

As an example, consider $u = x^2 - y^2$, which satisfies $\nabla^2 u = 0$. Hence

$$v = \int_0^x -2y\,dx - \int_0^y 2 \cdot 0\,dy + c = -2xy + c$$

Problems

1. Find the harmonic conjugate of $x^3 - 3xy^2$, of $e^x \cos y$, of $x/(x^2 + y^2)$.

2. Show that $u(x, y) = \sin x \cosh y$ and $v(x, y) = \cos x \sinh y$ are conjugate harmonics and that the curves $u(x, y) =$ constant, $v(x, y) =$ constant are orthogonal. What do the straight lines $y =$ constant transform into?

3. If $u(x, y)$, $v(x, y)$ are conjugate harmonics, show that the angle between any two curves in the x-y plane remains invariant under the transformation $u = u(x, y)$, $v = v(x, y)$, that is, the transformed curves have the same angle of intersection.

CHAPTER 5

STATIC AND DYNAMIC ELECTRICITY

57. Electrostatic Forces. We assume that the reader is familiar with the methods of generating electrostatic charges. It is found by experiment that the repulsion of two like point charges is inversely proportional to the square of the distance between the charges and directly proportional to the product of their charges. The forces act along the line joining the two charges. We define the electrostatic unit of charge (e.s.u.) as that charge which produces a force of one dyne on a like charge situated one centimeter from it when both are placed in a vacuum. The electrostatic intensity at a point P is the force that would act on a unit charge placed at P as a result of the rest of the charges, provided that the unit test charge does not affect the original distribution of charges. For a single charge q placed at the origin of our coordinate system, the electric intensity, or field, is given by $\mathbf{E} = (q/r^3)\mathbf{r}$. For many charges the field at P is given by

$$\mathbf{E} = -\sum_{\alpha=1}^{n} \frac{q_\alpha}{r_\alpha^{\,3}}\mathbf{r}_\alpha \tag{196}$$

where $\mathbf{r}_\alpha$ represents the vector from P to the charge q_α.

We have seen in Example 25 that $\nabla \cdot (\mathbf{r}/r^3) = 0$. Consequently,

$$\boxed{\nabla \cdot \mathbf{E} = 0} \tag{197}$$

so that the divergence of the electrostatic-field vector is zero at any point in space where no charge exists. Hence $\mathbf{E}$ is solenoidal except where charges exist, for there $\mathbf{E}$ is discontinuous. If the coordinates of P are ξ, η, ζ and the coordinates of q_i are x_i, y_i, z_i, then $r_i = [(\xi - x_i)^2 + (\eta - y_i)^2 + (\zeta - z_i)^2]^{\frac{1}{2}}$, and

$$\nabla \frac{1}{r_i} = \frac{\partial}{\partial \xi}\left(\frac{1}{r_i}\right)\mathbf{i} + \frac{\partial}{\partial \eta}\left(\frac{1}{r_i}\right)\mathbf{j} + \frac{\partial}{\partial \zeta}\left(\frac{1}{r_i}\right)\mathbf{k} = \frac{\mathbf{r}_i}{r_i^3}$$

$$\mathbf{r}_i = (x_i - \xi)\mathbf{i} + (y_i - \eta)\mathbf{j} + (z_i - \zeta)\mathbf{k}$$

so that (196) reduces to

$$\mathbf{E} = -\nabla\varphi \tag{198}$$

where $\varphi = \sum_{\alpha=1}^{n} q_\alpha/r_\alpha$. We call φ the electrostatic potential. For any closed path which does not pass through a point charge, we have $\oint \mathbf{E} \cdot d\mathbf{r} = -\oint \nabla\varphi \cdot d\mathbf{r} = -\oint d\varphi = 0$. Thus $\mathbf{E}$ is also irrotational, and

$$\nabla \times \mathbf{E} = 0 \tag{199}$$

We also note that $\int_P^\infty \mathbf{E} \cdot d\mathbf{r} = \varphi(P) - \varphi(\infty) = \varphi(P)$, since $\varphi(\infty) = 0$. Hence the work done by the field in taking a unit charge from P to ∞ is equal to the potential at P.

58. Gauss's Law. Let S be an imaginary closed surface that does not intersect any charges. In Example 66 we saw that $\iint_S (q\mathbf{r}/r^3) \cdot d\boldsymbol{\sigma} = 4\pi q$. This is true for each charge q_i inside S. Hence

$$\iint_S \sum_{\alpha=1}^{n} \frac{q_\alpha \mathbf{r}_\alpha \cdot d\boldsymbol{\sigma}}{r_\alpha{}^3} = 4\pi \sum_{\alpha=1}^{n} q_\alpha \tag{200}$$

For a charge outside S,

$$\iint_S \frac{q\mathbf{r}}{r^3} \cdot d\boldsymbol{\sigma} = \iiint_V \nabla \cdot \left(\frac{q\mathbf{r}}{r^3}\right) d\tau = 0 \tag{201}$$

since there is no discontinuity in $q\mathbf{r}/r^3$, $r > 0$. Adding (200) and (201), we obtain Gauss's law,

$$\iint_S \mathbf{E} \cdot d\boldsymbol{\sigma} = 4\pi Q \tag{202}$$

where Q is the total charge inside S. The theorem in words is that the total electric flux over any closed surface equals 4π times the total charge inside the surface.

Example 75. We define a *conductor* as a body with no electric field in its interior, for otherwise the "free" electrons would move and the field would not be static. The charge on a conductor must reside on the surface, for consider any small volume contained in the conductor and apply Gauss's theorem.

$$\iint \mathbf{E} \cdot d\mathbf{\sigma} = 4\pi q$$

and since $\mathbf{E} = 0$, we must have $q = 0$. This is true for arbitrarily small volumes, so that no excess of positive charges over negative charges exists. Hence the total charge must exist on the surface of the conductor.

If a body has the property that a charge placed on it continues to reside where placed in the absence of an external electric field,

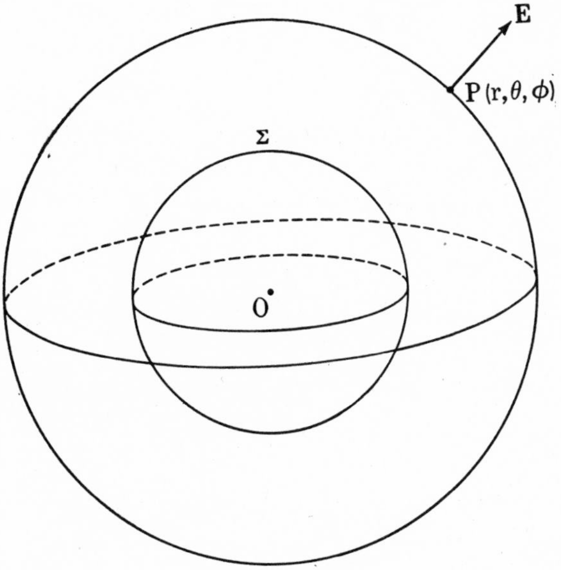

FIG. 54.

we call the body an *insulator*. Actually there is no sharp line of demarcation between conductors and insulators. Every body possesses some ability in conducting electrons.

At the surface of a conductor the field is normal to the surface, for any component of the field tangent to the surface would cause a flow of current in the conductor, this again being contrary to the assumption that the field is static (no large-scale motion of electrons occurring). Such a surface is called an equipotential surface. The field is everywhere normal to an equipotential surface, for the vector $\mathbf{E} = -\nabla\varphi$ is normal everywhere to the surface $\varphi(x, y, z) =$ constant.

Example 76. Consider a uniformly charged hollow sphere Σ. We shall show that the field outside the sphere is the same as if the total charge were concentrated at the center of the sphere and that in the interior of the sphere there is no field.

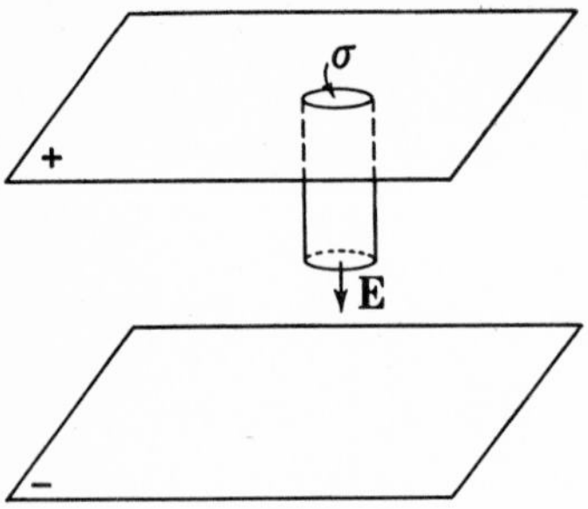

FIG. 55.

Let P be any point outside the sphere with spherical coordinates r, θ, φ. Construct an imaginary sphere through P concentric with the sphere Σ (see Fig. 54). From symmetry it is obvious that the intensity at any point of the sphere is the same as that at P. Moreover, the field is radial. Applying Gauss's law, we have $\iint \mathbf{E} \cdot d\boldsymbol{\sigma} = 4\pi Q$, or

$$\iint E\, dS = E \iint dS = 4\pi r^2 E = 4\pi Q$$

so that

$$E = \frac{Q}{r^2} \qquad \text{and} \qquad \mathbf{E} = \frac{Q}{r^3}\mathbf{r} \tag{203}$$

We leave it to the reader to show that $\mathbf{E} = 0$ inside Σ.

Example 77. *Field within a parallel-plate condenser.* Consider two infinite parallel plates with surface densities σ and $-\sigma$.

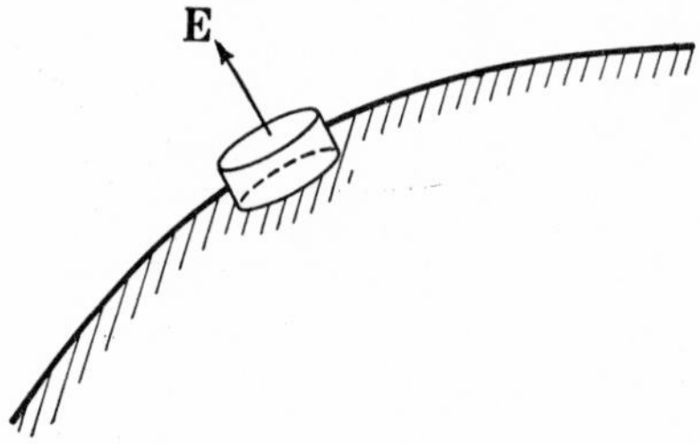

FIG. 56.

From symmetry the field is normal to the plates. We apply Gauss's law to the surface in Fig. 55 with unit cross-sectional area.

$$\iint \mathbf{E} \cdot d\boldsymbol{\sigma} = E = 4\pi\sigma \tag{204}$$

so that the field is uniform.

Example 78. We now determine the field in the neighborhood of a conductor. We consider the cylindrical pillbox of Fig. 56 and apply Gauss's law to obtain

$$EA = 4\pi\sigma A$$

or

$$\mathbf{E} = 4\pi\sigma\mathbf{N} \tag{205}$$

where σ is the charge per unit area and **N** is the unit normal vector to the surface of the conductor.

Example 79. *Force on the surface of a conductor.* We consider a small area on the surface of the conductor. The field at a point outside this area is due to (1) charges distributed on the rest of the conductor (call this field $\mathbf{E}_1$), and (2) the field due to the charge resting on the area in question, say $\mathbf{E}_2$ (see Fig. 57). From Example 78, $E_1 + E_2 = 4\pi\sigma$. Now the field inside the

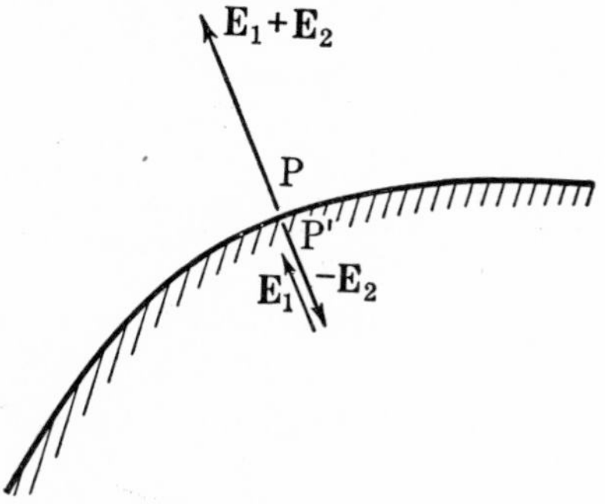

FIG. 57.

conductor at the point P' situated symmetrically opposite P is zero from Example 75. The field at P' is $E_1 - E_2 = 0$. Thus $E_1 = 2\pi\sigma$ per unit charge. For an area dS the force is

$$dE = (2\pi\sigma)(\sigma\, dS) = 2\pi\sigma^2\, dS \tag{206}$$

This force is normal to the surface. A charged soap film thus tends to expand.

Problems

1. Two hollow concentric spheres have equal and opposite charges Q and $-Q$. Find the work done in taking a unit test charge from the sphere of radius a to the sphere of radius b, $b > a$. The outer sphere is negatively charged.
2. Find the field due to any infinite uniformly charged cylinder.
3. Solve Prob. 1 for two infinite concentric cylinders.

4. Let $q_1, q_2, \ldots, q_n$ be a set of collinear electric charges residing on the line L. Let C be a circle whose plane is normal to L and whose center lies on L. Show that the electric flux through this circle is $N = \sum_{\alpha=1}^{n} 2\pi q_\alpha(1 - \cos \beta_\alpha)$, where β_α is the angle between L and any line from q_α to the circumference of C.

5. Let the line L of Prob. 4 be the x axis, and rotate a line of force Γ in the x-y plane about the x axis (see Fig. 58). If no

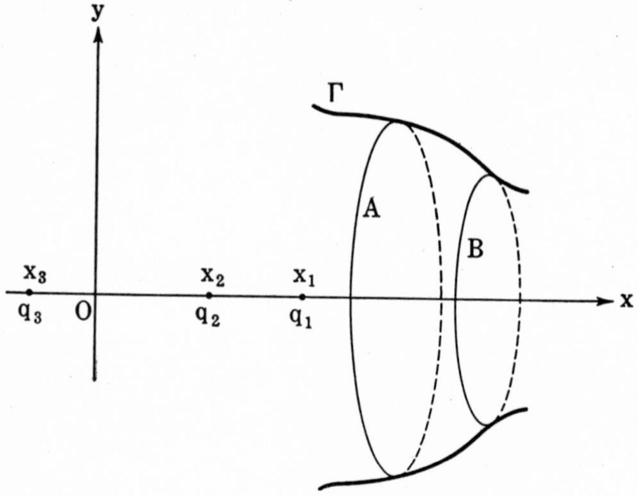

FIG. 58.

charges exist between the planes $x = A$, $x = B$, show that the equation of a line of force is

$$\sum_{\alpha=1}^{n} q_\alpha(x - x_\alpha)[(x - x_\alpha)^2 + y^2]^{-\frac{1}{2}} = \text{constant}$$

6. Point charges $+q$, $-q$ are placed at the points A, B. The line of force that leaves A making an angle α with AB meets the plane that bisects AB at right angles in P. Show that

$$\sin \frac{\alpha}{2} = \sqrt{2} \sin \left(\tfrac{1}{2} \measuredangle PAB\right)$$

59. Poisson's Formula. In Sec. 57, we saw that

$$\mathbf{E} = -\nabla \left[\sum_{\alpha=1}^{n} \frac{q_\alpha}{r_\alpha} \right] = -\nabla \varphi$$

For a continuous distribution of charge density ρ, we postulate that the potential is

$$\varphi = \iiint\limits_{\infty} \frac{\rho\, d\tau}{r} \tag{207}$$

where the integration exists over all of space. At any point P where no charges exist, $r > 0$, and we need not worry about the convergence of the integral. Now let us consider what happens at a point P where charges exist, that is, $r = 0$. Let us surround the point P by a small sphere R of radius ϵ. The integral $\iiint\limits_{V-R} (\rho\, d\tau/r)$ exists if ρ is continuous. We define φ at P as $\lim\limits_{\epsilon \to 0} \iiint\limits_{V-R} (\rho\, d\tau/r)$. This limit exists, for using spherical coordinates,

$$\left|\iiint\limits_{R} \frac{\rho\, d\tau}{r}\right| = \left|\int_0^{2\pi}\int_0^{\pi}\int_0^{\epsilon} \rho r \sin\theta\, dr\, d\theta\, d\varphi\right| < M\pi^2\epsilon^2$$

where M is the bound of ρ in the neighborhood of P. Thus

$$\left|\iiint\limits_{V-R} \frac{\rho\, d\tau}{r} - \iiint\limits_{V-R'} \frac{\rho\, d\tau}{r}\right| < M\pi^2(\epsilon^2 + \epsilon'^2)$$

where ϵ' is the radius of the sphere R' surrounding P. The Cauchy criterion holds, so that the limit exists. In much the same way we can show that

$$\mathbf{E} = \iiint\limits_{\infty} \frac{\rho \mathbf{r}}{r^3}\, d\tau \tag{208}$$

and that at a point P where a charge exists

$$\mathbf{E}(P) = \lim_{\epsilon \to 0} \iiint\limits_{V-R} \frac{\rho \mathbf{r}}{r^3}\, d\tau$$

converges.

Now from Gauss's law

$$\iint\limits_{S} \mathbf{E} \cdot d\mathbf{\delta} = 4\pi Q = 4\pi \iiint\limits_{V} \rho\, d\tau$$

In order to apply the divergence theorem to the surface integral, we must be sure that $\nabla \cdot \mathbf{E}$ is continuous at points where ρ is continuous. We assume this to be true, and the reader is referred to Kellogg's "Foundations of Potential Theory" for the proof of this. Thus

$$\iiint_V (\nabla \cdot \mathbf{E})\, d\tau = 4\pi \iiint_V \rho\, d\tau \tag{209}$$

Since (209) is true for all volumes, it is easy to see that

$$\boxed{\nabla \cdot \mathbf{E} = 4\pi\rho} \tag{210}$$

provided $\nabla \cdot \mathbf{E}$ and ρ are continuous. Since $\mathbf{E} = -\nabla\varphi$, we have Poisson's equation

$$\boxed{\nabla^2\varphi = -4\pi\rho} \tag{211}$$

and at places where no charges exist, $\rho = 0$, so that Laplace's equation, $\nabla^2\varphi = 0$, holds.

Example 80. In cylindrical coordinates

$$\nabla^2\varphi = \frac{1}{r}\left[\frac{\partial}{\partial r}\left(r\frac{\partial\varphi}{\partial r}\right) + \frac{\partial}{\partial\theta}\left(\frac{1}{r}\frac{\partial\varphi}{\partial\theta}\right) + \frac{\partial}{\partial z}\left(r\frac{\partial\varphi}{\partial z}\right)\right]$$

Consider an infinite cylinder of radius a and charge q per unit length. At points where no charge exists, we have $\nabla^2\varphi = 0$. Moreover, from symmetry, φ depends only on r. Thus

$$\nabla^2\varphi = \frac{1}{r}\frac{d}{dr}\left(r\frac{d\varphi}{dr}\right) = 0$$

$$r\frac{d\varphi}{dr} = \text{constant} = A$$

$$\varphi = A \log r + B$$

$$\mathbf{E} = -\nabla\varphi = -\frac{A}{r^2}\mathbf{r}, \qquad \mathbf{r} = x\mathbf{i} + y\mathbf{j}$$

Also $4\pi\sigma = (E_r)_{r=a} = -A/a$, so that $q = 2\pi a\sigma = -A/2$, and

$$\mathbf{E} = \frac{2q}{r^2}\mathbf{r} \tag{212}$$

Example 81. To prove that the potential is constant inside a conductor. From Green's formula we have

$$\iint_S \varphi \nabla\varphi \cdot d\mathbf{d} = \iiint_V (\nabla\varphi)^2 \, d\tau + \iiint_V \varphi \nabla^2\varphi \, d\tau$$

Inside the conductor no charge exists so that $\nabla^2\varphi = 0$. Moreover, for any surface inside the conductor, $\mathbf{E} = -\nabla\varphi = 0$ so that $\iiint_V (\nabla\varphi)^2 \, d\tau = 0$ for all volumes V inside the conductor. Therefore $(\nabla\varphi)^2 \equiv 0$, and $\frac{\partial\varphi}{\partial x} = \frac{\partial\varphi}{\partial y} = \frac{\partial\varphi}{\partial z} = 0$, so that $\varphi \equiv$ constant inside the conductor.

Problems

1. Solve Laplace's equation in spherical coordinates assuming the potential $V = V(r)$.
2. Find the field due to a two-dimensional infinite slab, of width $2a$, uniformly charged. Here we have $\varphi = \varphi(x)$ and must solve Laplace's equation and Poisson's equation separately for free space and for the slab, and we must satisfy the boundary condition for the potential at the edge of the slab. The space occupied by the slab is given by $-a \leqq x \leqq a$, $-\infty < y < \infty$.
3. Solve Laplace's equation for two concentric spheres of radii a, b, with $b > a$, with charges q, Q, and find the field.
4. Solve Laplace's equation and find the field due to an infinite uniformly charged plane.
5. Prove that two-dimensional lines of force also satisfy Laplace's equation.
6. Show that $\varphi = (A \cos nx + B \sin nx)(Ce^{ny} + De^{-ny})$ satisfies $\frac{\partial^2\varphi}{\partial x^2} + \frac{\partial^2\varphi}{\partial y^2} = 0$.
7. If φ_1 and φ_2 satisfy Laplace's equation, show that $\varphi_1 + \varphi_2$ and $\varphi_1 - \varphi_2$ satisfy Laplace's equation. Does $\varphi_1\varphi_2$ satisfy Laplace's equation?
8. If φ_1 satisfies Laplace's equation and φ_2 satisfies Poisson's equation, show that $\varphi_1 + \varphi_2$ satisfies Poisson's equation.

60. Dielectrics. If charges reside in a medium other than a vacuum, it is found that the inverse-square force needs readjustment. That this is reasonable can be seen from the following considerations. We consider a parallel-plate condenser separated by glass (Fig. 59). Assuming that the molecular structure of glass consists of positive and negative particles, the electrons being bound to the nucleus, we see that the field due to the oppositely charged plates might well cause a displacement of the electrons away from the negative plate and toward the positive plate. This tends to weaken the field, so that $E = 4\pi\sigma/\kappa$, where $\kappa > 1$. κ is called the dielectric constant.

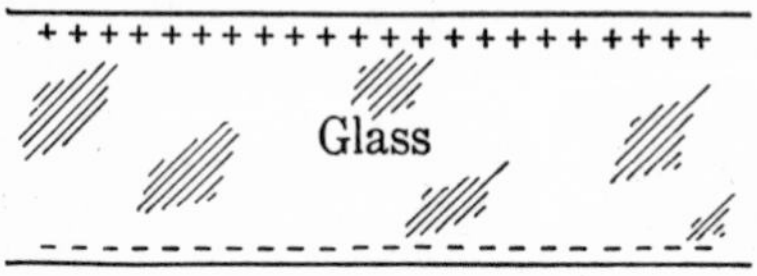

FIG. 59

It is found experimentally that $\mathbf{E} = (qq'/\kappa r^3)\mathbf{r}$ for charges in a dielectric. Applying this force, we see that Gauss's law is modified to read

$$\iint \mathbf{E} \cdot d\mathbf{\delta} = \frac{4\pi Q}{\kappa} \tag{213}$$

and if κ is a constant,

$$\iint \kappa\mathbf{E} \cdot d\mathbf{\delta} = \iint \mathbf{D} \cdot d\mathbf{\delta} = 4\pi Q \tag{214}$$

where $\mathbf{D}$ is defined as the displacement vector, $\mathbf{D} = \kappa\mathbf{E} = -\kappa \nabla\varphi$. Poisson's equation becomes $\nabla \cdot \mathbf{D} = -\nabla \cdot (\kappa \nabla\varphi) = 4\pi\rho$, and for constant κ

$$\nabla^2\varphi = -\frac{4\pi\rho}{\kappa} \tag{215}$$

For $\rho = 0$ we still have Laplace's equation $\nabla^2\varphi = 0$.

In the most general case, we have

$$D_i = \sum_{j=1}^{3} \kappa_{ij}E_j, \qquad i = 1, 2, 3$$

where $\mathbf{D} = D_1\mathbf{i} + D_2\mathbf{j} + D_3\mathbf{k}$, $\mathbf{E} = E_1\mathbf{i} + E_2\mathbf{j} + E_3\mathbf{k}$, and $\kappa_{ij} = \kappa_{ji}$.

61. Energy of the Electrostatic Field. Let us bring charges $q_1, q_2, \ldots, q_n$ from infinity to positions $P_1, P_2, \ldots, P_n$, and calculate the work done in bringing about this distribution. It takes no work to bring q_1 to P_1, since there is no field. To bring q_2 to P_2, work must be done against the field set up by q_1. This amount of work is q_1q_2/r_{12}, where r_{12} is the distance between P_1 and P_2. In bringing q_3 to P_3, we do work against the separate fields due to q_1 and q_2. This work is q_1q_3/r_{13} and q_2q_3/r_{23}. We continue this process and obtain for the total work

$$W = \frac{1}{2}\sum_{j=1}^{n}\sum_{i=1}^{n}\frac{q_iq_j}{r_{ij}} \tag{216}$$

The $\frac{1}{2}$ occurs because q_1q_2/r_{12} occurs twice in the summation process, once as q_1q_2/r_{12} and again as q_2q_1/r_{21}. The quantity W is called the electrostatic energy of the field. Since $\varphi_i = \sum_{j=1}^{n} q_j/r_{ij}$, we have $W = \frac{1}{2}\sum_{i=1}^{n} q_i\varphi_i$. For a continuous distribution of charge, we replace the summation by an integral, so that

$$W = \frac{1}{2}\iiint_{\infty} \rho\varphi\, d\tau \tag{217}$$

Now assume that all the charges are contained in some finite sphere. We have $\nabla \cdot \mathbf{D} = 4\pi\rho$ so that

$$W = \frac{1}{8\pi}\iiint_V \varphi\nabla\cdot\mathbf{D}\, d\tau = \frac{1}{8\pi}\iiint_V \nabla\cdot(\varphi\mathbf{D})\, d\tau - \frac{1}{8\pi}\iiint_V \nabla\varphi\cdot\mathbf{D}\, d\tau$$

Applying the divergence theorem,

$$W = \frac{1}{8\pi}\iint_S \varphi\mathbf{D}\cdot d\boldsymbol{\sigma} - \frac{1}{8\pi}\iiint_V \nabla\varphi\cdot\mathbf{D}\, d\tau$$

Now φ is of the order of $1/r$ for large r, and $\mathbf{D}$ is of the order of $1/r^2$, while $d\sigma$ is of the order of r^2. We may take our volume of integration as large as we please, since $\rho = 0$ outside a fixed sphere. Hence $\lim_{r\to\infty}\iint_S \varphi\mathbf{D}\cdot d\boldsymbol{\sigma} = 0$, so that

$$W = \frac{1}{8\pi}\iiint_{\infty} (\mathbf{E}\cdot\mathbf{D})\, d\tau \tag{218}$$

The energy density is $w = (1/8\pi)\mathbf{E} \cdot \mathbf{D}$. For an isotropic medium, $\mathbf{D} = \kappa\mathbf{E}$ and $W = (1/8\pi) \iiint_\infty \kappa \mathbf{E}^2 \, d\tau$.

Example 82. Let us compute the energy if our space contains a charge q distributed uniformly over the surface of a sphere of radius a. We have

$$\mathbf{D} = \mathbf{E} = \frac{q}{r^3}\mathbf{r},\ r \geqq a \qquad \text{and} \qquad \mathbf{D} = \mathbf{E} = 0,\ r < a$$

The total energy is

$$W = \frac{q^2}{8\pi} \int_0^{2\pi} \int_0^{\pi} \int_a^{\infty} \frac{1}{r^2} \sin\theta \, dr \, d\theta \, d\varphi$$
$$= \frac{q^2}{2a}$$

62. Discontinuities of D and E at the Boundary of Two Dielectrics. Let S be the surface of discontinuity between two media with dielectric constant κ_1 and κ_2. We apply Gauss's law to a pillbox with a face in each medium (Fig. 60). Assuming no charges exist on the surface of discontinuity, we have

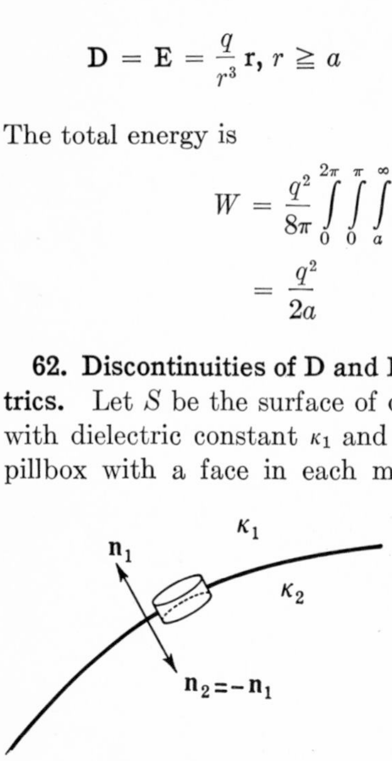

FIG. 60.

$$\iint \mathbf{D} \cdot d\boldsymbol{\sigma} = 0$$

so that $\mathbf{D}_1 \cdot \mathbf{n}_1 + \mathbf{D}_2 \cdot \mathbf{n}_2 = 0$. Since $\mathbf{n}_1 = -\mathbf{n}_2$, we have

$$\underline{\overline{D_{N_1} = D_{N_2}}} \tag{219}$$

We have taken the pillbox very flat so that the sides contribute a negligible amount to the flux. Equation (219) states that the normal component of the displacement vector $\mathbf{D}$ is continuous across a surface of discontinuity containing no charges.

We next consider a closed curve Γ with sides parallel to the surface of discontinuity and ends negligible in size (Fig. 61).

Since the field is conservative,

$$\oint \mathbf{E} \cdot d\mathbf{r} = 0 \qquad \text{or} \qquad \underline{\overline{E_{T_1} = E_{T_2}}} \tag{220}$$

In other words, the tangential component of the electric vector **E** is continuous across a surface of discontinuity. Combining (219) and (220), we have

$$\frac{D_{N_1}}{E_{T_1}} = \frac{D_{N_2}}{E_{T_2}} \quad \text{or} \quad \frac{\kappa_1 E_{N_1}}{E_{T_1}} = \frac{\kappa_2 E_{N_2}}{E_{T_2}}$$

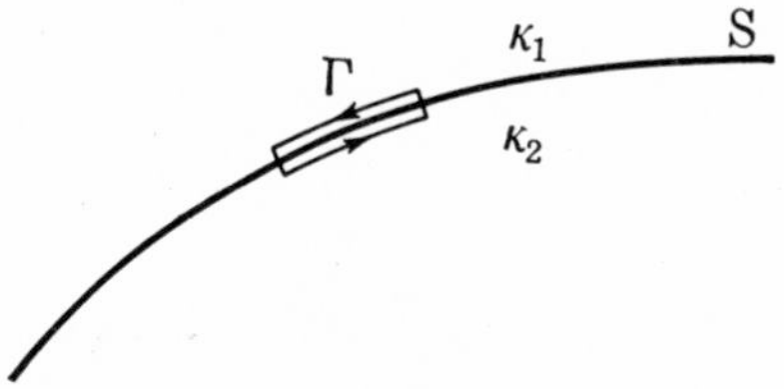

FIG. 61.

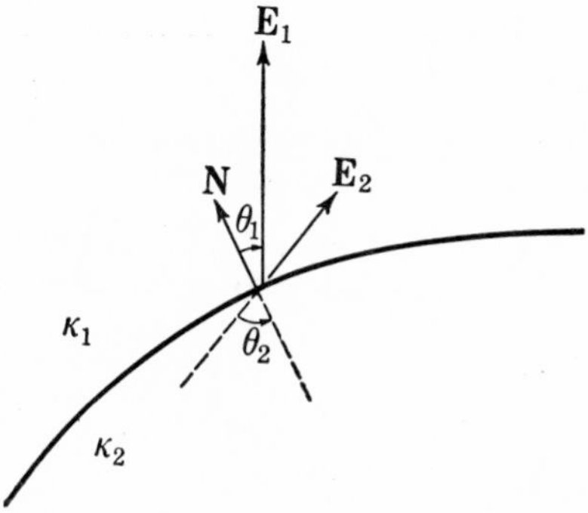

FIG. 62.

for isotropic media. Hence

$$\frac{\tan \theta_1}{\tan \theta_2} = \frac{\kappa_1}{\kappa_2} \tag{221}$$

which is the law of refraction (see Fig. 62).

63. Green's Reciprocity Theorem. Let us consider any distribution of volume and surface charges, the surfaces being conductors. Let ρ be the volume density and σ the surface density. If φ is the potential function for this distribution of charges, then $\nabla^2\varphi = -4\pi\rho$. We shall make use of the fact that $\mathbf{E} = -\nabla\varphi$ and that at the surface $E_n = 4\pi\sigma$, or $\mathbf{E} \cdot d\mathbf{\delta} = -4\pi\sigma\, dS$.

A new distribution of charges would yield a new potential function φ' such that $\nabla^2\varphi' = -4\pi\rho'$. Our problem is to find a relationship between the fundamental quantities ρ, σ, φ of the old distribution and ρ', σ', φ' of the new distribution. To do so, we apply Green's formula

$$\iiint\limits_V (\varphi\,\nabla^2\varphi' - \varphi'\,\nabla^2\varphi)\,d\tau = \iint\limits_S (\varphi\,\nabla\varphi' - \varphi'\,\nabla\varphi)\cdot d\mathbf{d}$$

which reduces to

$$-4\pi \iiint\limits_V (\varphi\rho' - \varphi'\rho)\,d\tau = 4\pi \iint\limits_S (\varphi\sigma' - \varphi'\sigma)\,dS$$

or

$$\iiint\limits_V \varphi\rho'\,d\tau + \iint\limits_S \varphi\sigma'\,dS = \iiint\limits_V \varphi'\rho\,d\tau + \iint\limits_S \varphi'\sigma\,dS \qquad (222)$$

This is Green's reciprocity theorem. It states that the potential φ of a given distribution when multiplied by the corresponding charge (ρ', σ') in the new distribution and then summed over all of the space is equal to the sum of the products of the potentials (φ') in the new distribution by the charges (ρ, σ) in the old distribution, that is, a reciprocal property prevails.

Example 83. Let a sphere of radius a be grounded, that is, its potential is zero, and place a charge q at a point P, b units from the center of the sphere, $b > a$. The charge q will induce a charge Q on the sphere. We desire to find Q. We construct a new distribution as follows: Place a unit charge on the sphere, and assume no other charges in space. The potential due to this charged sphere is $\varphi' = 1/r$. For the sphere we have initially $\varphi = 0$, $Q = ?$, and afterward, $\varphi' = 1/a$, $q' = 1$. For the point P we have initially $\varphi_P = ?$, $q = q$, and afterward, $\varphi = 1/b$, $q' = 0$. Applying the reciprocity theorem, we have

$$0 \cdot 1 + \varphi_P \cdot 0 = Q \cdot \frac{1}{a} + q \cdot \frac{1}{b}$$

so that $Q = -(a/b)q$. This is the total charge induced on the sphere when it is grounded. Note that this method does not tell us the surface distribution of the induced charge.

Problems

1. A conducting sphere of radius a is embedded in the center of a sphere of radius b and dielectric constant κ. The conductor

is grounded, and a point charge q is placed at a distance r from its center, $r > b > a$. Show that the charge induced on the sphere is $Q = -\kappa abq\{r[b + (\kappa - 1)a]\}^{-1}$.

2. A pair of concentric conductors of radii a and b are connected by a wire. A point charge q is detached from the inner one and moved radially with uniform speed V to the outer one. Show that the rate of transfer of the induced charge (due to q) from the inner to the outer sphere is

$$\frac{dQ}{dt} = -qab(b - a)^{-1}V(a + Vt)^{-2}$$

3. A spherical condenser with inner radius a and outer radius b is filled with two spherical layers of dielectrics κ_1 and κ_2, the boundary between being given by $r = \frac{1}{2}(a + b)$. If, when both shells are earthed, a point charge on the dielectric boundary induces equal charges on the inner and outer shells, show that $\kappa_1/\kappa_2 = b/a$.

4. A conductor has a charge e, and V_1, V_2 are the potentials of two equipotential surfaces which completely surround it ($V_1 > V_2$). The space between these two surfaces is now filled with a dielectric of inductive capacity κ. Show that the change in the energy of the system is $\frac{1}{2}e(V_1 - V_2)(\kappa - 1)\kappa^{-1}$.

5. The inner sphere of a spherical condenser (radii a, b) has a constant charge E, and the outer conductor is at zero potential. Under the internal forces, the outer conductor contracts from radius b to radius b_1. Prove that the work done by the electric forces is $\frac{1}{2}E^2(b - b_1)b^{-1}b_1^{-1}$.

64. Method of Images. We consider a charge q placed at a point $P(b, 0, 0)$ and ask if it is possible to find a point $Q(z, 0, 0)$ such that a certain charge q' at Q will cause the potential over the sphere $x^2 + y^2 + z^2 = a^2$, $a < b$, to vanish. The answer is "Yes"! We proceed as follows: From Fig. 63 we have

$$s^2 = z^2 + a^2 - 2az \cos \theta$$
$$t^2 = b^2 + a^2 - 2ab \cos \theta$$

We choose z so that $zb = a^2$, and call $Q(a^2/b, 0, 0)$ the image point of $P(b, 0, 0)$ with respect to the sphere. Thus

$$s^2 = \frac{a^2}{b^2}(a^2 + b^2 - 2ab \cos \theta) = \frac{a^2}{b^2}t^2$$

and

$$s = \frac{a}{b} t$$

The potential at S due to charges q and q' at P and Q is

$$\varphi = \frac{q'}{s} + \frac{q}{t} = \frac{1}{t}\left(\frac{b}{a} q' + q\right)$$

and $\varphi = 0$ if we choose $q' = -(a/b)q$.

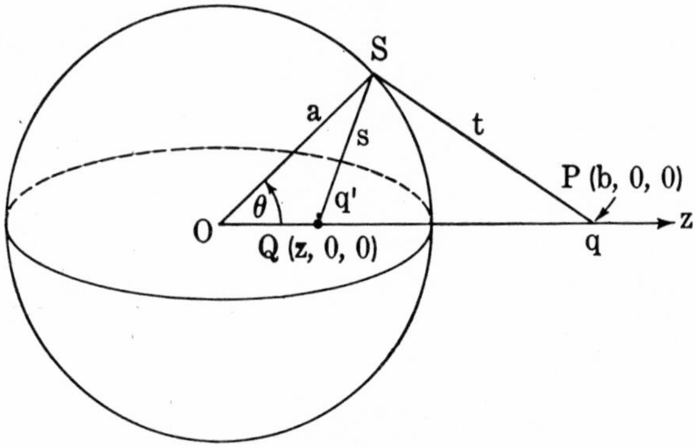

Fig. 63.

The potential at any point R with spherical coordinates r, θ, φ is

$$\Phi = \frac{q}{(r^2 + b^2 - 2rb \cos \theta)^{\frac{1}{2}}} - \frac{(a/b)q}{[r^2 + (a^4/b^2) - (2a^2/b)r \cos \theta]^{\frac{1}{2}}} \tag{223}$$

with $\Phi = 0$ on S and $\nabla^2\Phi = 0$ where no charges exist.

Now let us consider the sphere of Example 83. The function of (223) satisfies Laplace's equation and is zero on the sphere. From the uniqueness theorem of Example 69, Φ of (223) is the potential function for the problem of Example 83. The radial field is given by

$$E_r = -\frac{\partial \Phi}{\partial r} = \frac{q(r - b \cos \theta)}{(r^2 + b^2 - 2rb \cos \theta)^{\frac{3}{2}}} - \frac{(a/b)q[r - (a^2/b) \cos \theta]}{[r^2 + (a^4/b^2) - (2a^2/b)r \cos \theta]^{\frac{3}{2}}}$$

and the surface distribution is given by

$$\sigma = \frac{(E_r)_{r=a}}{4\pi} = -\frac{q}{4\pi}\,\frac{b^2 - a^2}{a(a^2 + b^2 - 2ab\cos\theta)^{\frac{3}{2}}}$$

Problems

1. A charge q is placed at a distance a from an infinite grounded plane. Find the image point, the field, and the induced surface density.

2. Two semiinfinite grounded planes intersect at right angles. A charge q is placed on the bisector of the planes. What distribution of charges is equivalent to this system? Find the field and the surface distribution induced on the planes.

3. An infinite plate with a hemispherical boss of radius a is at zero potential under the influence of a point charge q on the axis of the boss at a distance f from the plate. Find the surface density at any point of the plate, and show that the charge is attracted toward the plate with a force

$$\frac{q^2}{4f^2} + \frac{4q^2a^3f^3}{(f^4 - a^4)^2}$$

65. Conjugate Harmonic Functions. If we are dealing with a two-dimensional problem in electrostatics, we look for a solution of Laplace's equation $\nabla^2 V = 0$. The curves $V(x, y) =$ constant represent the equipotential lines. We know that these curves are orthogonal to the lines of force, so that the conjugate function $U(x, y)$ (see Sec. 56) will represent the lines of force. We know that $\nabla^2 V = \nabla^2 U = 0$.

We now give an example of the use of conjugate harmonic functions. In Example 73 we saw that

$$\begin{aligned} U(x, y) &= A \tan^{-1} \frac{y}{x} \\ V(x, y) &= \frac{A}{2} \log (x^2 + y^2) \end{aligned} \tag{224}$$

are conjugate functions satisfying Laplace's equation. If we take $V(x, y)$ as the potential function, then the equipotentials are the circles $(A/2) \log (x^2 + y^2) = C$, or $x^2 + y^2 = e^{2C/A}$.

Hence the potential due to an infinite charged conducting cylinder is

$$V(x, y) = \frac{A}{2} \log (x^2 + y^2) = \frac{A}{2} \log r^2 = A \log r, \quad r^2 = x^2 + y^2$$

since $A \log r$ satisfies Laplace's equation and satisfies the boundary condition that $V =$ constant for $r = a$, the radius of the charged cylinder. If q is the charge per unit length, then

$$\sigma = \frac{q}{2\pi a} = \frac{(E_r)_{r=a}}{4\pi} = \frac{-\left.\frac{\partial V}{\partial r}\right|_{r=a}}{4\pi} = -\frac{A}{4\pi a}$$

so that $A = -2q$ and $V = -2q \log r$.

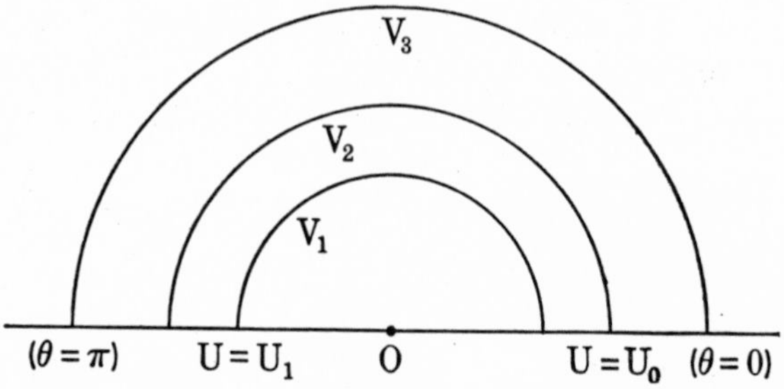

FIG. 64.

If we choose $U(x, y) = A \tan^{-1} (y/x)$ as our potential function, then the equipotentials $U =$ constant are the straight lines $A \tan^{-1} (y/x) = C$, or $y = x \tan (C/A)$. As a special case we may take the straight lines $\theta = 0$, $\theta = \pi$ as conducting planes raised to different potentials (see Fig. 64). The lines of force are the circles $(A/2) \log (x^2 + y^2) = V$.

The theory of conjugate functions belongs properly to the theory of functions of a complex variable. With the aid of the Schwarz transformation it is possible to find the conjugate functions associated with more difficult problems involving the two-dimensional Laplace equation.

Problems

1. By considering Example 72, find the potential function and lines of force for two semiinfinite planes intersecting at right angles.

2. What physical problems can be solved by the transformation $x = a \cosh U \cos V$, $y = a \sinh U \sin V$? Show that

$$\nabla^2 U = \nabla^2 V = 0$$

66. Integration of Laplace's Equation. Let S be the surface of a region R for which $\nabla^2\varphi = 0$. Let P be any point of R, and let r be the distance from P to any point of the surface S. We make use of Green's formula

$$\iiint_R (\varphi\,\nabla^2\psi - \psi\,\nabla^2\varphi)\,d\tau = \iint_S (\varphi\,\nabla\psi - \psi\,\nabla\varphi)\cdot d\boldsymbol{\sigma}$$

We choose $\psi = 1/r$, and this produces a discontinuity inside R, namely, at P, where $r = 0$. In order to overcome this difficulty, we proceed as in Example 66. Surround P by a sphere Σ of radius ϵ. Using the fact that $\nabla^2\varphi = \nabla^2\psi = 0$ inside R' (R minus the Σ sphere), we obtain

$$0 = \iint_S \left(\varphi\,\nabla\frac{1}{r} - \frac{1}{r}\nabla\varphi\right)\cdot d\boldsymbol{\sigma} + \iint_\Sigma \left(\varphi\,\nabla\frac{1}{r} - \frac{1}{r}\nabla\varphi\right)\cdot d\boldsymbol{\sigma} \qquad (225)$$

Now $\nabla(1/r) = -\mathbf{r}/r^3$, and on the sphere Σ,

$$\nabla\frac{1}{r}\cdot d\boldsymbol{\sigma} = -\frac{\mathbf{r}}{r^3}\cdot\frac{-\mathbf{r}}{r}\,dS = \frac{dS}{\epsilon^2}$$

and $(1/r)\,\nabla\varphi\cdot d\boldsymbol{\sigma}$ is of the order $\epsilon|\nabla\varphi|$, so that by letting $\epsilon \to 0$, (225) reduces to

$$\varphi(P) = \frac{1}{4\pi}\iint_S \left(\frac{1}{r}\nabla\varphi - \varphi\,\nabla\frac{1}{r}\right)\cdot d\boldsymbol{\sigma} \qquad (226)$$

This remarkable formula states that the value of φ at any point P is determined by the value of φ and $\nabla\varphi$ on the surface S.

Problems

1. If P is any point outside the closed surface S, show that $\iint_S [(1/r)\nabla\varphi - \varphi\,\nabla(1/r)]\cdot d\boldsymbol{\sigma} = 0$, where $\nabla^2\varphi = 0$ inside S and r is the distance from P to any point of S.

2. Let φ satisfy $\nabla^2\varphi = \dfrac{\partial^2\varphi}{\partial x^2} + \dfrac{\partial^2\varphi}{\partial y^2} = 0$. Let Γ be the closed boundary of a simply connected region in the x-y plane. If P is an interior point of Γ, show that

$$\varphi(P) = \frac{1}{2\pi} \oint \left\{ \left(\log \frac{1}{r}\right) \frac{\partial \varphi}{\partial n} - \varphi \frac{\partial[\log (1/r)]}{\partial n} \right\} ds$$

where use is made of the fact that

$$\iint_A (u \nabla^2 v - v \nabla^2 u)\, dA = \oint \left(u \frac{\partial v}{\partial n} - v \frac{\partial u}{\partial n} \right) ds$$

n being the normal to the curve.

3. Let φ be harmonic outside the closed surface S and assume that $\varphi \to 0$ and $r|\nabla\varphi| \to 0$ as $r \to \infty$. If P is a point outside S, show that

$$\varphi(P) = \frac{1}{4\pi} \iint_S \left(\frac{1}{r} \nabla\varphi - \varphi \nabla \frac{1}{r} \right) \cdot d\boldsymbol{\sigma}$$

where the normal $d\boldsymbol{\sigma}$ is inward on S.

4. Let φ be harmonic and regular inside sphere Σ. Show that the value of φ at the center of Σ is the average of its values over the surface of the sphere. Use (226).

67. Solution of Laplace's Equation in Spherical Coordinates. From Sec. 23, Prob. 1,

$$\nabla^2 V = \frac{1}{r^2 \sin\theta} \left[\frac{\partial}{\partial r}\left(r^2 \sin\theta \frac{\partial V}{\partial r} \right) + \frac{\partial}{\partial \theta}\left(\sin\theta \frac{\partial V}{\partial \theta} \right) + \frac{\partial}{\partial \varphi}\left(\frac{1}{\sin\theta} \frac{\partial V}{\partial \varphi} \right) \right] = 0 \quad (227)$$

To solve (227), we assume a solution of the form

$$V(r, \theta, \varphi) = R(r)\Theta(\theta)\Phi(\varphi) \quad (228)$$

Substituting (228) into (227) and dividing by V, we obtain

$$\frac{\sin\theta}{R} \frac{d}{dr}\left(r^2 \frac{dR}{dr} \right) + \frac{1}{\Theta} \frac{d}{d\theta}\left(\sin\theta \frac{d\Theta}{d\theta} \right) + \frac{1}{\Phi \sin\theta} \frac{d^2\Phi}{d\varphi^2} = 0$$

Consequently

$$\frac{1}{R}\frac{d}{dr}\left(r^2\frac{dR}{dr}\right) = -\frac{1}{\Theta\sin\theta}\frac{d}{d\theta}\left(\sin\theta\frac{d\Theta}{d\theta}\right) - \frac{1}{\sin^2\theta\,\Phi}\frac{d^2\Phi}{d\varphi^2} \tag{229}$$

The left-hand side of (229) depends only on r, while the right-hand side of (229) depends on θ and φ. This is possible only if both quantities are constant, for on differentiating (229) with respect to r, we obtain $\frac{d}{dr}\left[\frac{1}{R}\frac{d}{dr}\left(r^2\frac{dR}{dr}\right)\right] = 0$. We choose as the constant of integration $c = -n(n+1)$, so that

$$\frac{1}{R}\frac{d}{dr}\left(r^2\frac{dR}{dr}\right) = -n(n+1)$$

or

$$r^2\frac{d^2R}{dr^2} + 2r\frac{dR}{dr} + n(n+1)R = 0 \tag{230}$$

It is easy to integrate (230), and we leave it to the reader to show that $R = Ar^n + Br^{-n-1}$ is the most general solution of (230). Returning to (229), we have

$$\frac{1}{\Phi}\frac{d^2\Phi}{d\varphi^2} = n(n+1)\sin^2\theta - \frac{\sin\theta}{\Theta}\frac{d}{d\theta}\left(\sin\theta\frac{d\Theta}{d\theta}\right) \tag{231}$$

Since we have again separated the variables, both sides of (231) are constant. We choose the constant to be negative, $-m^2$, m being an integer. This choice guarantees that the solution of

$$\frac{d^2\Phi}{d\varphi^2} + m^2\Phi = 0 \tag{232}$$

is single-valued when φ is increased by 2π. The solution of (232) is $\Phi = A\cos m\varphi + B\sin m\varphi$.

Finally, we obtain that $\Theta(\theta)$ satisfies

$$\sin\theta\frac{d}{d\theta}\left(\sin\theta\frac{d\Theta}{d\theta}\right) + [n(n+1)\sin^2\theta - m^2]\Theta = 0 \tag{233}$$

We make a change of variable by letting $\mu = \cos\theta$,

$$d\mu = -\sin\theta\,d\theta$$

so that (233) becomes

$$(1 - \mu^2)\frac{d}{d\mu}\left[(1 - \mu^2)\frac{d\Theta}{d\mu}\right] + [(1 - \mu^2)n(n + 1) - m^2]\Theta = 0 \quad (234)$$

If we assume that V is independent of φ (symmetry about the z axis), we have $m = 0$, so that (234) becomes

$$\frac{d}{d\mu}\left[(1 - \mu^2)\frac{d\Theta}{d\mu}\right] + n(n + 1)\Theta = 0 \quad (235)$$

This is Legendre's differential equation.

By the method of series solution, it can be shown that

$$\Theta = P_n(\mu) \equiv \frac{1}{2^n n!}\frac{d^n(\mu^2 - 1)^n}{d\mu^n}$$

satisfies (235); the $P_n(\mu)$ are called Legendre polynomials.

Two important properties of Legendre polynomials are the following:

$$\int_{-1}^{1} P_m(\mu)P_n(\mu)\,d\mu = 0 \quad \text{if } m \neq n \quad (236)$$

$$\int_{-1}^{1} P_n^2(\mu)\,d\mu = \frac{2}{2n + 1} \quad (237)$$

$= \left(\frac{2}{2n+1}\right)\delta_{mn}$

We give a proof of (236). P_n and P_m satisfy

$$\frac{d}{d\mu}\left[(1 - \mu^2)\frac{dP_n}{d\mu}\right] + n(n + 1)P_n = 0 \quad (238)$$

$$\frac{d}{d\mu}\left[(1 - \mu^2)\frac{dP_m}{d\mu}\right] + m(m + 1)P_m = 0 \quad (239)$$

Multiplying (238) by P_m and (239) by P_n and subtracting, we obtain

$$P_m\frac{d}{d\mu}\left[(1 - \mu^2)\frac{dP_n}{d\mu}\right] - P_n\frac{d}{d\mu}\left[(1 - \mu^2)\frac{dP_m}{d\mu}\right] + [n(n + 1) - m(m + 1)P_nP_m = 0$$

or

$$\frac{d}{d\mu}\left[(1-\mu^2)\left(P_m\frac{dP_n}{d\mu}-P_n\frac{dP_m}{d\mu}\right)\right] + [n(n+1)-m(m+1)]P_nP_m = 0 \quad (240)$$

Integrating between the limits -1 and $+1$, we obtain

$$[n(n+1)-m(m+1)]\int_{-1}^{1} P_mP_n\,d\mu = 0$$

and if $m \neq n$,

$$\int_{-1}^{1} P_mP_n\,d\mu = 0$$

A particular solution of (227) which is independent of φ, that is, $\dfrac{\partial V}{\partial \varphi} = 0$, is given by $V(r, \theta) = (A_nr^n + B_nr^{-n-1})P_n(\cos\theta)$. Now it is easy to show that any sum of solutions of (227) is also a solution, since (227) is linear in V. Consequently a more general solution is

$$V = \sum_{n=0}^{\infty}(A_nr^n + B_nr^{-n-1})P_n(\cos\theta) \quad (241)$$

provided that the series converges.

If we wish to solve a problem involving $\nabla^2V = 0$ with spherical boundaries, we try (241) as our solution. If we can find the constants A_n, B_n so that the boundary conditions are fulfilled, then (241) will represent the only solution, from our previous uniqueness theorems involving Laplace's equation.

We list a few Legendre polynomials:

$$\begin{aligned}
P_0(\mu) &= 1\\
P_1(\mu) &= \mu\\
P_2(\mu) &= \tfrac{1}{2}(3\mu^2 - 1)\\
P_3(\mu) &= \tfrac{1}{2}(5\mu^3 - 3\mu)\\
P_4(\mu) &= \tfrac{1}{8}(35\mu^4 - 30\mu^2 + 3)
\end{aligned} \quad (242)$$

$$\begin{aligned}
P_n(0) &= 0,\ n \text{ odd}\\
P_n(0) &= (-1)^{n/2}\frac{1\cdot 3\cdot 5\cdots(n-1)}{2\cdot 4\cdot 6\cdots n},\ n \text{ even}\\
P_n(1) &= 1\\
P_n(-\mu) &= (-1)^nP_n(\mu)
\end{aligned}$$

Problems

1. Prove (237).

2. Solve $\nabla^2 V = 0$ for rectangular coordinates by the method of Sec. 67, assuming $V = X(x)Y(y)Z(z)$.

3. Investigate the solution of $\nabla^2 V = 0$ in cylindrical coordinates.

68. Applications

Example 84. A dielectric sphere of radius a is placed in a uniform field $\mathbf{E}_0 = E_0\mathbf{k}$. We calculate the field inside the sphere. The potential due to the uniform field is $\varphi = -E_0 z = -E_0 r \cos\theta$. There will be an additional potential due to the presence of the dielectric sphere. Assume it to be of the form $ArP_1 \equiv Ar\cos\theta$ inside the sphere and $Br^{-2}P_1 \equiv Br^{-2}\cos\theta$ outside the sphere. We cannot have a term of the type $Cr^{-2}\cos\theta$ inside the sphere, for at the origin we would have an infinite field caused by the presence of the dielectric. Similarly, if a term of the type $Dr\cos\theta$ occurred outside the sphere, we would have an infinite field at infinity due to the presence of the sphere. If we let V_{I} be the potential inside and V_{II} the potential outside the sphere, we have

$$\begin{aligned} V_{\mathrm{I}} &= -E_0 r\cos\theta + Ar\cos\theta \\ V_{\mathrm{II}} &= -E_0 r\cos\theta + \frac{B}{r^2}\cos\theta \end{aligned} \tag{243}$$

Notice that V_{I} and V_{II} are special cases of (241). We have two unknown constants, A, B, and two boundary conditions,

$$\begin{aligned} &V_{\mathrm{I}} = V_{\mathrm{II}} \text{ at } r = a \\ D_{N_1} = D_{N_2} \qquad \text{or} \qquad &\kappa\frac{\partial V_{\mathrm{I}}}{\partial r} = \frac{\partial V_{\mathrm{II}}}{\partial r} \text{ at } r = a \end{aligned} \tag{244}$$

(see Sec. 62). From (243) and (244) we obtain

$$A = \frac{\kappa - 1}{\kappa + 2}E_0, \qquad B = a^3\frac{\kappa - 1}{\kappa + 2}E_0$$

so that

$$V_{\mathrm{I}} = -\frac{3}{\kappa + 2}E_0 r\cos\theta = -\frac{3}{\kappa + 2}E_0 z \tag{245}$$

We see that the field inside the dielectric sphere is

$$\mathbf{E} = -\nabla V_{\mathrm{I}} = \frac{3}{\kappa + 2}\mathbf{E}_0$$

and $\mathbf{E}$ is uniform of intensity less than $\mathbf{E}_0$ since $\kappa > 1$. Outside the sphere

$$V_{\mathrm{II}} = -E_0 r \cos\theta + \frac{\kappa - 1}{\kappa + 2}\frac{a^3}{r^2}E_0 \cos\theta \tag{246}$$

The radial field outside the sphere is given by

$$E_r = -\frac{\partial V_{\mathrm{II}}}{\partial r} = E_0 \cos\theta + 2\frac{\kappa - 1}{\kappa + 2}\frac{a^3 E_0}{r^3}\cos\theta$$

For a given r the maximum E_r is found at $\theta = 0$.

Example 85. A conducting sphere of radius a and charge Q is surrounded by a spherical dielectric layer up to $r = b$ (Fig. 65). Let us calculate the potential distribution. From spherical symmetry $V = V(r)$, so that we try

$$V_{\mathrm{I}} = \frac{A}{r}$$

$$V_{\mathrm{II}} = \frac{B}{r} + C$$

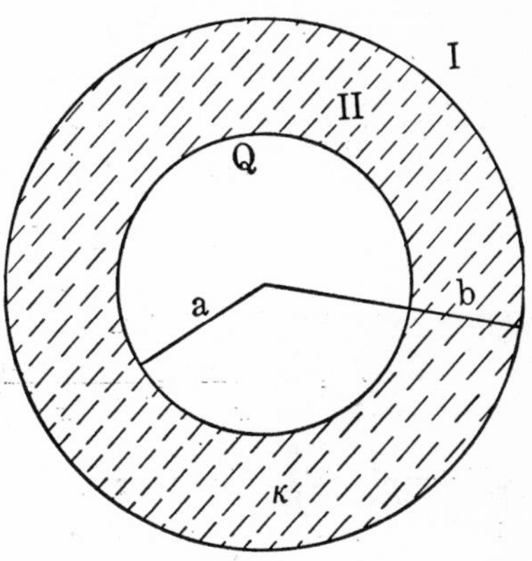

FIG. 65.

The boundary conditions are

(i) $$V_{\mathrm{I}} = V_{\mathrm{II}} \text{ at } r = b$$

(ii) $$\frac{\partial V_{\mathrm{I}}}{\partial r} = \kappa\frac{\partial V_{\mathrm{II}}}{\partial r} \text{ at } r = b$$

(iii) $$Q = \frac{1}{4\pi}\iint_S \mathbf{D}\cdot d\mathbf{\delta} = -\frac{\kappa}{4\pi}\int_0^{2\pi}\int_0^{\pi}\left(\frac{\partial V_{\mathrm{II}}}{\partial r}\right)_{r=a} a^2 \sin\theta\, d\theta\, d\varphi$$

From (i) $A/b = (B/b) + C$; from (ii) $-A/b^2 = -\kappa B/b^2$; from (iii) $Q = (\kappa/4\pi)(B/a^2)\iint dS = \kappa B$. Hence

$$V_{\text{I}} = \frac{Q}{r}, \qquad V_{\text{II}} = \frac{Q}{\kappa r} + \frac{Q}{b}\frac{\kappa - 1}{\kappa} \tag{247}$$

Example 86. A conducting sphere of radius a and charge Q is placed in a uniform field. We calculate the potential and the distribution of charge on the sphere. We assume a solution

$$V = -E_0 r \cos\theta + \frac{B}{r}$$

The boundary condition is

$$Q = \frac{1}{4\pi}\iint_S -\left.\frac{\partial V}{\partial r}\right|_{r=a} dS$$

so that

$$Q = \frac{1}{4\pi}\int_0^{2\pi}\int_0^{\pi}\left(E_0\cos\theta + \frac{B}{a^2}\right)a^2 \sin\theta\, d\theta\, d\varphi = B$$

and

$$V = -E_0 r\cos\theta + \frac{Q}{r}$$

For the charge distribution

$$\sigma = \frac{(E_r)_{r=a}}{4\pi} = \frac{-\left.\dfrac{\partial V}{\partial r}\right|_{r=a}}{4\pi} = \frac{1}{4\pi}\left(E_0\cos\theta + \frac{Q}{a^2}\right)$$

Example 87. Consider a charge q placed at $A(b, 0, 0)$. Let us compute the potential at any point $P(r, \theta, \varphi)$ (see Fig. 66). The potential at P is

$$V = \frac{q}{s} = q(r^2 + b^2 - 2rb\cos\theta)^{-\frac{1}{2}}$$

There are two cases to consider:

(*a*) $r < b$. Let $\mu = \cos\theta$, $x = r/b$, so that

$$V = \frac{q}{b}(1 - 2\mu x + x^2)^{-\frac{1}{2}}$$

Now $(1 - 2\mu x + x^2)^{-\frac{1}{2}}$ can be expanded in a Maclaurin series in powers of x, yielding

$$V = \frac{q}{b} \sum_{n=0}^{\infty} P_n(\mu) x^n = \frac{q}{b} \sum_{n=0}^{\infty} P_n(\mu) \left(\frac{r}{b}\right)^n \tag{248}$$

The proof is omitted here that the $P_n(\mu)$ are actually the Legendre polynomials. However, we might expect this, since V satisfies Laplace's equation and $P_n(\mu)r^n$ is a solution of $\nabla^2 V = 0$.

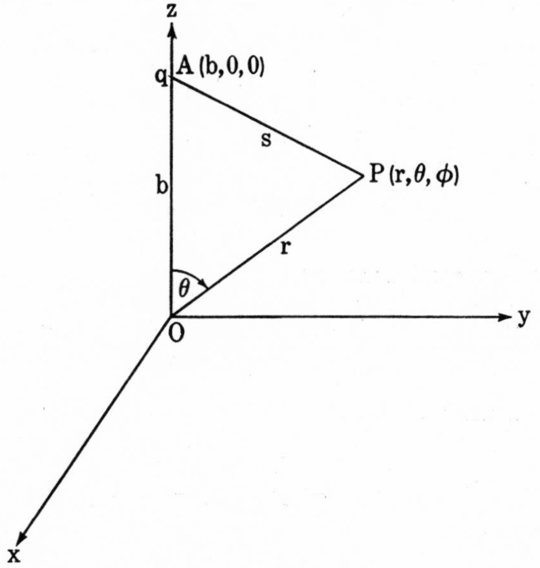

FIG. 66.

(b) $r > b$. In this case

$$V = \frac{q}{r} \sum_{n=0}^{\infty} P_n(\mu) \left(\frac{b}{r}\right)^n \tag{249}$$

Notice that each term is of the form $P_n(\mu)r^{-n-1}$, which satisfies Laplace's equation.

Example 88. A point charge $+q$ is placed at a distance b from the center of two concentric, earthed, conducting spheres of radii a and c, $a < b < c$. We find the potential at a point P for $a < r < b$.

For $r > b$, $V = (q/r) \sum_{0}^{\infty} (b/r)^n P_n(\cos \theta)$ due to the charge q; and for $r < b$, $V = (q/b) \sum_{0}^{\infty} (r/b)^n P_n(\cos \theta)$. Moreover, we have an induced potential of the form

$$V = q \sum_{0}^{\infty} (A_n r^n + B_n r^{-n-1}) P_n(\cos\theta) \tag{250}$$

which is due to the spheres, the A_n and B_n undetermined as yet. Hence

$$\left.\begin{aligned} &\text{For } r > b: \\ &\qquad V_1 = q \sum_{0}^{\infty} [A_n r^n + (B_n + b^n) r^{-n-1}] P_n(\cos\theta) \\ &\text{For } r < b: \\ &\qquad V_2 = q \sum_{0}^{\infty} [(A_n + b^{-n-1}) r^n + B_n r^{-n-1}] P_n(\cos\theta) \end{aligned}\right\} \tag{251}$$

The boundary conditions are

$$\begin{aligned} &\text{(i)} \qquad V_1 = 0 \text{ at } r = c \\ &\text{(ii)} \qquad V_2 = 0 \text{ at } r = a \end{aligned} \tag{252}$$

These yield the equations

$$\begin{aligned} &\text{(i)} \qquad A_n c^n + (B_n + b^n) c^{-n-1} = 0 \\ &\text{(ii)} \qquad (A_n + b^{-n-1}) a^n + B_n a^{-n-1} = 0 \end{aligned}$$

so that

$$B_n = \frac{a^{2n+1}}{b^{n+1}} \frac{(c^{2n+1} - b^{2n+1})}{(a^{2n+1} - c^{2n+1})},$$

$$A_n + b^{-(n+1)} = -\frac{b^{-n-1}(c^{2n+1} - b^{2n+1})}{a^{2n+1} - c^{2n+1}}$$

Hence

$$V_2(P) = q \sum_{n=0}^{\infty} \frac{b^{2n+1} - c^{2n+1}}{b^{n+1}(a^{2n+1} - c^{2n+1})} \left(r^n - \frac{a^{2n+1}}{r^{n+1}} \right) P_n(\cos\theta) \tag{253}$$

Problems

1. Show that the force acting on the sphere of Example 86 is $\mathbf{F} = \frac{1}{3} Q E_0 \mathbf{k}$.

2. A charge q is placed at a distance c from the center of a spherical hollow of radius a in an infinite dielectric of constant κ. Show that the force acting on the charge is

$$\frac{(\kappa - 1)q^2}{c^2} \sum_{n=0}^{\infty} \frac{n(n+1)}{n + \kappa(n+1)} \left(\frac{c}{a}\right)^{2n+1}$$

3. A point charge q is placed a distance c from the center of an earthed conducting sphere of radius a, on which a dielectric layer of outer radius b and constant κ exists. Show that the potential of this layer is

$$V = \frac{q}{c} \sum_{n=0}^{\infty} \frac{(2n+1)b^{2n+1}(r^n - a^{2n+1}r^{-n-1})}{c^n\{[(\kappa+1)n+1]b^{2n+1} + (n+1)(\kappa-1)a^{2n+1}\}} P_n(\cos\theta)$$

4. Show that the potential inside a dielectric shell of internal and external radii a and b, placed in a uniform field of strength E, is

$$V = \frac{9\kappa E}{9\kappa - 2(1-\kappa^2)[(b/a)^3 - 1]} r \cos\theta$$

5. The walls of an earthed rectangular conducting tube of infinite length are given by $x = 0$, $x = a$, $y = 0$, $y = b$. A point charge is placed at $x = x_0$, $y = y_0$, $z = z_0$ inside the tube. Show that the potential is given by

$$V = 8q \sum_{n=1}^{\infty} \sum_{m=1}^{\infty} (m^2a^2 + n^2b^2)^{-\frac{1}{2}} e^{-a^{-1}b^{-1}(m^2a^2+n^2b^2)^{\frac{1}{2}}\pi(z-z_0)} \sin\frac{n\pi x_0}{a} \sin\frac{n\pi x}{a} \sin\frac{m\pi y_0}{b} \sin\frac{m\pi y}{b}$$

69. Integration of Poisson's Equation. Instead of assuming that φ is harmonic, let us consider that φ satisfies $\nabla^2\varphi = -4\pi\rho$. By applying Green's formula as in Sec. 66, we immediately obtain

$$\varphi(P) = \iiint_R \frac{\rho}{r}\, d\tau + \iint_S \left(\frac{1}{r}\nabla\varphi - \varphi\nabla\frac{1}{r}\right)\cdot d\mathbf{d} \tag{254}$$

If we make the further assumption that φ is of the order of $1/r$ for large r and that $|\nabla\varphi| \sim 1/r^2$, we see that by pushing S out to infinity the surface integral will tend to zero. Our assumption is valid, for if we assume the charge distribution to be bounded by some sphere, then at large distances the potential will be of the order of $1/r$, since we may consider all the charges as essentially concentrated at a point. Thus

$$\varphi(P) = \iiint\limits_{\infty} \frac{\rho}{r}\, d\tau \tag{255}$$

70. Decomposition of a Vector into the Sum of Solenoidal and Irrotational Vectors. In Example 70, we saw that if $|\mathbf{f}|$ tends to zero like $1/r^2$ as $r \to \infty$, then $\mathbf{f}$ as uniquely determined by its curl and divergence.

We now proceed to write $\mathbf{f}$ as the sum of irrotational and solenoidal vectors. Let

$$\mathbf{W}(P) = \iiint\limits_{\infty} \frac{\mathbf{f}\, d\tau}{r} \tag{256}$$

where r is the distance from P to the element of integration $d\tau$. If we write $\mathbf{f} = f_1\mathbf{i} + f_2\mathbf{j} + f_3\mathbf{k}$, $\mathbf{W} = W_1\mathbf{i} + W_2\mathbf{j} + W_3\mathbf{k}$, then

$$\begin{aligned} W_1 &= \iiint\limits_{\infty} \frac{f_1}{r}\, d\tau \\ W_2 &= \iiint\limits_{\infty} \frac{f_2}{r}\, d\tau \\ W_3 &= \iiint\limits_{\infty} \frac{f_3}{r}\, d\tau \end{aligned} \tag{257}$$

We assume that the components of $\mathbf{f}$ are such that the integrals of (257) converge and that $|\mathbf{W}| \sim 1/r$, $|\nabla W_n| \sim 1/r^2$, $n = 1, 2, 3$. From Sec. 69, $\nabla^2 W_n = -4\pi f_n$, so that

$$\nabla^2 \mathbf{W} = -4\pi \mathbf{f} \tag{258}$$

From (256)

$$\begin{aligned} \nabla \cdot \mathbf{W} &= -\iiint\limits_{\infty} \mathbf{f} \cdot \nabla \frac{1}{r}\, d\tau \\ \nabla \times \mathbf{W} &= \iiint\limits_{\infty} \mathbf{f} \times \nabla \frac{1}{r}\, d\tau \end{aligned} \tag{259}$$

Now

$$\nabla \times (\nabla \times \mathbf{W}) = \nabla(\nabla \cdot \mathbf{W}) - \nabla^2 \mathbf{W}$$

so that

$$\mathbf{f} = \frac{1}{4\pi} \nabla \times (\nabla \times \mathbf{W}) - \frac{1}{4\pi} \nabla(\nabla \cdot \mathbf{W})$$

and hence

$$\mathbf{f} = \nabla \times \mathbf{A} + \nabla \varphi \tag{260}$$

where

$$\mathbf{A} = \frac{1}{4\pi} \nabla \times \mathbf{W}, \qquad \varphi = -\frac{1}{4\pi} \nabla \cdot \mathbf{W}$$

Problems

1. Show that (256) is a special case of (254).

2. Find an expression for $\varphi(P)$ if $\nabla^2 \varphi = -4\pi\rho$ inside S and if P is on the surface S.

3. $\mathbf{f} = yz\mathbf{i} + xz\mathbf{j} + (xy - xz)\mathbf{k}$. Express $\mathbf{f}$ as the sum of an irrotational and a solenoidal vector.

71. Dipoles. Let us consider two neighboring charges $-q$ and $+q$ situated at $P(x, y, z)$ and $Q(x + dx, y, z)$. The potential at the origin $O(0, 0, 0)$ due to $-q$ is $-q/r$, and that due to $+q$ is $q/(r + dr)$, where $r = (x^2 + y^2 + z^2)^{\frac{1}{2}}$ and

$$r + dr = [(x + dx)^2 + y^2 + z^2]^{\frac{1}{2}}$$

The potential at $O(0, 0, 0)$ due to both charges is

$$\varphi = \frac{q}{r + dr} - \frac{q}{r} \approx \frac{q}{r^2}\, dr$$

Now $dr = x\,dx/r$, so that $\varphi \approx qx\,dx/r^3$. If we now let $q \to \infty$ and $dx \to 0$ in such a way that $q\,dx$ remains finite, we have formed what is known as a dipole. Let $\mathbf{r}$ be the position vector from the origin to the dipole, and let $\mathbf{M} = q\,d\mathbf{r}$, where $d\mathbf{r}$ is the vector from the negative charge to the positive charge; $|d\mathbf{r}| = dx$. We have $(\mathbf{M} \cdot \mathbf{r})/r^3 = (q\mathbf{r} \cdot d\mathbf{r})/r^3 = (qx\,dx)/r^3$, so that

$$\varphi = \frac{\mathbf{M} \cdot \mathbf{r}}{r^3} \tag{261}$$

$\mathbf{M}$ is called the strength or moment of the dipole. For more than one dipole, the potential at a point P is given by

$$\varphi = \sum_{i=1}^{n} \frac{(\mathbf{M}_i \cdot \mathbf{r}_i)}{r_i^3}$$

where $\mathbf{r}_i$ is the vector from P to the dipole having strength $\mathbf{M}_i$.

Example 89. The field strength due to a dipole is $\mathbf{E} = -\nabla\varphi$ so that

$$\mathbf{E} = -\nabla\left(\frac{\mathbf{r}\cdot\mathbf{M}}{r^3}\right) = \frac{3(\mathbf{r}\cdot\mathbf{M})}{r^5}\mathbf{r} - \frac{\mathbf{M}}{r^3} \tag{262}$$

Example 90. *Potential energy of a dipole in a field of potential* φ. Let φ_1 be the potential at the charge q and φ_2 the potential at the charge $-q$. The energy of the dipole is

$$W = \varphi_1 q + \varphi_2(-q) = q(\varphi_1 - \varphi_2) = q\frac{\partial\varphi}{\partial s}\,ds = M\frac{\partial\varphi}{\partial s}$$

where ds is the distance between the charges. Now

$$d\varphi = d\mathbf{s}\cdot\nabla\varphi$$

so that $W = M\dfrac{\partial\mathbf{s}}{\partial s}\cdot\nabla\varphi = \mathbf{M}\cdot\nabla\varphi$.

72. Electric Polarization. Let us consider a volume filled with dipoles. The potential due to any single dipole is given by (261). If we let $\mathbf{P}$ be the dipole moment per unit volume, that is, $\mathbf{P} = \lim_{\Delta\tau\to 0} (\Delta\mathbf{M}/\Delta\tau)$, then the total potential due to the dipoles is

$$\varphi = \iiint_R \frac{\mathbf{r}\cdot\mathbf{P}}{r^3}\,d\tau \tag{263}$$

Now $\nabla\cdot(\mathbf{P}/r) = (1/r)\nabla\cdot\mathbf{P} - [(\mathbf{P}\cdot\mathbf{r})/r^3]$. The reason that we have taken $\nabla(1/r) = \mathbf{r}/r^3$ instead of $-\mathbf{r}/r^3$ is that

$$r = [(\xi - x)^2 + (\eta - y)^2 + (\zeta - z)^2]^{\frac{1}{2}}$$

and ∇ performs the differentiations with respect to x, y, z, the coordinates of the point P at which φ is being evaluated. The coordinates ξ, η, ζ belong to the region R and are the variables of integration, and $\mathbf{r} = (\xi - x)\mathbf{i} + (\eta - y)\mathbf{j} + (\zeta - z)\mathbf{k}$. Hence (263) becomes

$$\varphi = \iiint_R \nabla\cdot\left(\frac{\mathbf{P}}{r}\right)d\tau - \iiint_R \frac{\nabla\cdot\mathbf{P}}{r}\,d\tau$$

and

$$\varphi = \iint_S \frac{\mathbf{P}}{r} \cdot d\boldsymbol{\sigma} - \iiint_R \frac{\nabla \cdot \mathbf{P}}{r}\, d\tau \tag{264}$$

by applying the divergence theorem.

Example 91. Let us find the electric intensity at the center of a uniformly polarized sphere. Here $\mathbf{P} = P_0\mathbf{k}$, so that $\nabla \cdot \mathbf{P} = 0$ inside R. Hence (264) becomes

$$\varphi(x, y, z) = \iint_S \frac{P_0\mathbf{k} \cdot d\boldsymbol{\sigma}}{[(\xi - x)^2 + (\eta - y)^2 + (\zeta - z)^2]^{\frac{1}{2}}}$$

and

$$\mathbf{E} = -\nabla\varphi = -\iint_S \frac{P_0(\mathbf{k} \cdot d\boldsymbol{\sigma})[(\xi - x)\mathbf{i} + (\eta - y)\mathbf{j} + (\zeta - z)\mathbf{k}]}{[(\xi - x)^2 + (\eta - y)^2 + (\zeta - z)^2]^{\frac{3}{2}}}$$

$$\mathbf{E}(0, 0, 0) = -\iint_S \frac{(\xi\mathbf{i} + \eta\mathbf{j} + \zeta\mathbf{k})P_0(\mathbf{k} \cdot d\boldsymbol{\sigma})}{(\xi^2 + \eta^2 + \zeta^2)^{\frac{3}{2}}} \tag{265}$$

Now for points on the sphere,

$$\xi^2 + \eta^2 + \zeta^2 \equiv a^2,$$

and letting $\xi = a \sin\theta \cos\varphi$, $\eta = a \sin\theta \sin\varphi$, $\zeta = a\cos\theta$, it is easily seen that (265) reduces to

$$\mathbf{E}(0, 0, 0) = -\tfrac{4}{3}\pi P_0\mathbf{k} \tag{266}$$

$\mathbf{E}$ is independent of the radius of the sphere. By superimposing (concentrically) a sphere with an equal but negative polarization, we see that the field at the center of a uniformly polarized shell is zero.

Problems

1. Prove (262).
2. Prove (266).
3. If $\mathbf{M}_1$ and $\mathbf{M}_2$ are the vector moments of two dipoles at A and B, and if $\mathbf{r}$ is the vector from A to B, show that the energy of the system is $W = \mathbf{M}_1 \cdot \mathbf{M}_2 r^{-3} - 3(\mathbf{M}_1 \cdot \mathbf{r})(\mathbf{M}_2 \cdot \mathbf{r})r^{-5}$.
4. The dipole-moment density is given by $P = \mathbf{r}$ over a sphere of radius a. Calculate the field at the center of the sphere.

73. Magnetostatics. The same laws that have held for electrostatics are true for magnetostatics with the exception that $\nabla^2\varphi_m = 0$ always, since we cannot isolate a magnetic charge. We make the following correspondences, since all the laws of electrostatics were derived on the assumption of the inverse-square force law, which applies equally well for stationary magnets.

Electrostatics		*Magnetostatics*
$\mathbf{E}$	$\longleftrightarrow$	$\mathbf{H}$
q	$\longleftrightarrow$	q_m
$\mathbf{D}$	$\longleftrightarrow$	$\mathbf{B}$ (magnetic induction)
κ	$\longleftrightarrow$	μ (permeability)
$\mathbf{D} = \kappa\mathbf{E}$	$\longleftrightarrow$	$\mathbf{B} = \mu\mathbf{H}$
$w = \frac{1}{2}\mathbf{E}\cdot\mathbf{D}$	$\longleftrightarrow$	$w_m = \frac{1}{2}\mathbf{H}\cdot\mathbf{B}$
$\nabla\cdot\mathbf{D} = 4\pi\rho$	$\longleftrightarrow$	$\nabla\cdot\mathbf{B} = 0$

(267)

74. Solid Angle. Let $\mathbf{r}$ be the position vector from a point P to a surface of area dS and unit normal $\mathbf{N}$, that is, $d\boldsymbol{\sigma} = \mathbf{N}\,dS$. We define the solid angle subtended at P by the surface dS to be (see Fig. 67)

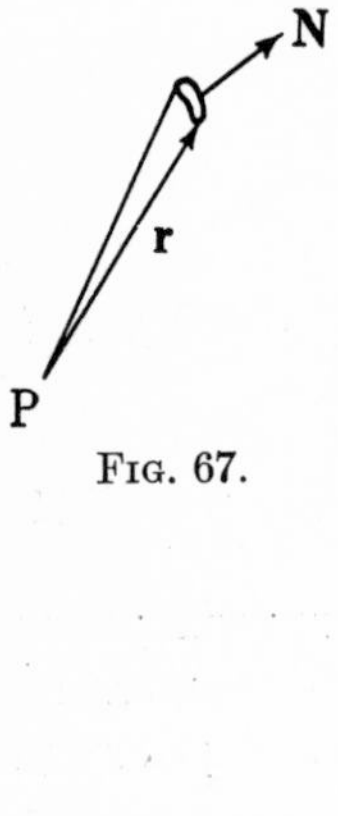

FIG. 67.

$$d\Omega = \frac{\mathbf{r}\cdot d\boldsymbol{\sigma}}{r^3}$$

The total solid angle of a surface is

$$\Omega = \iint_S \frac{\mathbf{r}\cdot d\boldsymbol{\sigma}}{r^3} \tag{268}$$

Example 92. Let S be a sphere and P the origin so that

$$\Omega(P) = \iint \frac{\mathbf{r}\cdot\mathbf{r}}{r^4}\,dS = 4\pi$$

Example 93. The magnetic dipole is the exact analogue of the electric dipole. We consider a magnetic shell, that is, a thin sheet magnetized uniformly in a direction normal to its surface (Fig. 68). Let β be the magnetic moment per unit area and assume β = constant. The potential at P is given by

$$\varphi = \iint_S \frac{\mathbf{r}\cdot\mathbf{M}}{r^3}\,dS = \beta \iint_S \frac{\mathbf{r}\cdot d\boldsymbol{\sigma}}{r^3} = \beta\Omega$$

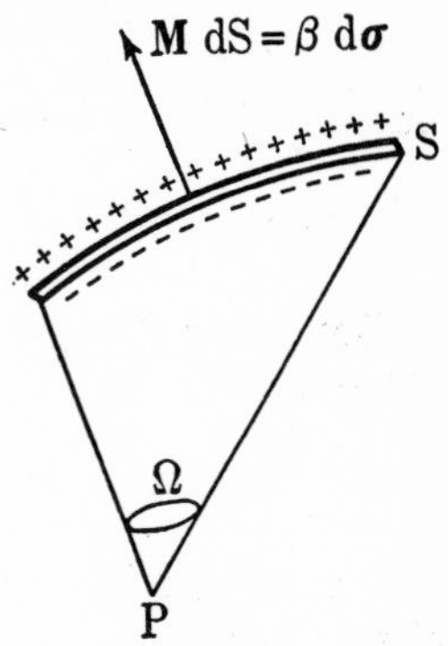

FIG. 68.

Now let P and Q be opposite points on the negative and positive sides of the surface S. We have

$$\varphi(Q) = -\beta(4\pi - \Omega)$$

so that the work done in taking a unit positive pole from a point P on the negative side of the shell to a point Q on the positive side of the shell is given by

$$W = \int_P^Q \mathbf{H}\cdot d\mathbf{r} = -\int_P^Q \nabla\varphi\cdot d\mathbf{r} = \varphi(P) - \varphi(Q)$$
$$= \beta\Omega + \beta(4\pi - \Omega)$$

$$\boxed{W = 4\pi\beta} \tag{269}$$

75. Moving Charges, or Currents. If two conductors at different potentials are joined together by a metal wire, it is found that certain phenomena occur (heating of the wire, magnetic field), so that one is led to believe that a flow of charge is taking place. Let $\mathbf{v}$ be the velocity of the charge and ρ the density of charge. We define current density by $\mathbf{j} = \rho\mathbf{v}$. The total charge passing through a surface per unit time is given by

$$\iint_S \rho\mathbf{v}\cdot d\boldsymbol{\sigma} = \iint_S \mathbf{j}\cdot d\boldsymbol{\sigma}$$

Now the total charge inside a closed surface S is $Q = \iiint_R \rho\,d\tau$. If there are no sources or sinks inside S, then the loss of charge per unit time is given by $-\dfrac{\partial Q}{\partial t} = -\iiint_R \dfrac{\partial\rho}{\partial t}\,d\tau$. Thus

$$\iint_S \rho\mathbf{v}\cdot d\boldsymbol{\sigma} = -\iiint_R \frac{\partial\rho}{\partial t}\,d\tau$$

Applying the divergence theorem, we have

$$\left.\begin{aligned} \nabla \cdot (\rho \mathbf{v}) + \frac{\partial \rho}{\partial t} &= 0 \\ \text{or} \qquad \nabla \cdot \mathbf{j} + \frac{\partial \rho}{\partial t} &= 0 \end{aligned}\right\} \tag{270}$$

Equations (270) are the statement of conservation of electric charge. We define a steady state as one for which ρ is independent of the time, $\dfrac{\partial \rho}{\partial t} = 0$, which implies $\nabla \cdot \mathbf{j} = 0$.

It has been found by experiment that if $\mathbf{E}$ is the electric field, then

$$\mathbf{j} = \lambda \mathbf{E} = -\lambda \nabla \varphi \tag{271}$$

where λ is the conductivity of the metal. This is Ohm's law. For the general case, $j_\alpha = \sum_{\beta=1}^{3} \lambda_{\alpha\beta} E_\beta$, and the simplest case, $\lambda \equiv$ constant, so that $\nabla^2 \varphi = 0$ for the steady state.

We now compute the work done on a charge q as it moves from a point of potential φ_1 to one of potential φ_2, $\varphi_1 > \varphi_2$. The energy at φ_1 is $q\varphi_1$ and at φ_2 is $q\varphi_2$. The loss in energy is

$$W = (\varphi_1 - \varphi_2)q$$

This loss in electrical energy does not go into mechanical energy, since the flow is assumed steady. Hence the electrical energy is converted into heat, $Q = (\varphi_1 - \varphi_2)q$. The power loss is $P = \dfrac{dQ}{dt} = (\varphi_1 - \varphi_2)\dfrac{dq}{dt}$, and since $\varphi_1 - \varphi_2 = RJ$ (another form of Ohm's law, where R is resistance and J current), we have

$$P = RJ^2 \tag{272}$$

76. Magnetic Effect of Currents (Oersted). Experiments show that electric currents produce magnetic fields. The mathematical expression for the magnetic field is given by

$$d\mathbf{H} = \frac{J\mathbf{r} \times d\mathbf{r}}{cr^3} \tag{273}$$

where $\mathbf{r}$ is the vector from the point P; P is the point at which we calculate the magnetic field $d\mathbf{H}$ due to the line current J in that portion of the wire $d\mathbf{r}$, and c is a constant, the ratio of the electrostatic to the electromagnetic unit of charge (see Fig. 69). Biot and Savart established this law for straight-line currents.

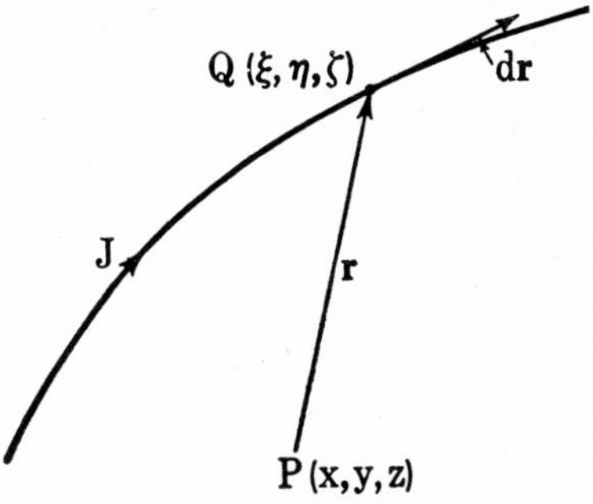

FIG. 69.

For a closed path

$$\mathbf{H} = \oint \frac{J\mathbf{r} \times d\mathbf{r}}{cr^3} \tag{274}$$

Now $r = [(\xi - x)^2 + (\eta - y)^2 + (\zeta - z)^2]^{\frac{1}{2}}$, and $\nabla(1/r) = \mathbf{r}/r^3$, where $\nabla = \mathbf{i}\dfrac{\partial}{\partial x} + \mathbf{j}\dfrac{\partial}{\partial y} + \mathbf{k}\dfrac{\partial}{\partial z}$. Hence

$$\mathbf{H} = \frac{1}{c}\oint J\,\nabla\frac{1}{r} \times d\mathbf{r} = \frac{1}{c}\oint \nabla \times \left(\frac{J}{r}\,d\mathbf{r}\right)$$

since ∇ does not operate on $d\mathbf{r}$ and J is a constant. Thus $\mathbf{H} = \nabla \times \mathbf{A}$, where $\mathbf{A} = (1/c)\oint J\,d\mathbf{r}/r = (1/c)\iiint_\infty \mathbf{j}\,d\tau/r$ is integrated over all space containing currents.

Now $\nabla \cdot \mathbf{A} = \iiint_R \nabla \cdot (\mathbf{j}/r)\,d\tau = \iint_S (\mathbf{j}/r) \cdot d\boldsymbol{\sigma}$, so that if all currents lie within a given sphere, we may push the boundary of R to infinity, since nothing new will be added to the integral yielding $\mathbf{A}$. But when S is expanded to a great distance, $\mathbf{j} = 0$ on S, so that that $\nabla \cdot \mathbf{A} = 0$. Also

$$\nabla \times \mathbf{H} = \nabla \times (\nabla \times \mathbf{A}) = \nabla(\nabla \cdot \mathbf{A}) - \nabla^2\mathbf{A}$$

so that $\nabla \times \mathbf{H} = -\nabla^2\mathbf{A}$. Now since $\mathbf{A} = (1/c)\iiint_\infty \mathbf{j}\,d\tau/r$, or

$$A_x = (1/c)\iiint_\infty j_x\,d\tau/r,\quad A_y = (1/c)\iiint_\infty j_y\,d\tau/r,$$

$$A_z = \frac{1}{c}\iiint_\infty \frac{j_z\,d\tau}{r}$$

we have from Sec. 69 that $\nabla^2\mathbf{A} = -(4\pi/c)\mathbf{j}$. Thus

$$\boxed{\nabla \times \mathbf{H} = \frac{4\pi}{c}\mathbf{j}} \tag{275}$$

Example 94. The work done in taking a unit magnetic pole around a closed path Γ in a magnetic field due to electric currents is

$$W = \oint \mathbf{H} \cdot d\mathbf{r} = \iint_S \nabla \times \mathbf{H} \cdot d\boldsymbol{\sigma} = \frac{4\pi}{c} \iint_S \mathbf{j} \cdot d\boldsymbol{\sigma}$$

For an electric current J in a wire that loops Γ, we have

$$\boxed{W = \frac{4\pi}{c} J} \tag{276}$$

Example 95. The magnetic field at a point P, r units away from an infinite straight-line wire carrying a current J, is obtained by use of Example 94.

$$\oint \mathbf{H} \cdot d\mathbf{r} = H(2\pi r) = \frac{4\pi}{c} J \qquad \text{so that} \qquad H = \frac{2J}{cr}$$

Example 96. We compute the dimensions of $c/\sqrt{\kappa\mu}$. Now $f_e = q_e q_e'/\kappa r^2$, and $f_m = q_m q_m'/\mu r^2$ so that

$$\frac{[M][L]}{[T]^2} = \frac{[q_e]^2}{[\kappa][L]^2} = \frac{[q_m]^2}{[\mu][L]^2}$$

and

$$\frac{[J]}{[c]} = \frac{[dq_e/dt]}{[c]} = \frac{[q_e]}{[c][T]} = \frac{[M]^{\frac{1}{2}}[L][\kappa]^{\frac{1}{2}}}{[c][T]^2}$$

From (276),

$$\frac{\text{Work}}{\text{Unit pole}} = \oint \mathbf{H} \cdot d\mathbf{r} = \frac{4\pi}{c} J$$

so that

$$\frac{[M][L]^2}{[q_m][T]^2} = \frac{[J]}{[c]}$$

and

$$\frac{[M]^{\frac{1}{2}}[L]^{\frac{1}{2}}}{[T][\mu]^{\frac{1}{2}}} = \frac{[M]^{\frac{1}{2}}[L]^{\frac{3}{2}}[\kappa]^{\frac{1}{2}}}{[c][T]^2}$$

yielding

$$\left[\frac{c}{\sqrt{\mu\kappa}}\right] = \left[\frac{L}{T}\right]$$

We see that $c/\sqrt{\mu\kappa}$ has the dimensions of speed. We shall soon see the significance of this.

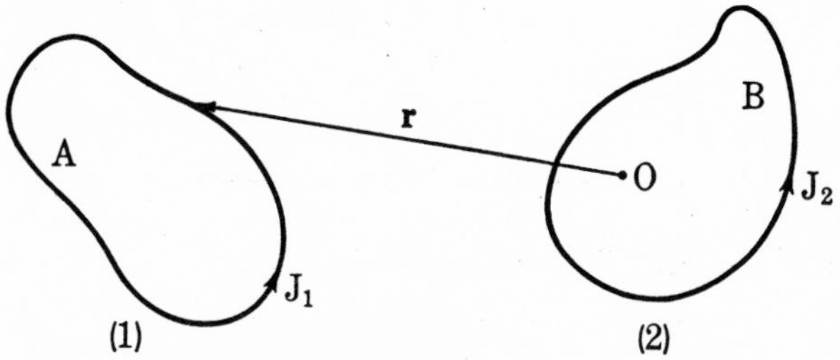

FIG. 70.

77. Mutual Induction and Action of Two Circuits. Consider two closed circuits with currents J_1 and J_2 (Fig. 70). The magnetic field at O due to J_1 is $\mathbf{H}_1 = \nabla \times \mathbf{A}_1$ where

$$\mathbf{A}_1 = \frac{J_1}{c} \int_{(1)} \frac{d\mathbf{r}}{r}$$

We define the mutual inductance of the two circuits as the magnetic flux through the surface B due to a unit current in (1). This is

$$M = \iint_B \mathbf{H}_1 \cdot d\boldsymbol{\sigma} = \iint_B \nabla \times \mathbf{A}_1 \cdot d\boldsymbol{\sigma}$$

$$= \int_{(2)} \mathbf{A}_1 \cdot d\mathbf{r} = \frac{1}{c} \int_{(2)} \left(\int_{(1)} \frac{d\mathbf{r}_1}{r} \right) \cdot d\mathbf{r}_2$$

Hence

$$M = \frac{1}{c} \int_{(2)} \int_{(1)} \frac{d\mathbf{r}_1 \cdot d\mathbf{r}_2}{r} \qquad (277)$$

The current element $J_2\, d\mathbf{r}_2$ sets up a magnetic field, so that from Newton's third law of action and reaction, any magnetic field will act on $J_2\, d\mathbf{r}_2$ with an equal and opposite force. Thus

$$df = J_2\, d\mathbf{r}_2 \times \mathbf{H}_1 = \frac{J_1 J_2}{c}\left(\int_{(1)} d\mathbf{r}_1 \times \nabla \frac{1}{r}\right) \times d\mathbf{r}_2 \tag{278}$$

and integrating over (2) we obtain

$$\mathbf{f} = \frac{J_1 J_2}{c} \int_{(2)} d\mathbf{r}_2 \times \int_{(1)} \nabla \frac{1}{r} \times d\mathbf{r}_1$$

Now

$$d\mathbf{r}_2 \times \left(\nabla \frac{1}{r} \times d\mathbf{r}_1\right) = \nabla \frac{1}{r}\,(d\mathbf{r}_1 \cdot d\mathbf{r}_2) - \left(d\mathbf{r}_2 \cdot \nabla \frac{1}{r}\right) d\mathbf{r}_1$$

and $\int_{(2)} [d\mathbf{r}_2 \cdot \nabla(1/r)]\, d\mathbf{r}_1 = \int_{(2)} d(1/r)\, d\mathbf{r}_1 = 0$, so that

$$\boxed{\mathbf{f} = \frac{J_1 J_2}{c} \int_{(2)} \int_{(1)} \left(\nabla \frac{1}{r}\right)(d\mathbf{r}_1 \cdot d\mathbf{r}_2)} \tag{279}$$

This is the force of loop (1) on loop (2). It is equal and opposite to the force of loop (2) on loop (1), this being immediately deducible from (279) when we keep in mind that

$$\nabla \frac{1}{r_{21}} = -\nabla \frac{1}{r_{12}}$$

In (279), $r = r_{21}$.

Example 97. We find the force per unit length between two long straight parallel wires carrying currents J_1 and J_2. We use (278) and the result of Example 95. We have $\mathbf{H}_1 = (2J_1/cd)\mathbf{i}$ at right angles to the plane containing the wires. Hence

$$d\mathbf{f} = J_2\, d\mathbf{r}_2 \times \frac{2J_1}{cd}\mathbf{i} = \frac{2J_1 J_2}{cd}\, d\mathbf{r}_2 \times \mathbf{i}$$

and the force per unit length is $F = 2J_1J_2/cd$. If the currents are parallel, **F** is an attractive force; if the currents are opposite, **F** is a repulsive force.

Problems

1. From (278) show that $\mathbf{f} = J_2 \iint_B d\boldsymbol{\sigma}_2 \times (\nabla \times \mathbf{H}_1)$.

2. Find the force between an infinite straight-line wire carrying a current J_1 and a square loop of side a with current J_2, the

extended plane of the loop containing the straight-line wire, and the shortest distance from the wire to the loop being d.

3. A current J flows around a circle of radius a, and a current J' flows in a very long straight wire in the same plane. Show that the mutual attraction is $4\pi JJ'/c(\sec \alpha - 1)$, where 2α is the angle subtended by the circle at the nearest point of the straight wire.

4. Show that $\mathbf{A} = (J/c) \iint_S d\boldsymbol{\sigma} \times \nabla(1/r)$ for a current J in a closed loop bounding the area S. For a small circular loop, show that $\mathbf{A} = (\mathbf{M} \times \mathbf{r}/r^3)$, where $\mathbf{r}$ is very much larger than the radius of the loop and is the vector to the center of the circle, and where

$$\mathbf{M} = \frac{J}{c} \iint d\boldsymbol{\sigma}$$

78. Law of Induction (Faraday). It has been found by experiment that a changing magnetic field produces an electromotive force in a circuit. If $\mathbf{B}$ is the magnetic inductance, the flux through a surface S with boundary curve Γ is given by $\iint_S \mathbf{B} \cdot d\boldsymbol{\sigma}$.

The law of induction states that

$$-\frac{1}{c}\frac{\partial}{\partial t} \iint \mathbf{B} \cdot d\boldsymbol{\sigma} = \oint_\Gamma \mathbf{E} \cdot d\mathbf{r}$$

Applying Stokes's theorem, we have

$$-\frac{1}{c}\frac{\partial \mathbf{B}}{\partial t} = \nabla \times \mathbf{E} \tag{280}$$

The time rate of change of magnetic inductance is proportional to the curl of the electric field. Equation (280) is a generalization of $\nabla \times \mathbf{E} = 0$, which is true for the electrostatic case in which $\mathbf{B} = 0$ and for the steady state for which $\dfrac{\partial \mathbf{B}}{\partial t} = 0$.

79. Maxwell's Equations. Up to the present we have, for an electrostatic field, $\nabla \times \mathbf{E} = 0$, $\nabla \cdot \mathbf{D} = 4\pi\rho$ and, for stationary currents, $\nabla \times \mathbf{H} = (4\pi/c)\mathbf{j}$, $\nabla \cdot \mathbf{B} = 0$.

Now $\nabla \times \mathbf{E} = -\frac{1}{c}\frac{\partial \mathbf{B}}{\partial t}$ is a generalization of $\nabla \times \mathbf{E} = 0$. Maxwell looked for a generalization of $\nabla \times \mathbf{H} = (4\pi/c)\mathbf{j}$. He decided to retain the two laws: (1) $\nabla \cdot \mathbf{D} = 4\pi\rho$ as the definition of charge, and (2) $\nabla \cdot \mathbf{j} + \frac{\partial \rho}{\partial t} = 0$ as the law of conservation of charge.

Let us assume

$$\nabla \times \mathbf{H} = \frac{4\pi}{c}(\mathbf{j} + \boldsymbol{\chi}) \tag{281}$$

as a generalization of $\nabla \times \mathbf{H} = (4\pi/c)\mathbf{j}$. We take the divergence of (281) and obtain

$$0 = \nabla \cdot \mathbf{j} + \nabla \cdot \boldsymbol{\chi} \tag{282}$$

so that

$$\begin{aligned}\nabla \cdot \boldsymbol{\chi} &= -\nabla \cdot \mathbf{j} = \frac{\partial \rho}{\partial t} = \frac{1}{4\pi}\frac{\partial}{\partial t}(\nabla \cdot \mathbf{D}) \\ &= \nabla \cdot \left(\frac{1}{4\pi}\frac{\partial \mathbf{D}}{\partial t}\right)\end{aligned}$$

We can choose $\boldsymbol{\chi} = \frac{1}{4\pi}\frac{\partial \mathbf{D}}{\partial t}$, so that

$$\boxed{\nabla \times \mathbf{H} = \frac{4\pi}{c}\left(\mathbf{j} + \frac{1}{4\pi}\frac{\partial \mathbf{D}}{\partial t}\right)} \tag{283}$$

We call $\frac{1}{4\pi}\frac{\partial \mathbf{D}}{\partial t}$ the displacement current.

We rewrite Maxwell's equations

$$\begin{aligned} &\text{(i)} && \nabla \cdot \mathbf{D} = 4\pi\rho \\ &\text{(ii)} && \nabla \cdot \mathbf{B} = 0 \\ &\text{(iii)} && \nabla \times \mathbf{E} = -\frac{1}{c}\frac{\partial \mathbf{B}}{\partial t} \\ &\text{(iv)} && \nabla \times \mathbf{H} = \frac{4\pi}{c}\left(\mathbf{j} + \frac{1}{4\pi}\frac{\partial \mathbf{D}}{\partial t}\right) \end{aligned} \tag{284}$$

We have in addition the equation

$$\text{(v)} \qquad \boxed{\mathbf{f} = \rho\left(\mathbf{E} + \frac{1}{c}\mathbf{v} \times \mathbf{B}\right)} \tag{285}$$

where $\mathbf{f}$ is the force on a charge ρ with velocity $\mathbf{v}$ moving in an electric field $\mathbf{E}$ and magnetic inductance $\mathbf{B}$. This result follows from Sec. 77.

Problems

1. Show that the equations of motion of a particle of mass m and charge e moving between the plates of a parallel-plate condenser producing a constant field E and subjected to a constant magnetic field H parallel to the plates are

$$m\frac{d^2x}{dt^2} = Ee - He\frac{dy}{dt}$$
$$m\frac{d^2y}{dt^2} = He\frac{dx}{dt}$$

Given that $\dfrac{dx}{dt} = \dfrac{dy}{dt} = x = y = 0$ when $t = 0$, show that $x = (E/\omega H)(1 - \cos \omega t)$, $y = (E/\omega H)(\omega t - \sin \omega t)$, where

$$\omega = \frac{He}{m}$$

2. From (iii) of (284) show that $\nabla \cdot \dfrac{\partial \mathbf{B}}{\partial t} = 0$.

3. From (i) and (iv) of (284) show that

$$\nabla \cdot \mathbf{j} = -\frac{1}{4\pi}\frac{\partial}{\partial t}(\nabla \cdot \mathbf{D}) = -\frac{\partial \rho}{\partial t}$$

4. Write down Maxwell's equations for a vacuum where $\mathbf{j} = \rho = 0$, $\mathbf{D} = \mathbf{E}$, $\mathbf{B} = \mathbf{H}$.

80. Solution of Maxwell's Equations for "Electrically" Free Space. We have $\rho = \mathbf{j} = 0$ and κ, μ are constants. Equations (284) become

$$
\begin{aligned}
&\text{(i)} & \nabla \cdot \mathbf{E} &= 0 \\
&\text{(ii)} & \nabla \cdot \mathbf{H} &= 0 \\
&\text{(iii)} & \nabla \times \mathbf{E} &= -\frac{\mu}{c}\frac{\partial \mathbf{H}}{\partial t} \\
&\text{(iv)} & \nabla \times \mathbf{H} &= \frac{\kappa}{c}\frac{\partial \mathbf{E}}{\partial t}
\end{aligned}
\tag{286}
$$

We take the curl of (iii) and obtain

$$\nabla \times (\nabla \times \mathbf{E}) = \nabla(\nabla \cdot \mathbf{E}) - \nabla^2\mathbf{E} = -\frac{\mu}{c}\nabla \times \frac{\partial \mathbf{H}}{\partial t}$$

or

$$\boxed{\nabla^2\mathbf{E} = \frac{\mu\kappa}{c^2}\frac{\partial^2\mathbf{E}}{\partial t^2}} \tag{287}$$

by making use of (i) and (iv). Similarly

$$\boxed{\nabla^2\mathbf{H} = \frac{\mu\kappa}{c^2}\frac{\partial^2\mathbf{H}}{\partial t^2}} \tag{287a}$$

Equation (287) represents a three-dimensional vector wave equation. To illustrate, consider a wave traveling down the x axis with velocity V and possessing the wave profile $y = f(x)$ at $t = 0$. At any time t it is easy to see that $y = f(x - Vt)$. From $y = f(x - Vt)$ we have $\frac{\partial^2 y}{\partial x^2} = f''(x - Vt)$ and

$$\frac{\partial^2 y}{\partial t^2} = V^2 f''(x - Vt)$$

so that

$$\frac{\partial^2 y}{\partial x^2} = \frac{1}{V^2}\frac{\partial^2 y}{\partial t^2} \tag{288}$$

Equation (287) represents three such equations, and $\mu\kappa/c^2$ plays the same role as $1/V^2$, so that $c/\sqrt{\mu\kappa}$ has the dimensions of a velocity (see Example 96). $c = 3 \times 10^{10}$ by experiment.

Example 98. We solve the wave equation $\nabla^2 f = \frac{1}{V^2}\frac{\partial^2 f}{\partial t^2}$ in spherical coordinates where $f = f(r, t)$.

$$\nabla f = f'(r, t)\,\nabla r = f'\,\frac{\mathbf{r}}{r}$$

$$\nabla^2 f = \nabla \cdot \left(\frac{f'\mathbf{r}}{r}\right) = \frac{f'}{r}\nabla \cdot \mathbf{r} + \nabla\left(\frac{f'}{r}\right) \cdot \mathbf{r}$$

$$= \frac{3f'}{r} + \frac{rf'' - f'}{r^2}\,\nabla r \cdot \mathbf{r} = f'' + \frac{2}{r}f'$$

Our wave equation is

$$\frac{\partial^2 f}{\partial r^2} + \frac{2}{r}\frac{\partial f}{\partial r} = \frac{1}{V^2}\frac{\partial^2 f}{\partial t^2} \tag{289}$$

Now let $u(r, t) = rf(r, t)$; then

$$\frac{\partial f}{\partial r} = \frac{1}{r}\frac{\partial u}{\partial r} - \frac{u}{r^2}, \qquad \frac{\partial^2 f}{\partial r^2} = \frac{1}{r}\frac{\partial^2 u}{\partial r^2} - \frac{2}{r^2}\frac{\partial u}{\partial r} + \frac{2}{r^3}u$$

and substituting into (289), we obtain

$$\frac{\partial^2 u}{\partial r^2} = \frac{1}{V^2}\frac{\partial^2 u}{\partial t^2}$$

of which the most general solution is

$$u(r, t) = g(r - Vt) + h(r + Vt)$$

and so

$$f(r, t) = \frac{1}{r}\,[g(r - Vt) + h(r + Vt)] \tag{290}$$

is the most general solution of (289).

Let us now try to determine a solution of Maxwell's equations for the case $\mathbf{E} = \mathbf{E}(x, t)$, $\mathbf{H} = \mathbf{H}(x, t)$.

Now $\nabla \cdot \mathbf{E} = 0$, so that $\dfrac{\partial E_x(x, t)}{\partial x} + \dfrac{\partial E_y(x, t)}{\partial y} + \dfrac{\partial E_z(x, t)}{\partial z} = 0$, which implies $\dfrac{\partial E_x(x, t)}{\partial x} = 0$. We are not interested in a uniform field in the x direction, so we choose $E_x = 0$. Hence

$$\mathbf{E} = E_y(x, t)\mathbf{j} + E_z(x, t)\mathbf{k}$$

and similarly

$$\mathbf{H} = H_y(x, t)\mathbf{j} + H_z(x, t)\mathbf{k}$$

Now we use Eq. (iii) of (286), $\nabla \times \mathbf{E} = -\frac{\mu}{c}\frac{\partial \mathbf{H}}{\partial t}$, so that

$$\begin{vmatrix} \mathbf{i} & \mathbf{j} & \mathbf{k} \\ \frac{\partial}{\partial x} & \frac{\partial}{\partial y} & \frac{\partial}{\partial z} \\ 0 & E_y & E_z \end{vmatrix} = -\frac{\mu}{c}\frac{\partial H_y}{\partial t}\mathbf{j} - \frac{\mu}{c}\frac{\partial H_z}{\partial t}\mathbf{k}$$

or

$$\begin{aligned} &\text{(i)} & \frac{\partial E_z}{\partial x} &= \frac{\mu}{c}\frac{\partial H_y}{\partial t} \\ &\text{(ii)} & \frac{\partial E_y}{\partial x} &= -\frac{\mu}{c}\frac{\partial H_z}{\partial t} \end{aligned} \tag{291}$$

Similarly, on using (iv) of (286), we obtain

$$\begin{aligned} &\text{(i)} & \frac{\partial H_z}{\partial x} &= -\frac{\kappa}{c}\frac{\partial E_y}{\partial t} \\ &\text{(ii)} & \frac{\partial H_y}{\partial x} &= \frac{\kappa}{c}\frac{\partial E_z}{\partial t} \end{aligned} \tag{292}$$

The four unknowns are E_y, E_z, H_y, H_z, which must satisfy (291) and (292). If we choose $H_y = E_z = 0$, we see that (i) of (291) and (ii) of (292) are satisfied. Differentiating (ii) of (291) with respect to x and (i) of (292) with respect to t, we obtain

$$\frac{\partial^2 E_y}{\partial x^2} = \frac{\mu\kappa}{c^2}\frac{\partial^2 E_y}{\partial t^2} \tag{293}$$

We leave it to the reader to show that

$$\frac{\partial^2 H_z}{\partial x^2} = \frac{\mu\kappa}{c^2}\frac{\partial^2 H_z}{\partial t^2} \tag{294}$$

These equations are of the type represented by (288). Hence a solution to Maxwell's equations is

$$\begin{aligned} \mathbf{E} &= [E_y^{(1)}(x - Vt) + E_y^{(2)}(x + Vt)]\mathbf{j} \\ \mathbf{H} &= [H_z^{(1)}(x - Vt) + H_z^{(2)}(x + Vt)]\mathbf{k} \end{aligned} \tag{295}$$

where $V = c(\mu\kappa)^{-\frac{1}{2}}$.

Both waves are transverse waves, that is, they travel down the x axis but have components perpendicular to the x axis.

Also note that $\mathbf{E} \cdot \mathbf{H} = 0$, so that $\mathbf{E}$ and $\mathbf{H}$ are always at right angles to each other.

By letting $H_z = E_y = 0$, we can obtain another solution, $\mathbf{E} = E_z(x, t)\mathbf{k}$, $\mathbf{H} = H_y(x, t)\mathbf{j}$. These two solutions are called the two states of polarization, the electric vector being always oriented 90° with the magnetic vector.

Example 99. We compute the energy density.

$$w_e = \frac{\mathbf{D} \cdot \mathbf{E}}{2} = \frac{\kappa \mathbf{E}^2}{2} = \frac{\kappa E_y^2}{2}$$

$$w_m = \frac{\mathbf{B} \cdot \mathbf{H}}{2} = \frac{\mu \mathbf{H}^2}{2} = \frac{\mu H_z^2}{2} = \frac{\kappa E_y^2}{2} = w_e$$

and $w = w_e + w_m = 2w_e = 2w_m = \kappa E_y^2$, and for both waves $w_T = \kappa(E_y^2 + E_z^2)$. We have here used the fact that

$$H_z = \sqrt{\frac{\kappa}{\mu}}\, E_y$$

(see Prob. 1).

Example 100. Maxwell's equations in a homogeneous conducting medium are

$$\text{(i)} \qquad \nabla \cdot \mathbf{E} = \frac{4\pi\rho}{\kappa}$$

$$\text{(ii)} \qquad \nabla \cdot \mathbf{H} = 0$$

$$\text{(iii)} \qquad \nabla \times \mathbf{E} = -\frac{\mu}{c}\frac{\partial \mathbf{H}}{\partial t}$$

$$\text{(iv)} \qquad \nabla \times \mathbf{H} = \frac{4\pi}{c}\left(\sigma \mathbf{E} + \frac{\kappa}{4\pi}\frac{\partial \mathbf{E}}{\partial t}\right)$$

Assume a periodic solution of the form

$$\mathbf{E} = \mathbf{E}_0(x, y, z)e^{-i\omega t}$$
$$\mathbf{H} = \mathbf{H}_0(x, y, z)e^{-i\omega t}$$

Substituting into (iv), we obtain

$$e^{-i\omega t}\, \nabla \times \mathbf{H}_0 = \frac{4\pi}{c}\left(\sigma e^{-i\omega t}\mathbf{E}_0 - \frac{i\kappa\omega}{4\pi}\mathbf{E}_0 e^{-i\omega t}\right)$$

or

$$\nabla \times \mathbf{H}_0 = \frac{4\pi}{c}\left(\sigma - \frac{i\kappa\omega}{4\pi}\right)\mathbf{E}_0 \qquad (296)$$

This equation is the same as that which occurs for "electrically" free space with a complex dielectric coefficient.

Problems

1. By letting $E_y = f(x - Vt)$, $H_z = F(x - Vt)$, $V = c/\sqrt{\kappa\mu}$, show that $H_z = \sqrt{\kappa/\mu}\, E_y$.

2. Derive (287*a*).

3. Let $r = x - Vt$, $s = x + Vt$, and show that $\dfrac{\partial^2 y}{\partial x^2} = \dfrac{1}{V^2}\dfrac{\partial^2 y}{\partial t^2}$ reduces to $\dfrac{\partial^2 y}{\partial r\, \partial s} = 0$. Integrate this equation and show that the general solution of (288) is $y = f(x - Vt) + F(x + Vt)$, where f and F are arbitrary functions.

4. Prove that Maxwell's equations for insulators ($\sigma = 0$) are

$$\nabla \times \mathbf{H} = \frac{\kappa}{c}\frac{\partial \mathbf{E}}{\partial t} \qquad \text{and} \qquad \nabla \times \mathbf{E} = -\frac{\mu}{c}\frac{\partial \mathbf{H}}{\partial t} \tag{297}$$

5. Show that the solution of (297) can be expressed in terms of a single vector **V**, the Hertzian vector, where

$$\mathbf{E} = \nabla(\nabla \cdot \mathbf{V}) - \frac{\kappa\mu}{c^2}\frac{\partial^2 \mathbf{V}}{\partial t^2}, \qquad \mathbf{H} = \frac{\kappa}{c}\nabla \times \frac{\partial \mathbf{V}}{\partial t}$$

and **V** satisfies $\nabla^2 \mathbf{V} = \dfrac{\kappa\mu}{c^2}\dfrac{\partial^2 \mathbf{V}}{\partial t^2}$.

6. Prove that $\mathbf{E} = \dfrac{\mu}{c}\nabla \times \dfrac{\partial \mathbf{W}}{\partial t}$, $\mathbf{H} = -\nabla(\nabla \cdot \mathbf{W}) + \dfrac{\kappa\mu}{c^2}\dfrac{\partial^2 \mathbf{W}}{\partial t^2}$ is a solution of (297), provided that **W** satisfies $\nabla^2 \mathbf{W} = \dfrac{\kappa\mu}{c^2}\dfrac{\partial^2 \mathbf{W}}{\partial t^2}$.

7. Derive (294).

8. Look up a proof of the laws of reflection and refraction.

9. By considering (i) and (iv) of Example 100, show that

$$\rho = \rho_0 e^{-4\pi\sigma t/\kappa}$$

81. Poynting's Theorem. Our starting point is Maxwell's equations. Dot Eq. (iii) of (284) with **H** and Eq. (iv) with **E** and subtract, obtaining

$$c(\mathbf{H} \cdot \nabla \times \mathbf{E} - \mathbf{E} \cdot \nabla \times \mathbf{H}) = -\mathbf{H} \cdot \frac{\partial \mathbf{B}}{\partial t} - 4\pi \mathbf{j} \cdot \mathbf{E} - \mathbf{E} \cdot \frac{\partial \mathbf{D}}{\partial t} \tag{298}$$

Now from (218)

$$\frac{\partial w_e}{\partial t} = \frac{1}{4\pi}\mathbf{E}\cdot\frac{\partial \mathbf{D}}{\partial t}, \qquad \frac{\partial w_m}{\partial t} = \frac{1}{4\pi}\mathbf{H}\cdot\frac{\partial \mathbf{B}}{\partial t}$$

Let us write $\mathbf{j} = \mathbf{j}_G + \mathbf{j}_c$ where $\mathbf{j}_G$ represents the galvanic current and $\mathbf{j}_c = \rho\mathbf{v}$, the conduction current. Now

$$\mathbf{E}\cdot\mathbf{j}_c = \mathbf{E}\cdot\rho\mathbf{v} = \rho\mathbf{E}\cdot\frac{d\mathbf{r}}{dt} = \frac{\partial w_{\text{mechanical}}}{\partial t} = \frac{\partial w_M}{\partial t}$$

and from Sec. 75 it is easy to prove that $\mathbf{E}\cdot\mathbf{j}_G$ is Joule's power loss $= \frac{\partial Q}{\partial t}\cdot$ Moreover, $\mathbf{H}\cdot\nabla\times\mathbf{E} - \mathbf{E}\cdot\nabla\times\mathbf{H} = \nabla\cdot(\mathbf{E}\times\mathbf{H})$, so that we rewrite (298) as

$$c\,\nabla\cdot(\mathbf{E}\times\mathbf{H}) = -4\pi\left(\frac{\partial w_e}{\partial t} + \frac{\partial w_m}{\partial t} + \frac{\partial w_M}{\partial t} + \frac{\partial Q}{\partial t}\right) \tag{299}$$

Integrating over a volume R and applying the divergence theorem, we obtain

$$\iiint_R \frac{\partial Q}{\partial t}\,d\tau + \frac{c}{4\pi}\iint_S \mathbf{E}\times\mathbf{H}\cdot d\mathbf{\sigma} = -\iiint_R \frac{\partial w}{\partial t}\,d\tau \tag{300}$$

where w is the total energy density.

We define $\mathbf{s} = (c/4\pi)\mathbf{E}\times\mathbf{H}$ as Poynting's vector. Equation (300) states that to determine the time rate of energy loss in a given volume V, we may find the flux through the boundary surface of the vector $\mathbf{s} = (c/4\pi)\mathbf{E}\times\mathbf{H}$ and add to this the rate of generation of heat within the volume. It is natural to interpret Poynting's vector as the density of energy flow.

Problems

1. Find the value of **E** and **H** on the surface of an infinite cylindrical wire carrying a current. Show that Poynting's vector represents a flow of energy into the wire, and show that this flow is just enough to supply the energy which appears as heat.

2. Find the Poynting vector around a uniformly charged sphere placed in a uniform magnetic field.

3. If **E** of Sec. 80 is sinusoidal, $\mathbf{E} = E_0 \sin \omega(x - Vt)\mathbf{k}$, find the energy density after finding the magnetic wave **H**.

82. Lorentz's Electron Theory. For charges moving with velocity $\mathbf{v}$, $\mathbf{j} = \rho\mathbf{v}$, and Maxwell's equations become

$$
\begin{aligned}
&\text{(i)} && \nabla \cdot \mathbf{D} = 4\pi\rho \\
&\text{(ii)} && \nabla \cdot \mathbf{B} = 0 \\
&\text{(iii)} && \nabla \times \mathbf{E} = -\frac{1}{c}\frac{\partial \mathbf{B}}{\partial t} \\
&\text{(iv)} && \nabla \times \mathbf{H} = \frac{4\pi}{c}\left(\rho\mathbf{v} + \frac{1}{4\pi}\frac{\partial \mathbf{D}}{\partial t}\right)
\end{aligned} \tag{301}
$$

These equations are due to Lorentz. From (ii) we can write $\mathbf{B} = \nabla \times (\mathbf{A}_0 + \nabla\chi) = \nabla \times \mathbf{A}$. Substitute this value of $\mathbf{B}$ into (iii) and obtain $\nabla \times \mathbf{E} = -\frac{1}{c}\nabla \times \frac{\partial \mathbf{A}}{\partial t}$, or its equivalent

$$\nabla \times \left(\mathbf{E} + \frac{1}{c}\frac{\partial \mathbf{A}}{\partial t}\right) = \mathbf{0}$$

Thus $\mathbf{E} + \frac{1}{c}\frac{\partial \mathbf{A}}{\partial t}$ is irrotational, so that $\mathbf{E} + \frac{1}{c}\frac{\partial \mathbf{A}}{\partial t} = -\nabla\varphi$.

Let $\mathbf{D} = \kappa\mathbf{E}$, $\mathbf{B} = \mu\mathbf{H}$, and substitute into (iv). We have

$$\frac{1}{\mu}\nabla \times (\nabla \times \mathbf{A}) = \frac{4\pi}{c}\left[\rho\mathbf{v} + \frac{\kappa}{4\pi}\left(-\frac{1}{c}\frac{\partial^2 \mathbf{A}}{\partial t^2} - \nabla\frac{\partial\varphi}{\partial t}\right)\right]$$

and since $\nabla \times (\nabla \times \mathbf{A}) = -\nabla^2\mathbf{A} + \nabla(\nabla \cdot \mathbf{A})$, we obtain

$$\nabla^2\mathbf{A} - \frac{\kappa\mu}{c^2}\frac{\partial^2 \mathbf{A}}{\partial t^2} = -\frac{4\pi\mu\rho}{c}\mathbf{v} + \nabla\psi \tag{302}$$

where

$$\psi = \nabla \cdot \mathbf{A} + \frac{\mu\kappa}{c^2}\frac{\partial\varphi}{\partial t}$$

Now

$$
\begin{aligned}
4\pi\rho = \nabla \cdot \mathbf{D} &= \kappa\,\nabla \cdot \left(-\nabla\varphi - \frac{1}{c}\frac{\partial \mathbf{A}}{\partial t}\right) \\
&= -\kappa\,\nabla^2\varphi - \frac{\kappa}{c}\left(\frac{\partial\psi}{\partial t} - \frac{\mu\kappa}{c}\frac{\partial^2\varphi}{\partial t^2}\right)
\end{aligned}
$$

so that

$$\nabla^2\varphi - \frac{\kappa\mu}{c^2}\frac{\partial^2\varphi}{\partial t^2} = -\frac{4\pi\rho}{\kappa} - \frac{1}{c}\frac{\partial\psi}{\partial t} \tag{303}$$

Equations (302) and (303) would be very much simplified if we could make $\psi \equiv \nabla \cdot \mathbf{A} + \frac{\kappa\mu}{c^2}\frac{\partial \varphi}{\partial t} = 0$. This is called the equation of gauge invariance. Let us see if this is possible.

Now $\mathbf{B} = \nabla \times \mathbf{A}_0$ and $\mathbf{E} + \frac{1}{c}\frac{\partial \mathbf{A}_0}{\partial t} = -\nabla\varphi_0$ so that

$$\mathbf{E} = -\frac{1}{c}\frac{\partial \mathbf{A}_0}{\partial t} - \nabla\varphi_0 = -\frac{1}{c}\frac{\partial \mathbf{A}}{\partial t} - \nabla\varphi$$

where $\mathbf{A} = \mathbf{A}_0 + \nabla\chi$. Thus

$$\frac{1}{c}\left(\frac{\partial \mathbf{A}}{\partial t} - \frac{\partial \mathbf{A}_0}{\partial t}\right) = \frac{1}{c}\frac{\partial}{\partial t}\nabla\chi = \nabla(\varphi_0 - \varphi)$$

and

$$\frac{1}{c}\frac{\partial \chi}{\partial t} = \varphi_0 - \varphi + \text{constant}$$

$$\frac{1}{c}\frac{\partial^2 \chi}{\partial t^2} = \frac{\partial \varphi_0}{\partial t} - \frac{\partial \varphi}{\partial t}$$

Now we desire

$$0 = \nabla \cdot \mathbf{A} + \frac{\kappa\mu}{c^2}\frac{\partial \varphi}{\partial t}$$

$$= \nabla \cdot \mathbf{A}_0 + \nabla^2\chi + \frac{\kappa\mu}{c}\left[\frac{\partial \varphi_0}{\partial t} - \frac{1}{c}\frac{\partial^2 \chi}{\partial t^2}\right]$$

or

$$\nabla^2\chi - \frac{\kappa\mu}{c^2}\frac{\partial^2 \chi}{\partial t^2} = -\nabla \cdot \mathbf{A}_0 - \frac{\kappa\mu}{c^2}\frac{\partial \varphi_0}{\partial t} \tag{304}$$

The right-hand side of (304) is a known function of x, y, z, t. This equation is called the inhomogeneous wave equation, and if the equation of gauge invariance is to hold, we must be able to solve it. If we can solve it, the Lorentz equations will reduce to four inhomogeneous wave equations and so will also be solvable. They are

$$\begin{aligned} \nabla^2\mathbf{A} - \frac{\kappa\mu}{c^2}\frac{\partial^2 \mathbf{A}}{\partial t^2} &= -\frac{4\pi\mu\rho}{c}\mathbf{v} \\ \nabla^2\varphi - \frac{\kappa\mu}{c^2}\frac{\partial^2 \varphi}{\partial t^2} &= -\frac{4\pi}{\kappa}\rho \end{aligned} \tag{305}$$

Problems

1. For the Lorentz transformations (see Prob. 11, Sec. 24), show that

$$\Box\varphi \equiv \frac{\partial^2\varphi}{\partial x^2} + \frac{\partial^2\varphi}{\partial y^2} + \frac{\partial^2\varphi}{\partial z^2} - \frac{1}{c^2}\frac{\partial^2\varphi}{\partial t^2} = \frac{\partial^2\varphi}{\partial \bar{x}^2} + \frac{\partial^2\varphi}{\partial \bar{y}^2} + \frac{\partial^2\varphi}{\partial \bar{z}^2} - \frac{1}{c^2}\frac{\partial^2\varphi}{\partial \bar{t}^2}$$

We call $\Box$ the D'Alembertian.

2. Consider the four-dimensional vector $C_i = (A_1, A_2, A_3, -\varphi)$, $i = 1, 2, 3, 4$, the A_i satisfying $\nabla^2\mathbf{A} = \frac{1}{c^2}\frac{\partial^2\mathbf{A}}{\partial t^2}$, while φ satisfies $\nabla^2\varphi = \frac{1}{c^2}\frac{\partial^2\varphi}{\partial t^2}$, with $\mathbf{H} = \nabla \times \mathbf{A}$, and $\mathbf{E} = -\frac{1}{c}\frac{\partial\mathbf{A}}{\partial t} - \nabla\varphi$. Let $x^1 = x$, $x^2 = y$, $x^3 = z$, $x^4 = ct$, and show that

$$F_{ij} \equiv \left(\frac{\partial C_i}{\partial x^j} - \frac{\partial C_j}{\partial x^i}\right) = \begin{pmatrix} 0 & -H_z & H_y & -E_x \\ H_z & 0 & -H_x & -E_y \\ -H_y & H_x & 0 & -E_z \\ E_x & E_y & E_z & 0 \end{pmatrix}$$

Show that $\sum_{j=1}^{4} \frac{\partial F_{ij}}{\partial x^j} = 0$, $i = 1, 2, 3, 4$, yields $\nabla \times \mathbf{H} = \frac{1}{c}\frac{\partial\mathbf{E}}{\partial t}$ and $\nabla \cdot \mathbf{E} = 0$. Also show that

$$\frac{\partial F_{\alpha\beta}}{\partial x^\gamma} + \frac{\partial F_{\gamma\alpha}}{\partial x^\beta} + \frac{\partial F_{\beta\gamma}}{\partial x^\alpha} = 0, \qquad \alpha, \beta, \gamma = 1, 2, 3, 4$$

yields $\nabla \times \mathbf{E} = -\frac{1}{c}\frac{\partial\mathbf{H}}{\partial t}$ and $\nabla \cdot \mathbf{H} = 0$.

3. If $\bar{F}_{ij} = \sum_{\alpha=1}^{4}\sum_{\beta=1}^{4} F_{\alpha\beta}\frac{\partial x^\alpha}{\partial \bar{x}^i}\frac{\partial x^\beta}{\partial \bar{x}^j}$, show that for the Lorentz transformations $\bar{F}_{12} \equiv -\bar{H}_{\bar{z}} = \frac{-H_z - (V/c)E_y}{[1 - (V^2/c^2)]^{\frac{1}{2}}}$. Complete the matrix $\bar{F}_{ij}$.

83. Retarded Potentials. Kirchhoff's Solution of

$$\nabla^2\varphi - \frac{1}{V^2}\frac{\partial^2\varphi}{\partial t^2} = -4\pi F(x, y, z, t)$$

To find the solution $\varphi(P, t)$ of the inhomogeneous wave equation at $t = 0$, we surround the point P by a small sphere Σ of

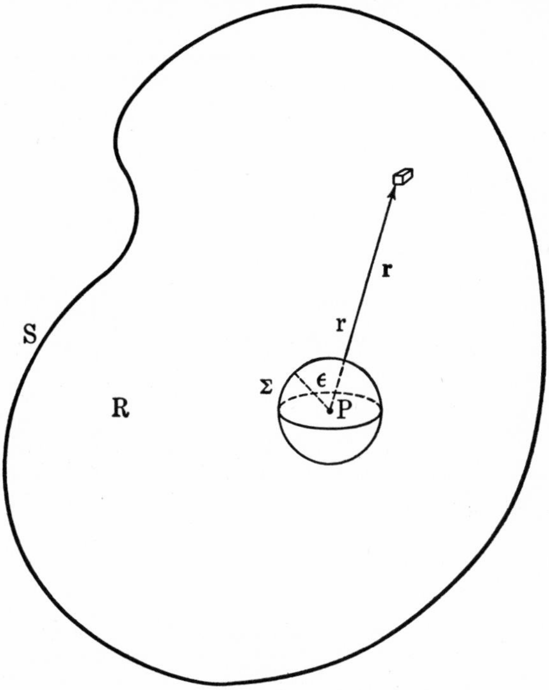

FIG. 71.

radius ϵ, and let S be the surface of a region R containing P (see Fig. 71). We apply Green's formula to this region.

$$\iiint_R (\psi \nabla^2\varphi - \varphi \nabla^2\psi)\, d\tau = \iint_\Sigma (\psi \nabla\varphi - \varphi \nabla\psi) \cdot d\boldsymbol{\sigma} + \iint_S (\psi \nabla\varphi - \varphi \nabla\psi) \cdot d\boldsymbol{\sigma} \tag{306}$$

We choose for ψ a solution of $\nabla^2\psi - \dfrac{1}{V^2}\dfrac{\partial^2\psi}{\partial t^2} = 0$. We know that $\psi = f(r + Vt)/r$ is one such solution, where f is arbitrary. Equation (306) now becomes

$$\frac{1}{V^2}\iiint_R \left(\psi \frac{\partial^2\varphi}{\partial t^2} - \varphi \frac{\partial^2\psi}{\partial t^2}\right) d\tau - 4\pi \iiint_R F\psi\, d\tau = \iint_\Sigma \cdots + \iint_S \cdots \tag{307}$$

Equation (307) is true for all values of t so that we may integrate (307) with respect to t between limits $t = t_1$ and $t = t_2$. We obtain

$$\frac{1}{V^2}\iiint\limits_R \left[\psi \frac{\partial \varphi}{\partial t} - \varphi \frac{\partial \psi}{\partial t}\right]_{t_1}^{t_2} d\tau - 4\pi \int_{t_1}^{t_2} dt \iiint\limits_R F\psi\, d\tau$$
$$= \int_{t_1}^{t_2} dt \left(\iint\limits_\Sigma \cdots + \iint\limits_S \cdots\right) \qquad (308)$$

Now on Σ, $\psi = (1/\epsilon)f(\epsilon + Vt)$ and

$$\nabla\psi \cdot d\mathbf{\delta} = -\left.\frac{\partial \psi}{\partial r}\right|_{r=\epsilon} = \frac{1}{\epsilon^2}[f(\epsilon + Vt) - \epsilon f'(\epsilon + Vt)]\, dS$$

so that (308) reduces to

$$\frac{1}{V^2}\iiint\limits_R \left[\frac{f(r+Vt)}{r}\frac{\partial \varphi}{\partial t} - \varphi \frac{Vf'(r+Vt)}{r}\right]_{t_1}^{t_2} d\tau$$
$$- 4\pi \iiint\limits_R \int_{t_1}^{t_2} \frac{Ff(r+Vt)}{r}\, dt\, d\tau = \int_{1}^{t_2} dt \left\{\iint\limits_\Sigma \left[\frac{f(\epsilon + Vt)}{\epsilon}\nabla\varphi\right.\right.$$
$$\left.\left. + \varphi\frac{\epsilon f'(\epsilon + Vt) - f(\epsilon + Vt)}{\epsilon^2}\mathbf{R}\right] \cdot d\mathbf{\delta} + \iint\limits_S \cdots\right\} \qquad (309)$$

Let us now return to a consideration of $f(r + Vt)$. Since f is arbitrary, let us choose $f \equiv 0$ for $|r + Vt| > \delta$, with the additional restriction that $\int_{-\delta}^{\delta} f(r + Vt)\, d(r + Vt) = 1$, where δ is arbitrary for the moment. Notice that $f' \equiv 0$ for $|r + Vt| > \delta$.

Now let us choose $t_2 > 0$ and t_1 negatively large, so that for all values of r in the region R, $|r + Vt| > \delta$. Hence

$$\left[\frac{f(r+Vt)}{r}\frac{\partial \varphi}{\partial t} - \varphi \frac{Vf'(r+Vt)}{r}\right]_{t_1}^{t_2} = 0$$

since $|r + Vt_2| > \delta$, $|r + Vt_1| > \delta$.
Moreover

$$\int_{t_1}^{t_2} \frac{F}{r} f(r+Vt)\, dt = \frac{1}{V}\int_{t_1}^{t_2} \frac{F}{r} f(r+Vt)\, d(r+Vt)$$
$$= \frac{1}{V}\int_{-\delta}^{\delta} \frac{F}{r} f(x)\, dx \qquad (310)$$

for a fixed r. Now if δ is chosen very small, the value of (310) reduces to approximately

$$\frac{1}{V}\left(\frac{F}{r}\right)_{x=0}\int_{-\delta}^{\delta} f(x)\,dx = \frac{1}{V}\left(\frac{F}{r}\right)_{t=-r/V}$$

Hence the left-hand side of (309) reduces to

$$-\frac{4\pi}{V}\iiint_R \left(\frac{F}{r}\right)_{t=-r/V} d\tau \tag{311}$$

Now considering the right-hand side of (309), we see that

$$\lim_{\epsilon\to 0}\iint_\Sigma \left[\frac{f(\epsilon + Vt)}{\epsilon}\nabla\varphi + \varphi\frac{f'(\epsilon + Vt)}{\epsilon}\mathbf{R}\right]\cdot d\boldsymbol{\sigma} = 0$$

since $d\boldsymbol{\sigma}$ is of the order ϵ^2, and f, f', φ, $\nabla\varphi$ are bounded for a fixed δ. We also have that

$$\lim_{\epsilon\to 0} - \int_{t_1}^{t_2} dt \iint_\Sigma \frac{\varphi f(\epsilon + Vt)}{\epsilon^2}\,dS = -\frac{1}{V}4\pi\varphi(P) \tag{312}$$

since $\iint_\Sigma dS = 4\pi\epsilon^2$, and for small δ,

$$\int_{t_1}^{t_2}\varphi f(\epsilon + Vt)\,d(\epsilon + Vt) = \int_{-\delta}^{\delta}\varphi f(x)\,dx \approx \varphi(P)$$

Finally,

$$\begin{aligned}
\iint_S \int_{t_1}^{t_2} dt&\left[\frac{f(r + Vt)}{r}\nabla\varphi - \varphi\left(\frac{rf' - f}{r^2}\right)\nabla r\right]\cdot d\boldsymbol{\sigma}\\
&= \iint_S \int_{t_1}^{t_2} dt\left[\frac{f(r + Vt)}{r}\nabla\varphi + \frac{\varphi f}{r^2}\nabla r\right]\cdot d\boldsymbol{\sigma}\\
-\iint_S \int_{t_1}^{t_2} dt\,\frac{\varphi}{r}f'(\nabla r\cdot d\boldsymbol{\sigma}) &= \iint_S \int_{t_1}^{t_2} dt\left[\frac{f(r + Vt)}{r}\nabla\varphi\right.\\
&\left.+\frac{\varphi f}{r^2}\nabla r\right]\cdot d\boldsymbol{\sigma} + \iint_S \int_{t_1}^{t_2}\frac{1}{rV}f\frac{\partial\varphi}{\partial t}\,dt\,(\nabla r\cdot d\boldsymbol{\sigma})
\end{aligned} \tag{313}$$

on integrating by parts and noticing that $f]_{t_1}^{t_2} = 0$. Finally the right-hand side of (313) becomes equivalent to

$$\frac{1}{V}\iint_S\left(\frac{1}{r}\nabla\varphi\bigg|_{t=-r/V} - \varphi\bigg|_{t=-r/V}\nabla\frac{1}{r} + \frac{1}{rV}\frac{\partial\varphi}{\partial t}\bigg|_{t=-r/V}\nabla r\right)\cdot d\boldsymbol{\sigma} \tag{314}$$

Combining (311), (312), and (314), we obtain

$$\varphi(P) = \iiint_R \frac{F}{r}\bigg|_{t=-r/V} d\tau - \frac{1}{4\pi}\iint_S \left(\frac{\nabla\varphi}{r} - \varphi\,\nabla\frac{1}{r} + \frac{1}{rV}\frac{\partial\varphi}{\partial t}\nabla r\right)_{t=-r/V} \cdot d\boldsymbol{\sigma} \tag{315}$$

Now let S recede to infinity and assume that φ, $\nabla\varphi$, $\dfrac{\partial\varphi}{\partial t}$, when evaluated at $t = -r/V$ on the surface S, have the value zero until a definite time T. For large r, $t = -r/V$ is negative and so is always less than T. Hence the surface element vanishes, and

$$\varphi(P) = \iiint_\infty \frac{F}{r}\bigg|_{t=-r/V} d\tau \tag{316}$$

The solutions to (305) are thus seen to be

$$\begin{aligned} \mathbf{A}(P, t) &= \iiint_\infty \frac{\mu\rho\mathbf{v}}{cr}\bigg|_{t-(r/V)} d\tau \\ \varphi(P, t) &= \iiint_\infty \frac{\rho}{\kappa r}\bigg|_{t-(r/V)} d\tau \end{aligned} \tag{317}$$

where $V = c/\sqrt{\mu\kappa}$.

Finally,

$$\begin{aligned} \mathbf{B} &= \nabla \times \mathbf{A} \\ \mathbf{E} &= -\frac{1}{c}\frac{\partial\mathbf{A}}{\partial t} - \nabla\varphi \end{aligned} \tag{318}$$

The physical interpretation of these results is simple. The values of the magnetic and electric intensities at any particular point P at any instant t are, in general, determined not by the state of the rest of the field (ρ, $\mathbf{v}$) at that particular instant, but by its previous history. The effects at P, due to elements at a distance r from P, depend on the state of the element at a previous time $t - (r/V)$. This is just the difference in time required for the waves to travel from the element to P with the velocity $V = c/\sqrt{\mu\kappa}$, hence the name retarded potential. Had we considered the function $f(r - Vt)$, we should have obtained a solution depending on the advanced potentials. Physically this is impossible, since future events cannot affect past events!

Problem

1. A short length of wire carries an alternating current, $\mathbf{j} = \rho\mathbf{v} = I_0 (\sin \omega t)\mathbf{k}, -l/2 \leqq z \leqq l/2.$

(a) At distances far removed from the wire, show that

$$\mathbf{A} = \frac{I_0 l}{cr} \sin \omega \left(t - \frac{r}{c}\right) \mathbf{k}$$

and that in spherical coordinates

$$A_r = \frac{I_0 l}{cr} \sin \omega \left(t - \frac{r}{c}\right) \cos \theta$$

$$A_\theta = -\frac{I_0 l}{cr} \sin \omega \left(t - \frac{r}{c}\right) \sin \theta$$

$$A_\varphi = 0$$

(b) Show that $H_r = H_\theta = 0$, and that

$$H_\varphi = \frac{I_0 l}{cr} \sin \theta \left[\frac{\omega}{c} \cos \omega \left(t - \frac{r}{c}\right) + \frac{1}{r} \sin \omega \left(t - \frac{r}{c}\right)\right]$$

(c) Find φ from the equation of gauge invariance, and then E_r, E_θ, E_φ from $\mathbf{E} + \dfrac{1}{c}\dfrac{\partial \mathbf{A}}{\partial t} = -\nabla\varphi.$

CHAPTER 6

MECHANICS

84. Kinematics of a Particle. We shall describe the motion of a particle relative to a cartesian coordinate system. The motion of any particle is known when $\mathbf{r} = x(t)\mathbf{i} + y(t)\mathbf{j} + z(t)\mathbf{k}$ is known, where t is the time. We have seen that the velocity and acceleration, relative to this frame of reference, will be given by

$$\mathbf{v} = \frac{dx}{dt}\mathbf{i} + \frac{dy}{dt}\mathbf{j} + \frac{dz}{dt}\mathbf{k}$$

$$\mathbf{a} = \frac{d^2x}{dt^2}\mathbf{i} + \frac{d^2y}{dt^2}\mathbf{j} + \frac{d^2z}{dt^2}\mathbf{k}$$

The velocity may also be given by $\mathbf{v} = v\mathbf{t}$, where v is the speed and $\mathbf{t}$ is the unit tangent vector to the curve $\mathbf{r} = \mathbf{r}(t)$. Differentiating, we obtain

$$\mathbf{a} = \frac{d\mathbf{v}}{dt} = \frac{dv}{dt}\mathbf{t} + v\frac{d\mathbf{t}}{ds}\frac{ds}{dt} = \frac{dv}{dt}\mathbf{t} + \kappa v^2\mathbf{n} \tag{319}$$

by making use of (95). Analyzing (319), we see that the acceleration of the particle can be resolved into two components: a tangential acceleration of magnitude $\frac{dv}{dt}$, and a normal acceleration of magnitude $v^2\kappa = v^2/\rho$. This latter acceleration is called centripetal acceleration and is due to the fact that the velocity vector is changing direction, and so we expect the curvature to play a role here.

For a particle moving in a plane, we have seen in Sec. 17, Example 18, that the acceleration may be given by

$$\mathbf{a} = \left[\frac{d^2r}{dt^2} - r\left(\frac{d\theta}{dt}\right)^2\right]\mathbf{R} + \frac{1}{r}\frac{d}{dt}\left(r^2\frac{d\theta}{dt}\right)\mathbf{P}$$

Example 101. Let us assume that a particle moves in a plane and that its acceleration is only radial. In this case we must have $\frac{1}{r}\frac{d}{dt}\left(r^2\frac{d\theta}{dt}\right) = 0$, and integrating, $\frac{1}{2}r^2\frac{d\theta}{dt} = h =$ constant. From the calculus we know that the sectoral area is given by $dA = \frac{1}{2}r^2\,d\theta$ (see Fig. 72). Thus $\frac{dA}{dt} =$ constant, so that equal areas are swept out in equal intervals of time.

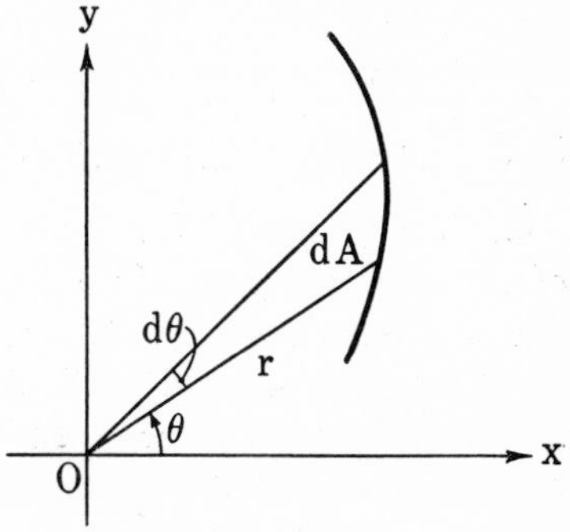

FIG. 72.

Example 102. For a particle moving around a circle $r = b$ with constant angular speed $\omega_0 = \frac{d\theta}{dt}$, we have $\frac{dr}{dt} = 0$ and $\frac{d}{dt}(r^2\omega_0) = 0$, so that $\mathbf{a} = -b\omega_0^2\mathbf{R}$.

Example 103. To find the tangential and normal components of the acceleration if the velocity and acceleration are known.

$$\mathbf{v} = v\mathbf{t}, \qquad \mathbf{a} = a_t\mathbf{t} + a_n\mathbf{n}$$

and

$$\mathbf{a}\cdot\mathbf{v} = va_t \qquad \text{so that} \qquad a_t = \frac{\mathbf{a}\cdot\mathbf{v}}{v}$$

Also

$$\mathbf{a}\times\mathbf{v} = va_n\mathbf{n}\times\mathbf{t} = -va_n\mathbf{b}$$

and

$$a_n = \frac{|\mathbf{a}\times\mathbf{v}|}{|\mathbf{v}|}$$

Problems

1. A particle moves in a plane with no radial acceleration and constant angular speed ω_0. Show that $r = Ae^{\omega_0 t} + Be^{-\omega_0 t}$.

2. A particle moves according to the law

$$\mathbf{r} = \cos t\,\mathbf{i} + \sin t\,\mathbf{j} + t^2\mathbf{k}$$

Find the tangential and normal components of the acceleration.

3. A particle describes the circle $r = a \cos \theta$ with constant speed. Show that the acceleration is constant in magnitude and directed toward the center of the circle.

4. A particle P moves in a plane with constant angular speed ω about O. If the rate of increase of its acceleration is parallel to OP, prove that $\dfrac{d^2r}{dt^2} = \frac{1}{3}r\omega^2$.

5. If the tangential and normal components of the acceleration of a particle moving in a plane are constant, show that the particle describes a spiral.

85. Motion about a Fixed Axis. In Sec. 10, Example 12, we saw that the velocity is given by $\mathbf{v} = \boldsymbol{\omega} \times \mathbf{r}$. Differentiating, we obtain

$$\mathbf{a} = \boldsymbol{\omega} \times \frac{d\mathbf{r}}{dt} + \frac{d\boldsymbol{\omega}}{dt} \times \mathbf{r}$$

$$\mathbf{a} = \boldsymbol{\omega} \times \mathbf{v} + \boldsymbol{\alpha} \times \mathbf{r} \tag{320}$$

where $\boldsymbol{\alpha}$ is the angular acceleration $\dfrac{d\boldsymbol{\omega}}{dt}$. Since $\mathbf{v} = \boldsymbol{\omega} \times \mathbf{r}$, we have also

$$\begin{aligned} \mathbf{a} &= \boldsymbol{\omega} \times (\boldsymbol{\omega} \times \mathbf{r}) + \boldsymbol{\alpha} \times \mathbf{r} \\ &= (\boldsymbol{\omega} \cdot \mathbf{r})\boldsymbol{\omega} - \omega^2\mathbf{r} + \boldsymbol{\alpha} \times \mathbf{r} \end{aligned}$$

If we take the origin on the line of $\boldsymbol{\omega}$ in the plane of the motion, then $\boldsymbol{\omega}$ is perpendicular to $\mathbf{r}$ or $\boldsymbol{\omega} \cdot \mathbf{r} = 0$, so that

$$\mathbf{a} = -\omega^2\mathbf{r} + \boldsymbol{\alpha} \times \mathbf{r}$$

$\boldsymbol{\alpha} \times \mathbf{r}$ is the tangential acceleration, and $\boldsymbol{\omega} \times (\boldsymbol{\omega} \times \mathbf{r})$ is the centripetal acceleration.

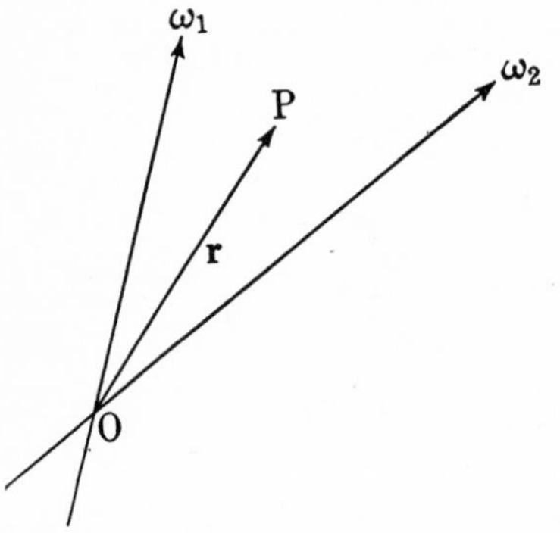

FIG. 73.

If we assume that a particle P is rotating about two intersecting lines simultaneously, with angular velocities $\boldsymbol{\omega}_1$, $\boldsymbol{\omega}_2$ (Fig. 73), we can choose our origin at the point of intersection so that

$$\mathbf{v}_1 = \boldsymbol{\omega}_1 \times \mathbf{r}, \qquad \mathbf{v}_2 = \boldsymbol{\omega}_2 \times \mathbf{r}$$

and the total velocity is

$$\mathbf{v} = \mathbf{v}_1 + \mathbf{v}_2 = (\boldsymbol{\omega}_1 + \boldsymbol{\omega}_2) \times \mathbf{r}$$

A particle on a spinning top that is also precessing experiences such motion.

86. Relative Motion. Let A and B be two particles traversing curves Γ_1 and Γ_2 (Fig. 74). $\mathbf{r}_1$ and $\mathbf{r}_2$ are the vectors from a point O to A and B, respectively.

$$\mathbf{r}_2 = \mathbf{r} + \mathbf{r}_1 \tag{321}$$

Definition: $\frac{d\mathbf{r}}{dt}$ is the relative velocity of B with respect to A, written $\mathbf{V}_A(B)$.

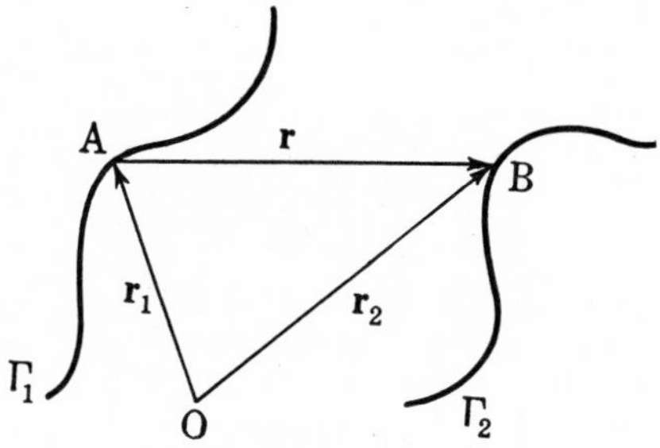

FIG. 74.

Differentiating (321), we have

$$\frac{d\mathbf{r}_2}{dt} = \frac{d\mathbf{r}}{dt} + \frac{d\mathbf{r}_1}{dt}$$

or

$$\boxed{\mathbf{V}_O(B) = \mathbf{V}_A(B) + \mathbf{V}_O(A)} \tag{322}$$

More generally, we have

$$\mathbf{V}_O(A) = \mathbf{V}_{A_1}(A) + \mathbf{V}_{A_2}(A_1) + \mathbf{V}_{A_3}(A_2) + \cdots + \mathbf{V}_{A_n}(A_{n-1}) + \mathbf{V}_O(A_n)$$

It is important to note that $\mathbf{V}_A(B) = -\mathbf{V}_B(A)$.

Example 104. A man walks eastward at 3 miles per hour, and the wind appears to come from the north. He then decreases his speed to 1 mile per hour and notices that the wind comes from the northwest (Fig. 75). What is the velocity of the wind?

We have

$$\mathbf{V}_G(W) = \mathbf{V}_M(W) + \mathbf{V}_G(M) \qquad G\text{(ground)}$$

In the first case

$$\mathbf{V}_M(W) = -k\mathbf{j}, \qquad \mathbf{V}_G(M) = 3\mathbf{i}$$

so that

$$\mathbf{V}_G(W) = -k\mathbf{j} + 3\mathbf{i}$$

In the second case,

$$\mathbf{V}_M(W) = h(\mathbf{i} - \mathbf{j}), \qquad \mathbf{V}_G(M) = \mathbf{i}$$

so that

$$\mathbf{V}_G(W) = h(\mathbf{i} - \mathbf{j}) + \mathbf{i} = (h + 1)\mathbf{i} - h\mathbf{j},$$

and

$$3 = h + 1, \qquad -k = -h, \qquad \mathbf{V}_G(W) = 3\mathbf{i} - 2\mathbf{j}$$

The speed of the wind is $\sqrt{13}$ miles per hour, and its direction makes an angle of $\tan^{-1} \frac{3}{2}$ with the south line.

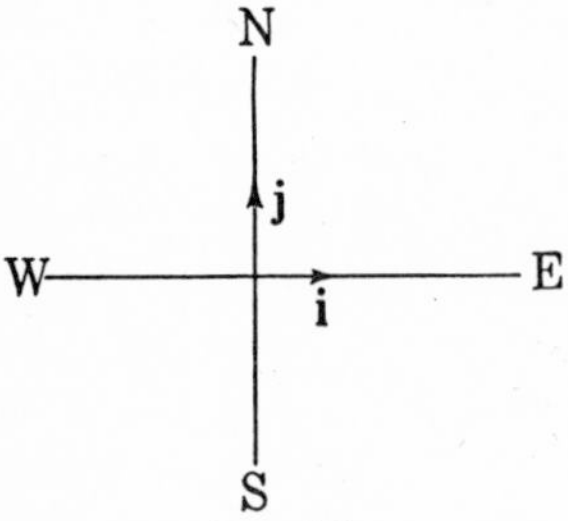

FIG. 75.

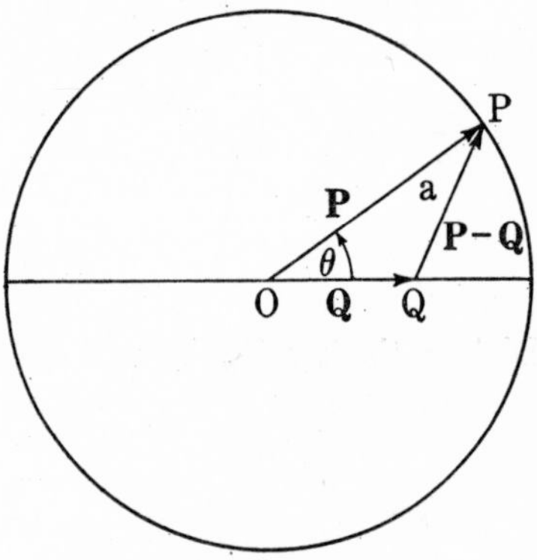

FIG. 76.

Example 105. To find the relative motion of two particles moving with the same speed v, one of which describes a circle of radius a while the other moves along the diameter (Fig. 76). We have

$$\mathbf{P} = a \cos \theta \, \mathbf{i} + a \sin \theta \, \mathbf{j}, \qquad a \frac{d\theta}{dt} = v$$

$$\mathbf{Q} = (a - vt)\mathbf{i}$$

This assumes that both particles started together.

$$\frac{d\mathbf{P}}{dt} - \frac{d\mathbf{Q}}{dt} = \left(-a \sin \theta \frac{d\theta}{dt} + v \right) \mathbf{i} + a \cos \theta \frac{d\theta}{dt} \mathbf{j}$$

$$\mathbf{V}_Q(P) = v(1 - \sin \theta)\mathbf{i} + v \cos \theta \, \mathbf{j}$$

The relative speed is

$$|\mathbf{V}_Q(P)| = [v^2(1 - \sin\theta)^2 + v^2\cos^2\theta]^{\frac{1}{2}} = 2^{\frac{1}{2}}v(1 - \sin\theta)^{\frac{1}{2}}$$

Maximum $|\mathbf{V}_Q(P)|$ occurs at $\theta = 3\pi/2$, minimum at $\theta = \pi/2$.

Problems

1. A man traveling east at 8 miles per hour finds that the wind seems to blow from the north. On doubling his speed, he finds that it appears to come from the northeast. Find the velocity of the wind.

2. A, B, C are on a straight line, B midway between A and C. It then takes A 4 minutes to catch C, and B catches C in 6 minutes. How long does it take A to catch up to B?

3. An airplane has a true course west and an air speed of 200 miles per hour. The wind speed is 50 miles per hour from 130°. Find the heading and ground speed of the plane.

87. Dynamics of a Particle. Up to the present, nothing has been said of the forces that produce or cause the motion of a particle. Experiment shows that for a particle to acquire an acceleration relative to certain types of reference frames, there must be a force acting on the particle. The types of forces encountered most frequently are (1) mechanical (push, pull), (2) gravitational, (3) electrical, (4) magnetic, (5) electromagnetic. We shall be chiefly concerned with forces of the types (1) and (2). For the present we shall assume Newton's laws of motion hold for motion relative to the earth. Afterward we shall modify this. Newton's laws are:

(*a*) A particle free from the action of forces will remain fixed or will continue to move in a straight line with constant speed.

(*b*) Force is proportional to time rate of change of momentum, that is, $\mathbf{f} = \dfrac{d}{dt}(m\mathbf{v})$. In general, m = constant, so that

$$\mathbf{f} = m\frac{d\mathbf{v}}{dt} = m\mathbf{a}$$

The factor m is found by experiment to be an invariant for a given particle and is called the mass of the particle. In the theory of relativity, m is not a constant. $m\mathbf{v}$ is called the *momentum*.

(*c*) If A exerts a force on B, then B exerts an equal and opposite force on A. This is the law of action and reaction: $\mathbf{f}_{AB} = -\mathbf{f}_{BA}$.

By a particle we mean a finite mass occupying a point in our Euclidean space. This is a purely mathematical concept, and physically we mean a mass occupying negligible volume as compared to the distance between masses. For example, the earth and sun may be thought of as particles in comparison to their distance apart, to a first approximation.

88. Equations of Motion for a Particle. Newton's second law may be written $\mathbf{f} = m\dfrac{d\mathbf{v}}{dt} = m\mathbf{a}$. We postulate that the forces acting on a particle behave as vectors. This is an experimental fact. Hence if $\mathbf{f}_1, \mathbf{f}_2, \ldots, \mathbf{f}_n$ act on m, its acceleration is given by

$$\mathbf{a} = \frac{1}{m}(\mathbf{f}_1 + \mathbf{f}_2 + \cdots + \mathbf{f}_n) = \frac{1}{m}\sum_{i=1}^{n} \mathbf{f}_i = \frac{\mathbf{f}}{m}$$

We may also write $\mathbf{f} = m\dfrac{d^2\mathbf{r}}{dt^2}$, where $\mathbf{r}$ is the position vector from the origin of our coordinate system to the particle. If the particle is at rest or is moving with constant velocity, then $\dfrac{d^2\mathbf{r}}{dt^2} = 0$, and so $\mathbf{f} = 0$, and conversely. Hence a necessary and sufficient condition that a particle be in static equilibrium is that the vector sum of the forces acting on it be zero.

A standard body is taken as the unit mass (pound mass). A poundal is the force required to accelerate a one-pound mass one foot per second per second. The mass of any other body can be compared with the unit mass by comparing the weights (force of gravity at mean sea level) of the two objects. This assumes the equivalence of gravitational mass and inertial mass.

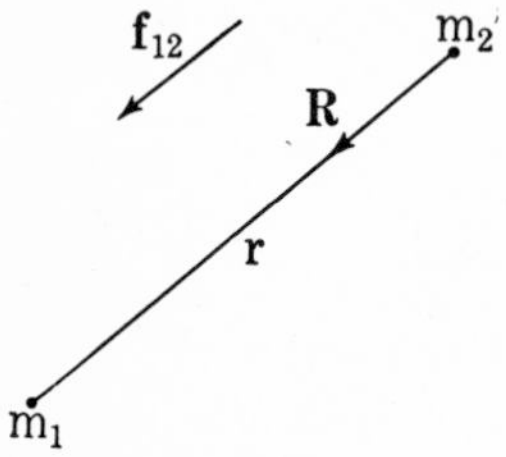

FIG. 77.

Example 106. Newton's law of gravitation for two particles is that every pair of particles in the universe exerts a mutual attraction with a force directed along the line joining the particles, the magnitude of the force being inversely proportional to the square of the distance between them and directly proportional to the product of their masses. $\mathbf{f}_{12} = (Gm_1m_2/r^2)\mathbf{R}$ (see Fig. 77). G is a universal constant. Let the mass of the sun be M and that of the earth be m. We shall assume that the sun is fixed at the origin of a given coordinate system (Fig. 78). The force acting on the earth due to the sun is

$$\mathbf{f} = -(GmM/r^3)\mathbf{r}$$

From the second law

$$-\frac{GmM}{r^3}\mathbf{r} = m\frac{d^2\mathbf{r}}{dt^2} = m\frac{d\mathbf{v}}{dt}$$

so that

$$\frac{d\mathbf{v}}{dt} = -\frac{GM}{r^3}\mathbf{r} \tag{323}$$

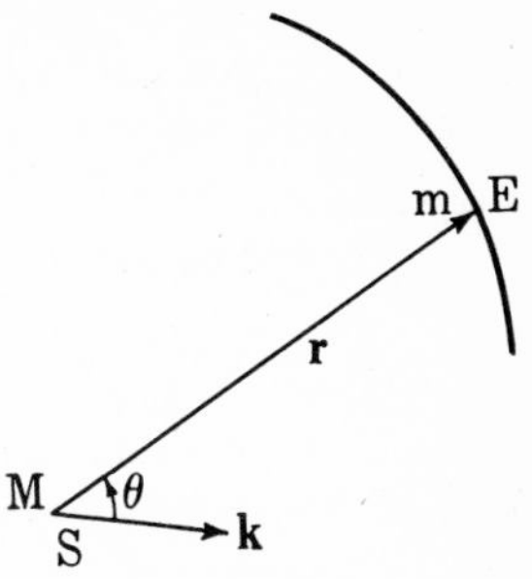

FIG. 78.

Now

$$\frac{d}{dt}(\mathbf{r}\times\mathbf{v}) = \mathbf{r}\times\frac{d\mathbf{v}}{dt}$$

and hence

$$\frac{d}{dt}(\mathbf{r}\times\mathbf{v}) = \mathbf{r}\times\left(-\frac{GM}{r^3}\mathbf{r}\right) = 0$$

This implies

$$\left.\begin{aligned} \mathbf{r}\times\mathbf{v} &= \mathbf{h} = \text{constant vector} \\ \text{or}\quad \mathbf{r}\times\frac{d\mathbf{r}}{dt} &= \mathbf{h} \end{aligned}\right\} \tag{324}$$

Since $|\mathbf{r}\times d\mathbf{r}|$ = twice sectoral area, we have $2\dfrac{dA}{dt} = |\mathbf{h}|$, or equal areas are swept out in equal intervals of time. This is Kepler's first law of planetary motion. Moreover, $\mathbf{r}\cdot\left[\mathbf{r}\times\dfrac{d\mathbf{r}}{dt}\right] = \mathbf{r}\cdot\mathbf{h} = 0$, so that $\mathbf{r}$ remains perpendicular to the fixed vector $\mathbf{h}$, and the motion is planar. Now

$$\frac{d\mathbf{v}}{dt}\times\mathbf{h} = -\frac{GM\mathbf{r}}{r^3}\times\mathbf{h} = -\frac{GM}{r^3}\mathbf{r}\times(\mathbf{r}\times\mathbf{v})$$

from (324), and $\dfrac{d}{dt}(\mathbf{v}\times\mathbf{h}) = \dfrac{d\mathbf{v}}{dt}\times\mathbf{h}$, so that

$$\frac{d}{dt}(\mathbf{v}\times\mathbf{h}) = -\frac{GM}{r^3}\mathbf{r}\times(\mathbf{r}\times\mathbf{v}) \tag{325}$$

Now $\mathbf{r} = r\mathbf{R}$, where $\mathbf{R}$ is a unit vector. Hence

$$\mathbf{v} = \frac{d\mathbf{r}}{dt} = r\frac{d\mathbf{R}}{dt} + \frac{dr}{dt}\mathbf{R}$$

so that (325) becomes

$$\begin{aligned}
\frac{d}{dt}(\mathbf{v} \times \mathbf{h}) &= -\frac{GM}{r^3}\mathbf{r} \times \left(\mathbf{r} \times r\frac{d\mathbf{R}}{dt}\right) \\
&= -GM\mathbf{R} \times \left(\mathbf{R} \times \frac{d\mathbf{R}}{dt}\right) \\
&= -GM\left[\left(\mathbf{R} \cdot \frac{d\mathbf{R}}{dt}\right)\mathbf{R} - \mathbf{R}^2\frac{d\mathbf{R}}{dt}\right] \\
&= GM\frac{d\mathbf{R}}{dt}
\end{aligned} \tag{326}$$

since $\mathbf{R}$ is a unit vector.

Integrating (326), we obtain

$$\mathbf{v} \times \mathbf{h} = GM\mathbf{R} + \mathbf{k}$$

and

$$\begin{aligned}
\mathbf{r} \cdot (\mathbf{v} \times \mathbf{h}) &= \mathbf{r} \cdot (GM\mathbf{R} + \mathbf{k}) \\
(\mathbf{r} \times \mathbf{v}) \cdot \mathbf{h} &= GMr + \mathbf{r} \cdot \mathbf{k} \\
h^2 &= GMr + \mathbf{r} \cdot \mathbf{k} \\
h^2 &= GMr + rk \cos (\mathbf{R}, \mathbf{k})
\end{aligned} \tag{327}$$

Thus

$$r = \frac{h^2/GM}{1 + (k/GM) \cos (\mathbf{r}, \mathbf{k})} \tag{328}$$

We choose the direction of the constant vector $\mathbf{k}$ as the polar axis, so that

$$r = \frac{h^2/GM}{1 + (k/GM) \cos \theta} \tag{329}$$

This is the polar equation of a conic section. For the planets these conic sections are closed curves, so that we obtain Kepler's second law, which states that the orbits of the planets are ellipses with the sun at one of the foci.

Let us now write the ellipse in the form

$$r = \frac{ep}{1 + e \cos \theta} \qquad \text{where } e = \frac{k}{GM},\ p = \frac{h^2}{k}$$

The curve crosses the polar axis at $\theta = 0$, $\theta = \pi$ so that the length of the major axis is

$$2a = \frac{ep}{1+e} + \frac{ep}{1-e} = \frac{2p}{1-e^2} = \frac{2h^2}{GM(1-e^2)}$$

For an ellipse, $b^2 = a^2 - c^2 = a^2 - e^2a^2$, or $b = a(1 - e^2)^{\frac{1}{2}}$. The area of the ellipse is $A = \pi ab = \pi a^2(1 - e^2)^{\frac{1}{2}}$, and since $\dfrac{dA}{dt} = \frac{1}{2}h$, the period for one complete revolution is

$$T = \frac{2A}{h} = \frac{2\pi a^2(1-e^2)^{\frac{1}{2}}}{a^{\frac{1}{2}}G^{\frac{1}{2}}M^{\frac{1}{2}}(1-e^2)^{\frac{1}{2}}} = \frac{2\pi a^{\frac{3}{2}}}{G^{\frac{1}{2}}M^{\frac{1}{2}}}$$

Thus

$$\frac{T^2}{a^3} = \frac{4\pi^2}{GM} \equiv \text{constant, for all planets} \tag{330}$$

This is Kepler's third law, which states that the squares of the periods of revolution of the planets are proportional to the cubes of the mean distances from the sun.

Problems

1. A particle of mass m is attracted toward the origin with the force $\mathbf{f} = -(k^2m/r^6)\mathbf{r}$. If it starts from the point $(a, 0)$ with the speed $v_0 = k/2^{\frac{1}{2}}a^2$ perpendicular to the x axis, show that the path is given by $r = a \cos \theta$.

2. A bead of mass m slides along a smooth rod which is rotating with constant angular speed ω, the rod always lying in a horizontal plane. Find the reaction between bead and rod.

3. A particle of mass m is attracted toward the origin with a force $-(mk^2/r^3)\mathbf{R}$. If it starts from the point $(a, 0)$ with velocity $v_0 > k/a$ perpendicular to the x axis, show that the equation of the path is

$$r = a \sec \left[\frac{(a^2v_0{}^2 - k^2)^{\frac{1}{2}}}{av_0}\,\theta\right]$$

4. In a uniform gravitational field (earth), a 16-pound shot leaves the putter's fingers 7 feet from the ground. At what angle should the shot leave to attain a maximum horizontal distance?

5. Assume a comet starts from infinity at rest and is attracted toward the sun. Let r_0 be its least distance to the sun. Show that the motion of the comet is given by $r = 2r_0/(1 + \cos\theta)$.

89. System of Particles. Let us consider a system consisting of a finite number of particles moving under the action of various forces. A given particle will be under the influence of two types of forces: (1) internal forces, that is, forces due to the interaction of the particle with the other particles of the system, and (2) all other forces acting on the particle, said forces being called external forces.

If $\mathbf{r}_j$ is the position vector to the particle of mass m_j, then we shall designate $\mathbf{f}_j^{(e)}$ as the sum of the external forces acting on the jth particle, and $\mathbf{f}_j^{(i)}$ as the sum of the internal forces acting on this particle. Newton's second law becomes for this particle

$$\mathbf{f}_j^{(e)} + \mathbf{f}_j^{(i)} = m_j \frac{d^2\mathbf{r}_j}{dt^2} \tag{331}$$

Unfortunately, we do not know, in general, $\mathbf{f}_j^{(i)}$, so that we shall not try to find the motion of each particle but shall look rather for the motion of the system as a whole. Since Eq. (331) is true for each j, we can sum up j for all the particles. This yields

$$\sum_{j=1}^{n} \mathbf{f}_j^{(e)} + \sum_{j=1}^{n} \mathbf{f}_j^{(i)} = \sum_{j=1}^{n} m_j \frac{d^2\mathbf{r}_j}{dt^2}$$

From Newton's third law we know that for every internal force there is an equal and opposite reaction, so that $\sum_{j=1}^{n} \mathbf{f}_j^{(i)} = 0$. This leaves

$$\sum_{j=1}^{n} \mathbf{f}_j^{(e)} = \sum_{j=1}^{n} m_j \frac{d^2\mathbf{r}_j}{dt^2} \tag{332}$$

We now define a new vector, called the center-of-mass vector, by the equation

$$\mathbf{r}_c = \bar{\mathbf{r}} = \frac{\sum_{j=1}^{n} m_j\mathbf{r}_j}{\sum_{j=1}^{n} m_j} = \frac{1}{M}\sum_{j=1}^{n} m_j\mathbf{r}_j \tag{333}$$

The end point of $\mathbf{r}_c$ is called the center of mass of the system. It is a geometric property and depends only on the position of the particles. Differentiating (333) twice with respect to time, we obtain

$$M \frac{d^2\mathbf{r}_c}{dt^2} = \sum_{j=1}^{n} m_j \frac{d^2\mathbf{r}_j}{dt^2}$$

so that (332) becomes

$$\mathbf{f} = \sum_{j=1}^{n} \mathbf{f}_j^{(e)} = M \frac{d^2\mathbf{r}_c}{dt^2} \tag{334}$$

Equation (334) states that the center of mass of the system accelerates as if the total mass were concentrated there and all the external forces acted at that point.

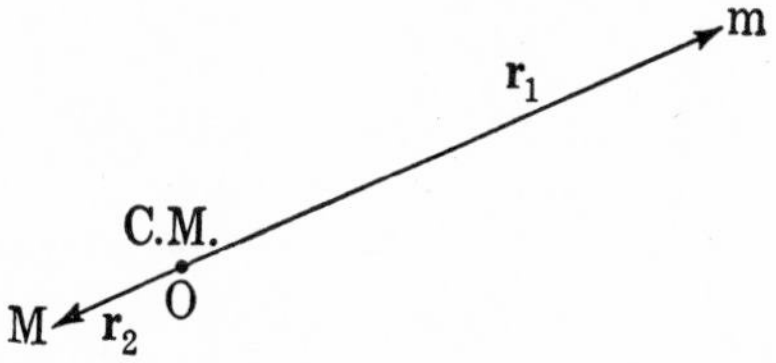

FIG. 79.

Example 107. If our system is composed of two particles in free space and if they are originally at rest, then the center of mass will always remain at rest, since $\mathbf{f} = 0$ so that $\dfrac{d^2\mathbf{r}_c}{dt^2} = 0$, and $\mathbf{r}_c \equiv$ *constant* satisfies the equation of motion and the initial condition $\dfrac{d\mathbf{r}_c}{dt} = 0$. For the earth and sun we may choose the center of mass as the origin of our coordinate system (Fig. 79). The equations of motion for earth and sun are

$$m \frac{d^2\mathbf{r}_1}{dt^2} = -\frac{GmM\mathbf{R}}{(\mathbf{r}_1 - \mathbf{r}_2)^2}, \qquad M \frac{d^2\mathbf{r}_2}{dt^2} = \frac{GmM\mathbf{R}}{(\mathbf{r}_1 - \mathbf{r}_2)^2}$$

Since $\mathbf{r}_c = 0$, we have $m\mathbf{r}_1 + M\mathbf{r}_2 = 0$, and

$$\frac{d^2\mathbf{r}_1}{dt^2} = \frac{-GM}{[1 + (m/M)]^2} \frac{\mathbf{r}_1}{r_1^3}$$

This shows that m is attracted toward the center of mass by an inverse-square force. The results of Example 106 hold by replacing M by $M[1 + (m/M)]^{-2}$.

Problems

1. Show that the center of mass is independent of the origin of our coordinate system.

2. Particles of masses 1, 2, 3, 4, 5, 6, 7, 8 are placed at the corners of a unit cube. Find the center of mass.

3. Find the center of mass of a uniform hemisphere.

4. Find the force of attraction of a hemisphere on another hemisphere, the two hemispheres forming a full sphere.

90. Momentum and Angular Momentum. The momentum of a particle of mass m and velocity $\mathbf{v}$ is defined as $\mathbf{M} = m\mathbf{v}$. The total momentum of a system of particles is given by $\mathbf{M} = \sum_{j=1}^{n} m_j\mathbf{v}_j$. We have at once that

$$\frac{d\mathbf{M}}{dt} = \sum_{j=1}^{n} m_j \frac{d\mathbf{v}_j}{dt} = \sum_{j=1}^{n} \mathbf{f}_j^{(e)} = \mathbf{f} \tag{335}$$

We emphasize again that the mass of each particle is assumed constant throughout the motion.

The vector quantity $\mathbf{r} \times m\mathbf{v}$ is defined as the angular momentum, or moment of momentum, of the particle about the origin O. The total angular momentum is given by

$$\mathbf{H} = \sum_{j=1}^{n} \mathbf{r}_j \times m_j\mathbf{v}_j \tag{336}$$

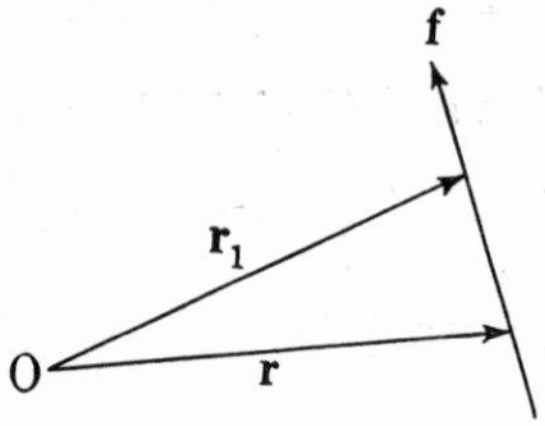

FIG. 80.

91. Torque, or Force Moment. Let $\mathbf{f}$ be a force acting in a given direction and let $\mathbf{r}$ be any vector from the origin whose end point lies on the line of action of the force (see Fig. 80). The vector quantity $\mathbf{r} \times \mathbf{f}$ is defined as the force moment, or torque, of $\mathbf{f}$ about O. For a system of forces,

$$\mathbf{L} = \sum_{j=1}^{n} \mathbf{r}_j \times \mathbf{f}_j \tag{337}$$

We immediately ask if the torque is different if we use a different vector $\mathbf{r}_1$ to the line of action of $\mathbf{f}$. The answer is in the negative, for

$$(\mathbf{r}_1 - \mathbf{r}) \times \mathbf{f} = 0$$

since $\mathbf{r}_1 - \mathbf{r}$ is parallel to $\mathbf{f}$. Hence $\mathbf{r}_1 \times \mathbf{f} = \mathbf{r} \times \mathbf{f}$.

What of the torque due to two equal and opposite forces both acting along the same line? It is zero, for

$$\mathbf{r}_1 \times \mathbf{f} + \mathbf{r}_1 \times (-\mathbf{f}) = \mathbf{r}_1 \times (\mathbf{f} - \mathbf{f}) \equiv 0$$

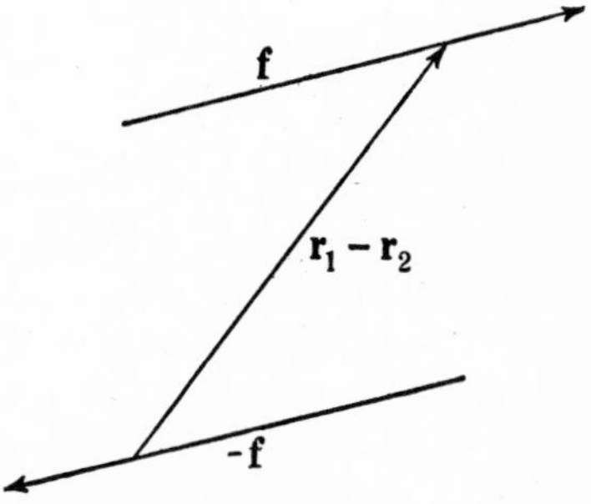

FIG. 81.

Two equal and opposite forces with different lines of action constitute a **couple** (see Fig. 81). Let $\mathbf{r}_1$ be a vector to $\mathbf{f}$ and $\mathbf{r}_2$ a vector to $-\mathbf{f}$. The torque due to this couple is

$$\begin{aligned} \mathbf{L} &= \mathbf{r}_1 \times \mathbf{f} + \mathbf{r}_2 \times (-\mathbf{f}) \\ &= (\mathbf{r}_1 - \mathbf{r}_2) \times \mathbf{f} \end{aligned}$$

The couple depends only on $\mathbf{f}$ and on any vector from the line of action of $-\mathbf{f}$ to the line of action of $\mathbf{f}$.

Problems

1. Show that if the resultant of a system of forces is zero, the total torque about one point is the same as that about any other point.

2. Show that the torques about two different points are equal, provided that the resultant of the forces is parallel to the vector joining the two origins.

3. Show that any set of forces acting on a body can be replaced by a single force, acting at an arbitrary point, plus a suitable couple. Prove this first for a single force.

4. Prove that the torque due to internal forces vanishes.

92. A Theorem Relating Angular Momentum with Torque. We are now in a position to prove that the time rate of change of angular momentum is equal to the sum of the external torques for a system of particles.

Since

$$\mathbf{H} = \sum_{j=1}^{n} \mathbf{r}_j \times m_j\mathbf{v}_j = \sum_{j=1}^{n} \mathbf{r}_j \times m_j\frac{d\mathbf{r}_j}{dt}$$

we have on differentiating

$$\frac{d\mathbf{H}}{dt} = \sum_{j=1}^{n} \mathbf{r}_j \times m_j \frac{d^2\mathbf{r}_j}{dt^2} + \sum_{j=1}^{n} \frac{d\mathbf{r}_j}{dt} \times m_j \frac{d\mathbf{r}_j}{dt}$$

$$= \sum_{j=1}^{n} \mathbf{r}_j \times (\mathbf{f}_j^{(e)} + \mathbf{f}_j^{(i)})$$

$$\frac{d\mathbf{H}}{dt} = \sum_{j=1}^{n} \mathbf{r}_j \times \mathbf{f}_j^{(e)} = \mathbf{L} \tag{338}$$

93. Moment of Momentum (*Continued*). It is occasionally more useful to choose a moving point Q as the origin of our

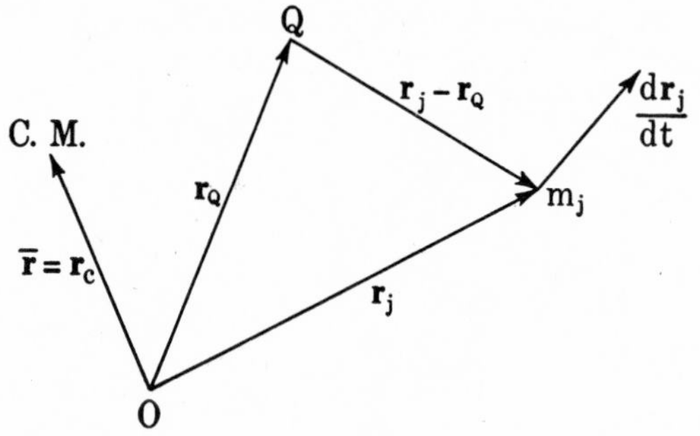

FIG. 82.

coordinate system. Let O be a fixed point and Q any point in space. We define

$$\mathbf{H}_Q{}^a \equiv \sum_{j=1}^{n} (\mathbf{r}_j - \mathbf{r}_Q) \times m_j \frac{d\mathbf{r}_j}{dt} \tag{339}$$

The superscript a stands for absolute momentum, that is, the velocity of m_j is taken relative to O, whereas the subscript Q stands for the fact that the lever arm is measured from Q to the particle m_j (Fig. 82). Differentiating (339), we obtain

$$\frac{d\mathbf{H}_Q{}^a}{dt} = \sum_{j=1}^{n} \left(\frac{d\mathbf{r}_j}{dt} - \frac{d\mathbf{r}_Q}{dt}\right) \times m_j \frac{d\mathbf{r}_j}{dt} + \sum_{j=1}^{n} (\mathbf{r}_j - \mathbf{r}_Q) \times m_j \frac{d^2\mathbf{r}_j}{dt^2}$$

$$= -\frac{d\mathbf{r}_Q}{dt} \times \sum_{j=1}^{n} m_j \frac{d\mathbf{r}_j}{dt} + \sum_{j=1}^{n} (\mathbf{r}_j - \mathbf{r}_Q) \times (\mathbf{f}_j{}^{(e)} + \mathbf{f}_j{}^{(i)})$$

Now $M\bar{\mathbf{r}} = \sum_{j=1}^{n} m_j\mathbf{r}_j$, so that $M \frac{d\bar{\mathbf{r}}}{dt} = \sum_{j=1}^{n} m_j \frac{d\mathbf{r}_j}{dt}$, and $\sum_{j=1}^{n} \mathbf{f}_j{}^{(i)} = 0$, $\sum_{j=1}^{n} \mathbf{r}_j \times \mathbf{f}_j{}^{(i)} = 0$ from Sec. 91, so that

$$\frac{d\mathbf{H}_Q{}^a}{dt} = -M \frac{d\mathbf{r}_Q}{dt} \times \frac{d\bar{\mathbf{r}}}{dt} + \sum_{j=1}^{n} (\mathbf{r}_j - \mathbf{r}_Q) \times \mathbf{f}_j{}^{(e)}$$

or

$$\frac{d\mathbf{H}_Q{}^a}{dt} = \mathbf{L}_Q{}^{(e)} - M \frac{d\mathbf{r}_Q}{dt} \times \frac{d\mathbf{r}_c}{dt} \tag{340}$$

We can simplify (340) under three conditions:

1. Q at rest, so that $\frac{d\mathbf{r}_Q}{dt} = 0$

2. Center of mass at rest, $\frac{d\mathbf{r}_c}{dt} = 0$

3. Velocity of Q is parallel to velocity of center of mass, $\frac{d\mathbf{r}_Q}{dt} \times \frac{d\mathbf{r}_c}{dt} = 0$

In all three cases

$$\frac{d\mathbf{H}_Q{}^a}{dt} = \mathbf{L}_Q{}^{(e)} \tag{341}$$

In particular, if $\mathbf{L}_Q{}^{(e)} = 0$, then $\mathbf{H}_Q{}^a =$ *constant*, and this is the law of conservation of angular momentum.

94. Moment of Relative Momentum about Q. In Sec. 93 we assumed that the absolute velocity of each particle was known. It is often more convenient to calculate the velocity of each particle relative to Q. This is $\frac{d\mathbf{r}_j}{dt} - \frac{d\mathbf{r}_Q}{dt}$. We now define relative moment of momentum about Q as

$$\mathbf{H}_Q{}^r = \sum_{j=1}^{n} (\mathbf{r}_j - \mathbf{r}_Q) \times m_j \frac{d}{dt} (\mathbf{r}_j - \mathbf{r}_Q) \tag{342}$$

Differentiating,

$$\frac{d\mathbf{H}_Q{}^r}{dt} = \sum_{j=1}^{n} (\mathbf{r}_j - \mathbf{r}_Q) \times m_j \left(\frac{d^2\mathbf{r}_j}{dt^2} - \frac{d^2\mathbf{r}_Q}{dt^2} \right)$$

$$= \sum_{j=1}^{n} (\mathbf{r}_j - \mathbf{r}_Q) \times (\mathbf{f}_j{}^{(e)} + \mathbf{f}_j{}^{(i)}) + \frac{d^2\mathbf{r}_Q}{dt^2} \times \sum_{j=1}^{n} m_j(\mathbf{r}_j - \mathbf{r}_Q)$$

We see that

$$\frac{d\mathbf{H}_Q{}^r}{dt} = \mathbf{L}_Q{}^{(e)} + \frac{d^2\mathbf{r}_Q}{dt^2} \times \sum_{j=1}^{n} m_j(\mathbf{r}_j - \mathbf{r}_Q) \tag{343}$$

Under what conditions does $\dfrac{d\mathbf{H}_Q{}^r}{dt} = \mathbf{L}_Q{}^{(e)}$? We need

$$\frac{d^2\mathbf{r}_Q}{dt^2} \times \sum_{j=1}^{n} m_j(\mathbf{r}_j - \mathbf{r}_Q) = 0$$

or

$$M \frac{d^2\mathbf{r}_Q}{dt^2} \times (\mathbf{r}_c - \mathbf{r}_Q) = 0 \tag{344}$$

Now (344) holds if

1. $\mathbf{r}_c = \mathbf{r}_Q$, or Q is at the center of mass.
2. Q moves with constant velocity, $\dfrac{d^2\mathbf{r}_Q}{dt^2} = 0$.
3. $\mathbf{r}_c - \mathbf{r}_Q$ is parallel to $\dfrac{d^2\mathbf{r}_Q}{dt^2}$.

Problems

1. Show that $\displaystyle\sum_{j=1}^{n} m_j(\mathbf{r}_j - \mathbf{r}_Q) \times \frac{d^2\mathbf{r}_Q}{dt^2} = M(\mathbf{r}_c - \mathbf{r}_Q) \times \frac{d^2\mathbf{r}_Q}{dt^2}$.

2. A system of particles lies in a plane, and each particle remains at a fixed distance from a point O in this plane, each

particle rotating about O with angular velocity $\boldsymbol{\omega}$. Show that $\mathbf{H}_O = I\boldsymbol{\omega}$, where $I = \sum_{j=1}^{n} m_j r_j^2$, and show that $\mathbf{L}_O = I \dfrac{d\boldsymbol{\omega}}{dt}$.

3. A hoop rolls down an inclined plane. What point can be taken as Q so that the equation of motion (343) would be simplified?

95. Kinetic Energy. We define the kinetic energy of a particle of mass m and velocity $\mathbf{v}$ as $T = \frac{1}{2}m\mathbf{v} \cdot \mathbf{v}$. For a system of particles,

$$T = \sum_{j=1}^{n} \frac{1}{2} m_j \mathbf{v}_j^2 = \sum_{j=1}^{n} \frac{1}{2} m_j \left(\frac{d\mathbf{r}_j}{dt}\right)^2 \tag{345}$$

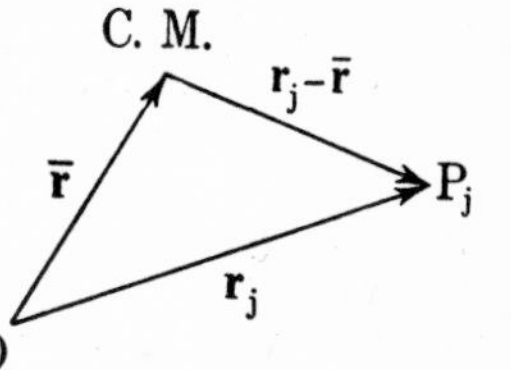

FIG. 83.

Now let $\mathbf{r}_c$ be the vector to the center of mass C (Fig. 83). It is obvious that

$$\mathbf{r}_j = \mathbf{r}_c + (\mathbf{r}_j - \mathbf{r}_c)$$

so that

$$\frac{d\mathbf{r}_j}{dt} = \frac{d\mathbf{r}_c}{dt} + \frac{d}{dt}(\mathbf{r}_j - \mathbf{r}_c)$$

$$\frac{d\mathbf{r}_j}{dt} \cdot \frac{d\mathbf{r}_j}{dt} = \left(\frac{d\mathbf{r}_c}{dt}\right)^2 + 2\frac{d\mathbf{r}_c}{dt} \cdot \frac{d}{dt}(\mathbf{r}_j - \mathbf{r}_c) + \left[\frac{d}{dt}(\mathbf{r}_j - \mathbf{r}_c)\right]^2$$

Hence

$$T = \frac{1}{2} M \left(\frac{d\mathbf{r}_c}{dt}\right)^2 + \frac{d\mathbf{r}_c}{dt} \cdot \sum_{j=1}^{n} m_j \frac{d}{dt}(\mathbf{r}_j - \mathbf{r}_c) + \sum_{j=1}^{n} \frac{1}{2} m_j \left[\frac{d}{dt}(\mathbf{r}_j - \mathbf{r}_c)\right]^2 \tag{346}$$

Now $M\mathbf{r}_c = \sum_{j=1}^{n} m_j \mathbf{r}_j = \left(\sum_{j=1}^{n} m_j\right) \mathbf{r}_c$, so that $\sum_{j=1}^{n} m_j \dfrac{d}{dt}(\mathbf{r}_j - \mathbf{r}_c) = 0$,

and (346) reduces to

$$T = \frac{1}{2} M \left(\frac{d\mathbf{r}_c}{dt}\right)^2 + \sum_{j=1}^{n} \frac{1}{2} m_j \left[\frac{d}{dt}(\mathbf{r}_j - \mathbf{r}_c)\right]^2 \tag{347}$$

This proves that the kinetic energy of a system of particles is equal to the kinetic energy of a particle having the total mass of the system and moving with the center of mass, plus the kinetic energy of the particles in their motion relative to the center of mass.

96. Work. If a particle moves along a curve Γ with velocity $\mathbf{v}$ under the action of a force $\mathbf{f}$, we define the work done by this force as

$$W = \int_\Gamma (\mathbf{f} \cdot \mathbf{v})\, dt = \int_\Gamma \mathbf{f} \cdot \frac{d\mathbf{r}}{dt}\, dt = \int_\Gamma \mathbf{f} \cdot d\mathbf{r}$$

$$= \int_\Gamma \mathbf{f} \cdot \frac{d\mathbf{r}}{ds}\, ds = \int_\Gamma \mathbf{f} \cdot \mathbf{t}\, ds \tag{348}$$

If $\mathbf{f}$ acts at right angles to the path, no work is done.

If the field is conservative, $\mathbf{f} = -\nabla\varphi$, the work done in taking the particle from a point A to a point B is independent of the path (see Sec. 52).

Now

$$m_j \frac{d\mathbf{v}_j}{dt} = \mathbf{f}_j^{(e)} + \mathbf{f}_j^{(i)}$$

$$m_j \mathbf{v}_j \cdot \frac{d\mathbf{v}_j}{dt} = \mathbf{f}_j^{(e)} \cdot \mathbf{v}_j + \mathbf{f}_j^{(i)} \cdot \mathbf{v}_j$$

and integrating and summing over all particles,

$$\sum_{j=1}^{n} \int_{t_0}^{t_1} m_j \mathbf{v}_j \cdot \frac{d\mathbf{v}_j}{dt}\, dt = \sum_{j=1}^{n} \int_{t_0}^{t_1} \mathbf{f}_j^{(e)} \cdot \mathbf{v}_j\, dt + \sum_{j=1}^{n} \int_{t_0}^{t_1} \mathbf{f}_j^{(i)} \cdot \mathbf{v}_j\, dt$$

or

$$\sum_{j=1}^{n} \tfrac{1}{2} m_j [\mathbf{v}_j^2(t_1) - \mathbf{v}_j^2(t_0)] = W^{(e)} + W^{(i)} \tag{349}$$

This is the principle of work and energy. The change in the kinetic energy of a system of particles is equal to the total work done by both the external and internal forces.

If the particles always remain at a constant distance apart, $(\mathbf{r}_j - \mathbf{r}_k)^2 \equiv$ constant, the internal forces do no work. Let $\mathbf{r}_1$ and $\mathbf{r}_2$ be the position vectors of two particles whose distance apart remains constant, and let $\mathbf{f}$ and $-\mathbf{f}$ be the internal forces of one particle on the other and conversely. Now

$$(\mathbf{r}_1 - \mathbf{r}_2) \cdot (\mathbf{r}_1 - \mathbf{r}_2) \equiv \text{constant}$$

so that

$$(\mathbf{r}_1 - \mathbf{r}_2) \cdot \left(\frac{d\mathbf{r}_1}{dt} - \frac{d\mathbf{r}_2}{dt}\right) = 0 \tag{350}$$

Also

$$\begin{aligned} W^{(i)} &= \int \mathbf{f} \cdot \mathbf{v}_1 \, dt + \int -\mathbf{f} \cdot \mathbf{v}_2 \, dt \\ &= \int \mathbf{f} \cdot (\mathbf{v}_1 - \mathbf{v}_2) \, dt \end{aligned}$$

Since $\mathbf{f}$ is parallel to $\mathbf{r}_1 - \mathbf{r}_2$, we have $\mathbf{f} = \alpha(\mathbf{r}_1 - \mathbf{r}_2)$ and

$$\mathbf{f} \cdot (\mathbf{v}_1 - \mathbf{v}_2) = \alpha(\mathbf{r}_1 - \mathbf{r}_2) \cdot (\mathbf{v}_1 - \mathbf{v}_2) \equiv 0$$

from (350). Thus $W^{(i)} = 0$.

Problems

1. A system of particles has an angular velocity $\boldsymbol{\omega}$. Show that $T = \sum_{j=1}^{n} \frac{1}{2} m_j |\boldsymbol{\omega} \times \mathbf{r}_j|^2$.

2. If $\boldsymbol{\omega}$ of Prob. 1 has a constant direction, show that $T = \frac{1}{2} I \omega^2$, where $I = \sum_{j=1}^{n} m_j d_j^2$, d_j being the shortest distance from m_j to line of $\boldsymbol{\omega}$.

3. Show that $\dfrac{dT}{dt} = \boldsymbol{\omega} \cdot \mathbf{L}$, by using the fact that $T = \sum_{j=1}^{n} \frac{1}{2} m_j v_j^2$ and that $\mathbf{v}_j = \boldsymbol{\omega} \times \mathbf{r}_j$.

4. Show that the kinetic energy of a system of rotating particles is constant if the system is subjected to no torques. What if $\mathbf{L}$ is perpendicular to $\boldsymbol{\omega}$?

5. A particle falls from infinity to the earth. Show that it strikes the earth with a speed of approximately 7.0 miles per second. Use the principle of work and energy.

97. Rigid Bodies. By a rigid body we mean a system of particles such that the relative distances between pairs of points remain constant during the discussion of our problem. Actually no such systems exist, but for practical purposes there do exist such rigid bodies, at least to a first approximation. Moreover, the rigid body may not consist of a finite number of particles, but rather will have a continuous distribution, at least to the unaided eye. We postulate that we can subdivide the body into a great many small parts so that we can apply our laws of motion for particles to this system, this postulate implying that we can use the integral calculus. Our laws of motion as derived above take the following form:

$$T = \iiint_R \tfrac{1}{2}\rho v^2 \, d\tau, \qquad \rho = \text{density}$$

$$\mathbf{r}_c = \frac{\iiint_R \rho \mathbf{r} \, d\tau}{\iiint_R \rho \, d\tau}$$

$$\iiint_R \mathbf{f}^{(e)} \, d\tau = M \frac{d^2\mathbf{r}_c}{dt^2} \tag{351}$$

$$\mathbf{H} = \iiint_R \rho \mathbf{r} \times \mathbf{v} \, d\tau$$

$$\frac{d\mathbf{H}}{dt} = \mathbf{L}^{(e)} = \iiint_R \mathbf{r} \times \mathbf{f}^{(e)} \, d\tau$$

where $\mathbf{f}^{(e)}$ is the external force per unit volume.

98. Kinematics of a Rigid Body. Let O be a point of a rigid body for which O happens to be fixed. It is easy to prove that the velocity of any other point P of the body must be perpendicular to the line joining O to P, for if $\mathbf{r}$ is the position vector from O to P, we have $\mathbf{r} \cdot \mathbf{r} \equiv$ constant throughout the motion so that $\mathbf{r} \cdot \frac{d\mathbf{r}}{dt} = 0$. Q.E.D.

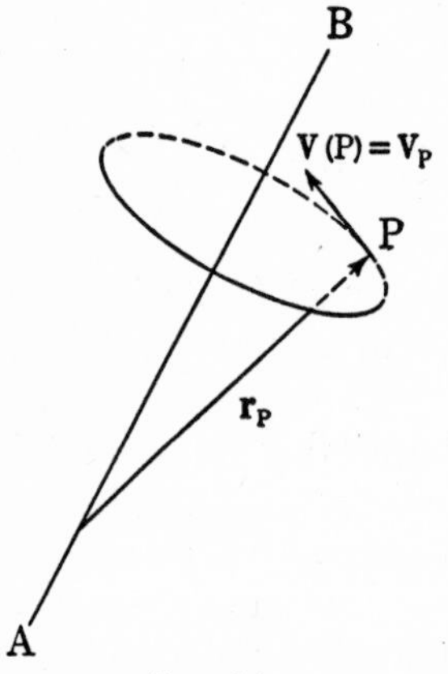

FIG. 84.

We next prove that if two points of a rigid body are fixed, then all other particles of the body are rotating around the line joining these two points. Let A and B be the fixed points and P any other point of the body. From above we have

$$\mathbf{v}(P) \cdot \overrightarrow{AP} = \mathbf{v}(P) \cdot \overrightarrow{BP} = 0$$

so that P is always moving perpendicular to the plane ABP. Moreover, since the body is rigid, the shortest distance from P to the line AB remains constant, so that P moves in a circle around AB (Fig. 84). We saw in Sec. 10 that the velocity of P could be written

$$\mathbf{v}_P = \boldsymbol{\omega} \times \mathbf{r}_P$$

Is $\boldsymbol{\omega}$ the same for all particles? Yes! Assume Q is rotating about AB with angular velocity $\boldsymbol{\omega}_1$, so that $\mathbf{v}_Q = \boldsymbol{\omega}_1 \times \mathbf{r}_Q$. Now $(\mathbf{r}_P - \mathbf{r}_Q)^2 \equiv$ constant, so that

$$(\mathbf{r}_P - \mathbf{r}_Q) \cdot (\mathbf{v}_P - \mathbf{v}_Q) = 0$$

or

$$(\mathbf{r}_P - \mathbf{r}_Q) \cdot (\boldsymbol{\omega} \times \mathbf{r}_P - \boldsymbol{\omega}_1 \times \mathbf{r}_Q) = 0$$

Thus

$$-\mathbf{r}_Q \cdot \boldsymbol{\omega} \times \mathbf{r}_P - \mathbf{r}_P \cdot \boldsymbol{\omega}_1 \times \mathbf{r}_Q = 0$$

and

$$\mathbf{r}_Q \times \mathbf{r}_P \cdot (\boldsymbol{\omega}_1 - \boldsymbol{\omega}) = 0$$

We leave it to the reader to conclude that $\boldsymbol{\omega}_1 = \boldsymbol{\omega}$.

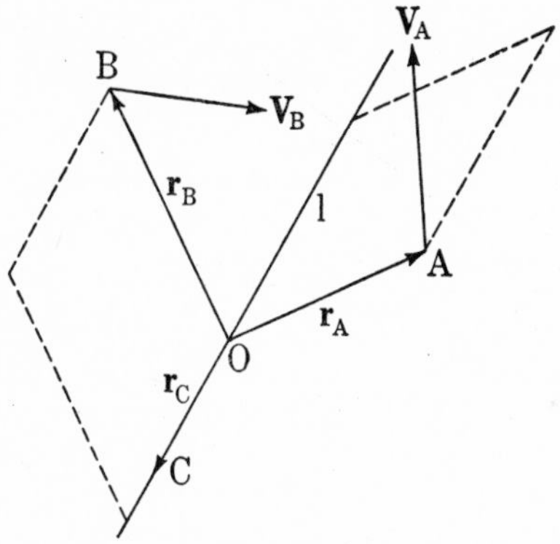

FIG. 85.

If one point of a rigid body is fixed, we cannot, in general, hope to find a fixed line about which the body is rotating. However, there does exist a moving line passing through the fixed point so that at any instant the body is actually rotating around this line. The proof proceeds as follows: Let O be the fixed point of our rigid body and let $\mathbf{r}_A$ be the position vector to a point A. From above we know that the velocity of A, $\mathbf{v}_A$, is perpendicular to $\mathbf{r}_A$. Construct the plane through O and A perpendicular to $\mathbf{v}_A$ (Fig. 85). Now choose a point B not in the plane. We also have that $\mathbf{v}_B \cdot \mathbf{r}_B = 0$, so that we can construct the plane through O and B perpendicular to $\mathbf{v}_B$. Both planes pass through O, so that their line of intersection, l, passes through O. Now consider any point C on this line. We have $\mathbf{v}_C \cdot \mathbf{r}_C = 0$. Moreover, $(\mathbf{r}_C - \mathbf{r}_A) \cdot (\mathbf{r}_C - \mathbf{r}_A) \equiv$ constant, so that

$$(\mathbf{r}_C - \mathbf{r}_A) \cdot (\mathbf{v}_C - \mathbf{v}_A) = 0$$

and $(\mathbf{r}_C - \mathbf{r}_A) \cdot \mathbf{v}_C = 0$, since $\mathbf{v}_A$ is perpendicular to $(\mathbf{r}_C - \mathbf{r}_A)$. Similarly $(\mathbf{r}_C - \mathbf{r}_B) \cdot \mathbf{v}_C = 0$. Hence the projections of $\mathbf{v}_C$ in three directions which are nonplanar are zero. This means that

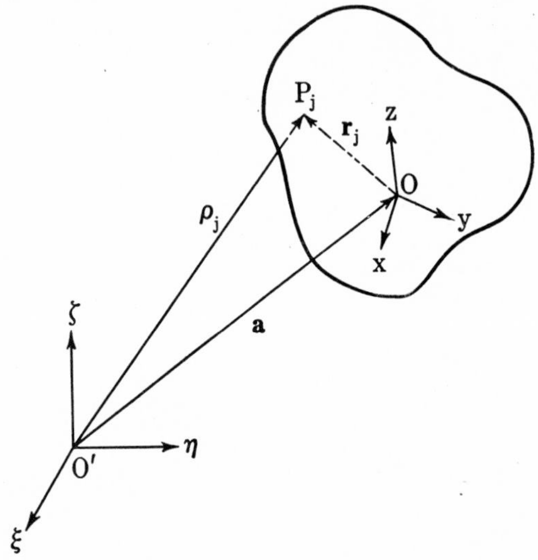

FIG. 86.

$\mathbf{v}_C \equiv 0$, so that we have two fixed points at this particular instant. Hence from the previous paragraph the motion is that of a rotation about the line l. If $\boldsymbol{\omega}$ is the angular-velocity vector, then $\mathbf{v}_j = \boldsymbol{\omega} \times \mathbf{r}_j$, where $\mathbf{r}_j$ is the vector from O to the jth particle.

Now let us consider the most general type of motion of a rigid body. Let O'-ξ-η-ζ represent a fixed coordinate system in space, and let O-x-y-z represent a coordinate system fixed in the rigid body (see Fig. 86). Let $\boldsymbol{\rho}_j$ and $\mathbf{r}_j$ represent the vectors from O' and O to the jth particle, and let $\mathbf{a}$ be the vector from O' to O. We have $\boldsymbol{\rho}_j = \mathbf{a} + \mathbf{r}_j$, and differentiating,

$$\frac{d\boldsymbol{\rho}_j}{dt} = \frac{d\mathbf{a}}{dt} + \frac{d\mathbf{r}_j}{dt}$$

Now $\frac{d\mathbf{r}_j}{dt}$ represents the velocity of P_j relative to O. This means O is fixed as far as P_j is concerned, and from above we know that $\frac{d\mathbf{r}_j}{dt} = \boldsymbol{\omega} \times \mathbf{r}_j$. Thus

$$\mathbf{v}_j = \frac{d\boldsymbol{\varrho}_j}{dt} = \frac{d\mathbf{a}}{dt} + \boldsymbol{\omega} \times \mathbf{r}_j \tag{352}$$

that is, the most general type of motion of a rigid body is that of a translation $\frac{d\mathbf{a}}{dt}$ plus a rotation $\boldsymbol{\omega} \times \mathbf{r}_j$.

We next ask the following question: If we change our origin from O to, say, O'' does $\boldsymbol{\omega}$ change? (Fig. 87.) The answer is "No"! Let $\mathbf{b}$ be the vector from O' to O''. Then

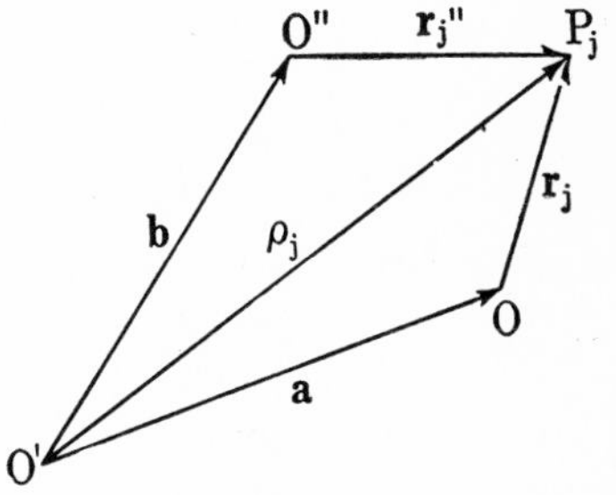

FIG. 87.

$$\mathbf{v}_j = \frac{d\boldsymbol{\varrho}_j}{dt} = \frac{d\mathbf{b}}{dt} + \boldsymbol{\omega}_1 \times \mathbf{r}_j''$$

But

$$\frac{d\mathbf{b}}{dt} = \frac{d\mathbf{a}}{dt} + \boldsymbol{\omega} \times (\mathbf{b} - \mathbf{a})$$

and

$$\mathbf{r}_j'' = (\mathbf{a} - \mathbf{b}) + \mathbf{r}_j$$

Thus

$$\mathbf{v}_j = \frac{d\mathbf{a}}{dt} + \boldsymbol{\omega} \times (\mathbf{b} - \mathbf{a}) + \boldsymbol{\omega}_1 \times (\mathbf{a} - \mathbf{b}) + \boldsymbol{\omega}_1 \times \mathbf{r}_j \tag{353}$$

Subtracting (352) from (353), we obtain

$$(\boldsymbol{\omega} - \boldsymbol{\omega}_1) \times (\mathbf{b} - \mathbf{a}) + (\boldsymbol{\omega}_1 - \boldsymbol{\omega}) \times \mathbf{r}_j = 0$$

or

$$(\boldsymbol{\omega} - \boldsymbol{\omega}_1) \times (\mathbf{b} - \mathbf{a} - \mathbf{r}_j) = -(\boldsymbol{\omega} - \boldsymbol{\omega}_1) \times \mathbf{r}_j'' = 0$$

We can certainly choose an $\mathbf{r}_j'' \neq 0$ and not parallel to the vector $\boldsymbol{\omega} - \boldsymbol{\omega}_1$, at any particular instant. Hence $\boldsymbol{\omega}_1 \equiv \boldsymbol{\omega}$.

Problems

1. Show that if $\mathbf{r}_1$ and $\mathbf{r}_2$ are two position vectors from the origin of the moving system of coordinates to two points in the rigid body, then $\mathbf{r}_1 \cdot \frac{d\mathbf{r}_2}{dt} + \mathbf{r}_2 \cdot \frac{d\mathbf{r}_1}{dt} = 0$.

2. A plane body is moving in its own plane. Find the point in the body which is instantaneously at rest.

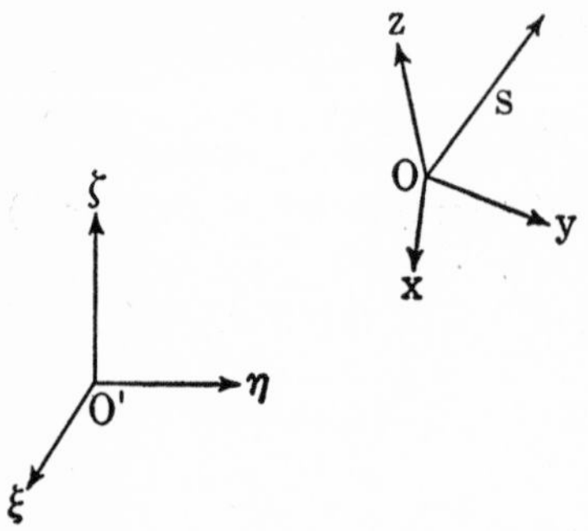

Fig. 88.

3. Show that the most general motion of a rigid body is a translation plus a rotation about a line parallel to the translation.

99. Relative Time Rate of Change of Vectors. Let **S** be any vector measured in the moving system of coordinates (Fig. 88).

$$\mathbf{S} = S_x\mathbf{i} + S_y\mathbf{j} + S_z\mathbf{k} \qquad (354)$$

To find out how **S** changes with time as measured by an observer at O', we differentiate (354),

$$\frac{d\mathbf{S}}{dt} = \frac{dS_x}{dt}\mathbf{i} + \frac{dS_y}{dt}\mathbf{j} + \frac{dS_z}{dt}\mathbf{k} + S_x\frac{d\mathbf{i}}{dt} + S_y\frac{d\mathbf{j}}{dt} + S_z\frac{d\mathbf{k}}{dt} \qquad (355)$$

We do not keep **i, j, k** fixed since **i, j, k** suffer motions relative to O'. But we do know that $\frac{d\mathbf{i}}{dt}$ is the velocity of a point one unit along the x axis, relative to O. Hence $\frac{d\mathbf{i}}{dt} = \boldsymbol{\omega} \times \mathbf{i}$, $\frac{d\mathbf{j}}{dt} = \boldsymbol{\omega} \times \mathbf{j}$, $\frac{d\mathbf{k}}{dt} = \boldsymbol{\omega} \times \mathbf{k}$. Hence (355) becomes

$$\frac{d\mathbf{S}}{dt} = \frac{dS_x}{dt}\mathbf{i} + \frac{dS_y}{dt}\mathbf{j} + \frac{dS_z}{dt}\mathbf{k} + \boldsymbol{\omega} \times (S_x\mathbf{i} + S_y\mathbf{j} + S_z\mathbf{k})$$

and

$$\boxed{\frac{d\mathbf{S}}{dt} = \frac{D\mathbf{S}}{dt} + \boldsymbol{\omega} \times \mathbf{S}} \qquad (356)$$

where $\frac{D\mathbf{S}}{dt}$ represents the time rate of change of **S** relative to the moving frame, for S_x is measured in the moving frame and so $\frac{dS_x}{dt}$ is the time rate of change of S_x as measured by an observer in the moving frame.

Intuitively, we expected the result of (356), for not only does **S** change relative to O, but to this change we must add the change in **S** because of the rotating frame. The reader might well ask, What of the motion of O itself? Will not this motion have to be considered? The answer is "No," for a translation of O only pulls **S** along, that is, **S** does not change length or direction if O is translated. It is the motion of **S** relative to the frame O-x-y-z and the rotation about O that produce changes in **S**.

Problems

1. Show that $\dfrac{d\boldsymbol{\omega}}{dt} = \dfrac{D\boldsymbol{\omega}}{dt}$.

2. For a pure translation show that $\dfrac{d\mathbf{S}}{dt} = \dfrac{D\mathbf{S}}{dt}$.

3. From (356) show that $\dfrac{d\mathbf{i}}{dt} = \boldsymbol{\omega} \times \mathbf{i}$.

100. Velocity. Let P be any point in space and let $\boldsymbol{\varrho}$ and $\mathbf{r}$ be the position vectors to P from O' and O, respectively (see Fig. 89). Obviously $\boldsymbol{\varrho} = \mathbf{a} + \mathbf{r}$, so that

$$\mathbf{v} = \frac{d\boldsymbol{\varrho}}{dt} = \frac{d\mathbf{a}}{dt} + \frac{d\mathbf{r}}{dt}$$

Now $\mathbf{r}$ is a vector measured in the O-x-y-z system, so that (356) applies to $\mathbf{r}$. This yields $\dfrac{d\mathbf{r}}{dt} = \dfrac{D\mathbf{r}}{dt} + \boldsymbol{\omega} \times \mathbf{r}$ and

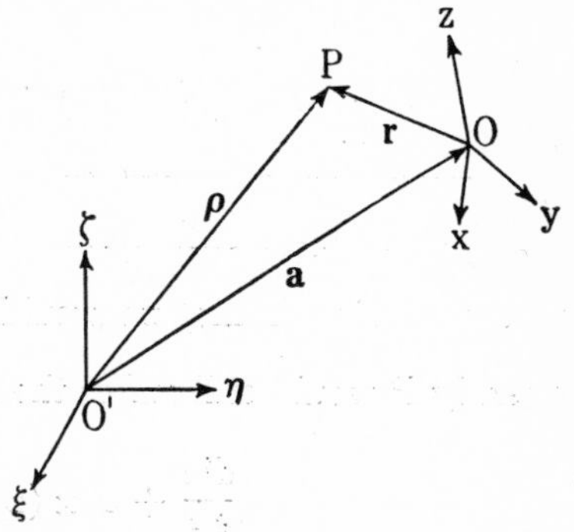

FIG. 89.

$$\mathbf{v} = \frac{d\boldsymbol{\varrho}}{dt} = \frac{d\mathbf{a}}{dt} + \boldsymbol{\omega} \times \mathbf{r} + \frac{D\mathbf{r}}{dt} \tag{357}$$

This result is expected. $\dfrac{d\mathbf{a}}{dt}$ is the drag velocity of P, $\boldsymbol{\omega} \times \mathbf{r}$ is the velocity due to the rotation of the O-x-y-z frame, and $\dfrac{D\mathbf{r}}{dt}$ is the velocity of P relative to the O-x-y-z frame. The vector sum is the velocity of P relative to the frame O'-ξ-η-ζ.

101. Acceleration. In Sec. 100 we saw that

$$\mathbf{v} = \frac{d\boldsymbol{\varrho}}{dt} = \frac{d\mathbf{a}}{dt} + \boldsymbol{\omega} \times \mathbf{r} + \frac{D\mathbf{r}}{dt}$$

To find the acceleration, we differentiate (357) and obtain

$$\frac{d\mathbf{v}}{dt} = \frac{d^2\boldsymbol{\varrho}}{dt^2} = \frac{d^2\mathbf{a}}{dt^2} + \frac{d}{dt}(\boldsymbol{\omega} \times \mathbf{r}) + \frac{d}{dt}\left(\frac{D\mathbf{r}}{dt}\right) \tag{358}$$

We apply (356) to $\boldsymbol{\omega} \times \mathbf{r}$ and obtain

$$\frac{d}{dt}(\boldsymbol{\omega} \times \mathbf{r}) = \boldsymbol{\omega} \times (\boldsymbol{\omega} \times \mathbf{r}) + \frac{D}{dt}(\boldsymbol{\omega} \times \mathbf{r})$$

Similarly

$$\frac{d}{dt}\left(\frac{D\mathbf{r}}{dt}\right) = \boldsymbol{\omega} \times \frac{D\mathbf{r}}{dt} + \frac{D}{dt}\left(\frac{D\mathbf{r}}{dt}\right)$$

so that (358) becomes

$$\frac{d^2\boldsymbol{\varrho}}{dt^2} = \frac{d^2\mathbf{a}}{dt^2} + \boldsymbol{\omega} \times (\boldsymbol{\omega} \times \mathbf{r}) + \frac{d\boldsymbol{\omega}}{dt} \times \mathbf{r} + 2\boldsymbol{\omega} \times \frac{D\mathbf{r}}{dt} + \frac{D^2\mathbf{r}}{dt^2} \tag{359}$$

Let us analyze each term of (359). If P were fixed relative to the moving frame, we would have $\frac{D\mathbf{r}}{dt} = \frac{D^2\mathbf{r}}{dt^2} = 0$ and consequently P would still suffer the acceleration

$$\frac{d^2\mathbf{a}}{dt^2} + \boldsymbol{\omega} \times (\boldsymbol{\omega} \times \mathbf{r}) + \frac{d\boldsymbol{\omega}}{dt} \times \mathbf{r}$$

This vector sum is appropriately called the drag acceleration of the particle. Now let us analyze each term of the drag acceleration. If the moving frame were not rotating, we would have $\boldsymbol{\omega} = 0$, and the drag acceleration reduces to the single term $\frac{d^2\mathbf{a}}{dt^2}$. This is the translational acceleration of O relative to O'. Now in Sec. 84 we saw that $\boldsymbol{\omega} \times (\boldsymbol{\omega} \times \mathbf{r})$ represented the centripetal acceleration due to rotation and $\frac{d\boldsymbol{\omega}}{dt} \times \mathbf{r}$ represented the tangential component of acceleration due to the angular-acceleration vector

$\dfrac{d\boldsymbol{\omega}}{dt}\cdot$ We easily explain the term $\dfrac{D^2\mathbf{r}}{dt^2}$ as the acceleration of P relative to the O-x-y-z frame. What, then, of the term $2\boldsymbol{\omega} \times \dfrac{D\mathbf{r}}{dt}$? This term is called the Coriolis acceleration, named after its discoverer. We do not try to give a geometrical or physical reason for its existence. Suffice to say, it occurs in Eq. (359) and must be considered when we discuss the motion of bodies moving over the earth's surface. Notice that the term disappears for particles at rest relative to the moving frame, for then $\dfrac{D\mathbf{r}}{dt} = 0$. It also does not exist for nonrotating frames.

Now Newton's second law states that force is proportional to the acceleration when the mass of the particle remains constant. It is found that the frame of reference for which this law holds best is that of the so-called "fixed stars." We call such a frame of reference an inertial frame. Any other coordinate system moving relative to an inertial frame with constant velocity is also an inertial frame, since from (359) we have $\dfrac{d^2\boldsymbol{\varrho}}{dt^2} = \dfrac{D^2\mathbf{r}}{dt^2}$ because $\boldsymbol{\omega} = 0$, $\dfrac{d\boldsymbol{\omega}}{dt} = 0$, $\dfrac{d\mathbf{a}}{dt} = constant$, $\dfrac{d^2\mathbf{a}}{dt^2} = 0$.

Let us now consider the motion of a particle relative to the earth. If $\mathbf{f}$ is the vector sum of the external forces (real forces, that is, gravitation, push, pull, etc.), then $\dfrac{d^2\boldsymbol{\varrho}}{dt^2} = \dfrac{\mathbf{f}}{m}$, and (359) becomes

$$\frac{D^2\mathbf{r}}{dt^2} = -\frac{d^2\mathbf{a}}{dt^2} - \boldsymbol{\omega} \times (\boldsymbol{\omega} \times \mathbf{r}) - \frac{d\boldsymbol{\omega}}{dt} \times \mathbf{r} - 2\boldsymbol{\omega} \times \frac{D\mathbf{r}}{dt} + \frac{\mathbf{f}}{m} \tag{360}$$

This is the differential equation of motion for a particle of mass m with external force $\mathbf{f}$ applied to it.

Example 108. Let us consider the earth as our rotating frame. The quantity $\boldsymbol{\omega} \times (\boldsymbol{\omega} \times \mathbf{r})$ is small, since $|\boldsymbol{\omega}| \approx 2\pi/86{,}164$ rad/sec, and for a particle near the earth's surface, $|\mathbf{r}| \approx (4{,}000)(5{,}280)$ feet. Also $\dfrac{d\boldsymbol{\omega}}{dt} \approx 0$ over a short time; $\dfrac{d^2\mathbf{a}}{dt^2} \approx 0$ over a short time;

so that (360) becomes

$$\frac{D^2\mathbf{r}}{dt^2} = -2\boldsymbol{\omega} \times \frac{D\mathbf{r}}{dt} + \frac{\mathbf{f}}{m} \tag{361}$$

Now consider a freely falling body starting from a point P at rest relative to the earth. Let the z axis be taken as the line joining the center of the earth to P, and let the x axis be taken perpendicular to the z axis in the eastward direction. We shall denote the latitude of the place by λ, assuming $\lambda > 0$. The equation of motion in the eastward direction is given by

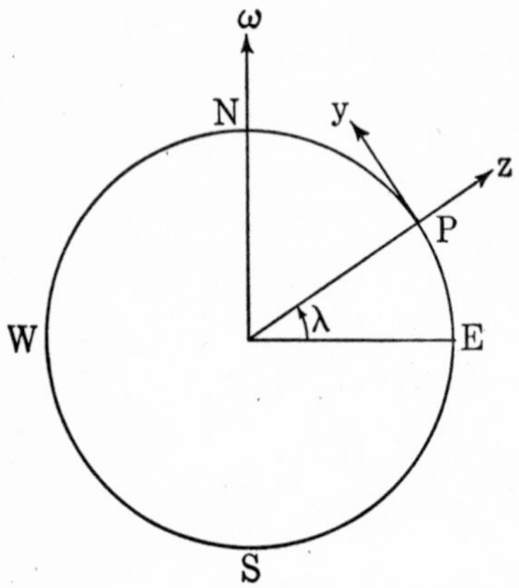

FIG. 90.

$$\frac{d^2x}{dt^2} = -2\left(\boldsymbol{\omega} \times \frac{D\mathbf{r}}{dt}\right)_e + \left(\frac{\mathbf{f}}{m}\right)_e$$

Now $\mathbf{f}$ (force of attraction) has no component eastward, so that $\left(\frac{\mathbf{f}}{m}\right)_e = 0$. We do not know $\frac{D\mathbf{r}}{dt}$, but to a first approximation it is $-gt\mathbf{k} + \frac{dx}{dt}\mathbf{i}$. Moreover, $\boldsymbol{\omega} = \omega \sin\lambda\,\mathbf{k} + \omega\cos\lambda\,\mathbf{j}$ (see Fig. 90). Hence $\left(\boldsymbol{\omega} \times \frac{D\mathbf{r}}{dt}\right)_e = -\omega gt\cos\lambda$, and

$$\frac{d^2x}{dt^2} = 2\omega gt\cos\lambda \tag{362}$$

If the particle remains in the vicinity of latitude λ, we can keep λ constant, so that on integrating (362), we obtain

$$\frac{dx}{dt} = \omega gt^2\cos\lambda$$

$$x = \frac{\omega gt^3}{3}\cos\lambda \tag{363}$$

(363) is to a first approximation the eastward deflection of a shot if it is dropped in the Northern Hemisphere. If h is the

distance the shot falls, then $h = \frac{1}{2}gt^2$ approximately, so that

$$x = \frac{2}{3}\omega h \cos\lambda \left(\frac{2h}{g}\right)^{\frac{1}{2}}$$

Problems

1. Show that the winds in the Northern Hemisphere have a horizontal deflecting Coriolis acceleration $2\omega v \sin\lambda$ at right angles to **v**.

2. A body is thrown vertically upward. Show that it strikes the ground $\frac{8}{3}\omega h \cos\lambda\,(2h/g)^{\frac{1}{2}}$ to the west.

3. Choose the x axis east, y axis south, z axis along the plumb line, and show that the equations of motion for a freely falling body are

$$\frac{d^2x}{dt^2} + 2\omega\sin\theta\frac{dz}{dt} - 2\omega\cos\theta\frac{dy}{dt} = 0$$

$$\frac{d^2y}{dt^2} + 2\omega\cos\theta\frac{dy}{dt} = 0$$

$$\frac{d^2z}{dt^2} - g - 2\omega\sin\theta\frac{dx}{dt} = 0$$

where θ is the colatitude.

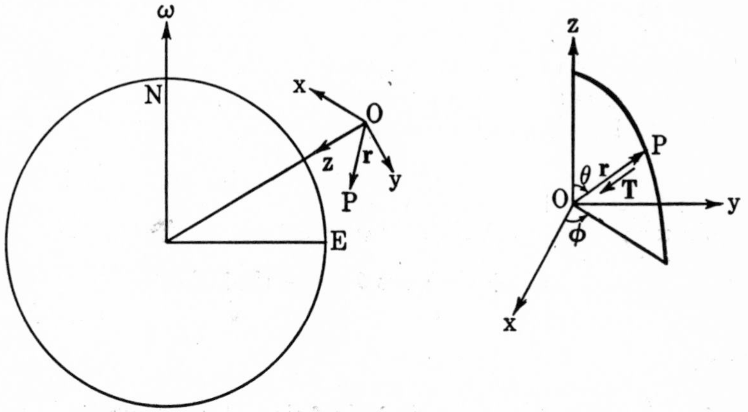

FIG. 91.

4. Using the coordinate system of Prob. 3, let us consider the motion of the Foucault pendulum (see Fig. 91).

Let $\mathbf{i}_1$, $\mathbf{i}_2$, $\mathbf{i}_3$ be the unit tangent vectors to the spherical curves r, θ, φ. We leave it to the reader to show that the acceleration along the $\mathbf{i}_3$ vector is $2 \cos \theta \, \dot{\varphi}\dot{\theta} + \sin \theta \, \ddot{\varphi}$ when the string is of unit length. The two external forces are $mg\mathbf{k}$ along the z axis and the tension in the string, $\mathbf{T} = -T\mathbf{r} = -T\mathbf{i}_1$. We wish to find the component of these forces along the $\mathbf{i}_3$ direction. $\mathbf{T}$ has no component in the $\mathbf{i}_3$ direction. Now $\mathbf{k} \cdot \mathbf{i}_3 = 0$, so that $mg\mathbf{k}$ has no component along the $\mathbf{i}_3$ direction. Finally, we must compute the $\mathbf{i}_3$ component of $-2\boldsymbol{\omega} \times \dfrac{D\mathbf{r}}{dt}$. The velocity vector is

$$\frac{D\mathbf{r}}{dt} = \dot{\theta}\mathbf{i}_2 + \sin \theta \, \dot{\varphi}\mathbf{i}_3$$

Also $\boldsymbol{\omega} = \omega(-\cos \lambda \, \mathbf{j} - \sin \lambda \, \mathbf{k})$, so that we must find the relationship between $\mathbf{i}_1, \mathbf{i}_2, \mathbf{i}_3$ and $\mathbf{i}, \mathbf{j}, \mathbf{k}$.

Now

$$\begin{aligned}
\mathbf{r} = \mathbf{i}_1 &= \sin \theta \cos \varphi \, \mathbf{i} + \sin \theta \sin \varphi \, \mathbf{j} + \cos \theta \, \mathbf{k} \\
\mathbf{i}_2 = \frac{\partial \mathbf{i}_1}{\partial \theta} &= \cos \theta \cos \varphi \, \mathbf{i} + \cos \theta \sin \varphi \, \mathbf{j} - \sin \theta \, \mathbf{k} \\
\mathbf{i}_3 = \frac{1}{\sin \theta} \frac{\partial \mathbf{i}_1}{\partial \varphi} &= -\sin \varphi \, \mathbf{i} + \cos \varphi \, \mathbf{j}
\end{aligned}$$

Thus

$$\begin{aligned}
\mathbf{i} &= (\mathbf{i} \cdot \mathbf{i}_1)\mathbf{i}_1 + (\mathbf{i} \cdot \mathbf{i}_2)\mathbf{i}_2 + (\mathbf{i} \cdot \mathbf{i}_3)\mathbf{i}_3 \\
&= \sin \theta \cos \varphi \, \mathbf{i}_1 + \cos \theta \cos \varphi \, \mathbf{i}_2 - \sin \varphi \, \mathbf{i}_3 \\
\mathbf{j} &= \sin \theta \sin \varphi \, \mathbf{i}_1 + \cos \theta \sin \varphi \, \mathbf{i}_2 + \cos \varphi \, \mathbf{i}_3 \\
\mathbf{k} &= \cos \theta \, \mathbf{i}_1 - \sin \theta \, \mathbf{i}_2
\end{aligned}$$

$$-2\boldsymbol{\omega} \times \frac{D\mathbf{r}}{dt} = 2\omega \begin{vmatrix} \mathbf{i}_1 & \mathbf{i}_2 & \mathbf{i}_3 \\ \cos \lambda \sin \theta \sin \varphi + \sin \lambda \cos \theta & \cos \lambda \cos \theta \sin \varphi - \sin \lambda \sin \theta & \cos \lambda \cos \varphi \\ 0 & \dot{\theta} & \sin \theta \, \dot{\varphi} \end{vmatrix}$$

and

$$\left(-2\boldsymbol{\omega} \times \frac{D\mathbf{r}}{dt}\right)_{\mathbf{i}_3} = \dot{\theta}(\sin \lambda \sin \theta \sin \varphi + \sin \lambda \cos \theta)$$

Equation (361) yields

$$2 \cos \theta \, \dot{\varphi}\dot{\theta} + \sin \theta \, \ddot{\varphi} = 2\omega(\dot{\theta} \sin \lambda \sin \theta \sin \varphi + \dot{\theta} \sin \lambda \cos \theta) \tag{364}$$

For small oscillations, $\sin\theta \approx 0$, and (364) reduces to

$$\dot{\varphi} = \omega \sin\lambda \tag{365}$$

Hence the pendulum rotates about the vertical in the clockwise sense when viewed from the point of suspension with an angular speed $\omega \sin\lambda$. At latitude 30° the time for one complete oscillation is 48 hours.

5. Find the equation of motion by considering the $\mathbf{i}_2$ components of (361) for the Foucault pendulum.

102. Motion of a Rigid Body with One Point Fixed. The motion of a rigid body with one point fixed will depend on the forces acting on the body. Let O-x-y-z be a coordinate system fixed in the moving body, and let O-ξ-η-ζ be the coordinate system fixed in space. O is the fixed point of the body. In Sec. 94 we saw that $\dfrac{d\mathbf{H}_O{}^r}{dt} = \mathbf{L}_O$. Now $\mathbf{H}_O{}^r = \iiint\limits_R \mathbf{r} \times \rho \dfrac{d\mathbf{r}}{dt}\, d\tau$. We can replace $\dfrac{d\mathbf{r}}{dt}$ by $\boldsymbol{\omega} \times \mathbf{r}$ ($\boldsymbol{\omega}$ unknown). Thus

$$\begin{aligned}\mathbf{H}_O{}^r &= \iiint\limits_R \rho\mathbf{r} \times (\boldsymbol{\omega} \times \mathbf{r})\, d\tau \\ &= \iiint\limits_R \rho[r^2\boldsymbol{\omega} - (\mathbf{r}\cdot\boldsymbol{\omega})\mathbf{r}]\, d\tau\end{aligned} \tag{366}$$

Let

$$\begin{aligned}\boldsymbol{\omega} &= \omega_x\mathbf{i} + \omega_y\mathbf{j} + \omega_z\mathbf{k} \\ \mathbf{r} &= x\mathbf{i} + y\mathbf{j} + z\mathbf{k}\end{aligned}$$

so that

$$\begin{aligned}r^2\boldsymbol{\omega} - (\mathbf{r}\cdot\boldsymbol{\omega})\mathbf{r} &= (x^2 + y^2 + z^2)(\omega_x\mathbf{i} + \omega_y\mathbf{j} + \omega_z\mathbf{k}) \\ &\quad + (x\omega_x + y\omega_y + z\omega_z)(x\mathbf{i} + y\mathbf{j} + z\mathbf{k}) \\ &= [(y^2 + z^2)\omega_x - xy\omega_y - xz\omega_z]\mathbf{i} \\ &\quad + [-xy\omega_x + (z^2 + x^2)\omega_y - yz\omega_z]\mathbf{j} \\ &\quad + [-xz\omega_x - yz\omega_y + (x^2 + y^2)\omega_z]\mathbf{k}\end{aligned}$$

We thus obtain

$$\begin{aligned}\mathbf{H}_O{}^r &= \mathbf{i}[\omega_x \textstyle\iiint \rho(y^2 + z^2)\, d\tau - \omega_y \iiint \rho xy\, d\tau - \omega_z \iiint \rho xz\, d\tau] \\ &\quad + \mathbf{j}[-\omega_x \textstyle\iiint \rho xy\, d\tau + \omega_y \iiint \rho(z^2 + x^2)\, d\tau - \omega_z \iiint \rho yz\, d\tau] \\ &\quad + \mathbf{k}[-\omega_x \textstyle\iiint \rho xz\, d\tau - \omega_y \iiint \rho yz\, d\tau + \omega_z \iiint \rho(x^2 + y^2)\, d\tau]\end{aligned} \tag{367}$$

The quantities

$$
\begin{aligned}
A &= \iiint \rho(y^2 + z^2)\, d\tau \\
B &= \iiint \rho(z^2 + x^2)\, d\tau \\
C &= \iiint \rho(x^2 + y^2)\, d\tau \\
D &= \iiint \rho yz\, d\tau \\
E &= \iiint \rho zx\, d\tau \\
F &= \iiint \rho xy\, d\tau
\end{aligned}
\tag{368}
$$

are independent of the motion and are constants of the body. That they are independent of the motion is seen from the fact that for a particle with coordinates x, y, z, the scalars x, y, z remain invariant because the O-x-y-z frame is fixed in the body. The quantities A, B, C are the moments of inertia about the x, y, z axes, and D, E, F are called the products of inertia. We assume the student has studied these integrals in the integral calculus.

Now from Sec. 99 we have $\dfrac{d\mathbf{H}_O^r}{dt} = \dfrac{D\mathbf{H}_O^r}{dt} + \boldsymbol{\omega} \times \mathbf{H}_O^r$ so that

$$\mathbf{L}_O = \frac{D\mathbf{H}_O^r}{dt} + \boldsymbol{\omega} \times \mathbf{H}_O^r$$

Hence

$$
\begin{aligned}
L_x\mathbf{i} + L_y\mathbf{j} + L_z\mathbf{k} = {} & \mathbf{i}\left(A\frac{d\omega_x}{dt} - F\frac{d\omega_y}{dt} - E\frac{d\omega_z}{dt}\right) \\
& + \mathbf{j}\left(-F\frac{d\omega_x}{dt} + B\frac{d\omega_y}{dt} - D\frac{d\omega_z}{dt}\right) \\
& + \mathbf{k}\left(-E\frac{d\omega_x}{dt} - D\frac{d\omega_y}{dt} + C\frac{d\omega_z}{dt}\right) \\
& + \begin{vmatrix} \mathbf{i} & \mathbf{j} & \mathbf{k} \\ \omega_x & \omega_y & \omega_z \\ A\omega_x - F\omega_y - E\omega_z, & -F\omega_x + B\omega_y - D\omega_z, & -E\omega_x - D\omega_y + C\omega_z \end{vmatrix}
\end{aligned}
\tag{369}
$$

In the special case when the axes are so chosen that the products of inertia vanish (see Sec. 107), we have Euler's celebrated equations of motion:

$$
\begin{aligned}
L_x &= A\frac{d\omega_x}{dt} + (C - B)\omega_y\omega_z \\
L_y &= B\frac{d\omega_y}{dt} + (A - C)\omega_z\omega_x \\
L_z &= C\frac{d\omega_z}{dt} + (B - A)\omega_x\omega_y
\end{aligned}
\tag{370}
$$

103. Applications. If no torques are applied to the body of Sec. 102, Euler's equations reduce to

$$
\begin{aligned}
&\text{(i)} \qquad & A\frac{d\omega_x}{dt} + (C - B)\omega_y\omega_z &= 0 \\
&\text{(ii)} \qquad & B\frac{d\omega_y}{dt} + (A - C)\omega_z\omega_x &= 0 \\
&\text{(iii)} \qquad & C\frac{d\omega_z}{dt} + (B - A)\omega_x\omega_y &= 0
\end{aligned}
\tag{371}
$$

Multiplying (i), (ii), (iii) by ω_x, ω_y, ω_z, respectively, and adding, we obtain

$$A\omega_x\frac{d\omega_x}{dt} + B\omega_y\frac{d\omega_y}{dt} + C\omega_z\frac{d\omega_z}{dt} = 0$$

Integrating yields

$$A\omega_x^2 + B\omega_y^2 + C\omega_z^2 = \text{constant} \tag{372}$$

This is one of the integrals of the motion. We obtain another integral by multiplying (i), (ii), (iii) by $A\omega_x$, $B\omega_y$, $C\omega_z$, and adding. This yields

$$A^2\omega_x\frac{d\omega_x}{dt} + B^2\omega_y\frac{d\omega_y}{dt} + C^2\omega_z\frac{d\omega_z}{dt} = 0$$

so that

$$A^2\omega_x^2 + B^2\omega_y^2 + C^2\omega_z^2 = \text{constant} \tag{373}$$

If originally the motion was that of a rotation of angular velocity $\boldsymbol{\omega}$ about a principal axis (x axis), then initially

$$\begin{aligned} \omega_x(0) &= \omega_0 \\ \omega_y(0) &= 0 \\ \omega_z(0) &= 0 \end{aligned} \tag{374}$$

and we notice that (371) and the boundary condition (374) are satisfied by

$$\begin{aligned} \omega_x(t) &\equiv \omega_0 \\ \omega_y(t) &\equiv 0 \\ \omega_z(t) &\equiv 0 \end{aligned}$$

so that the motion continues to be one of constant angular velocity about the x axis. Here we have used a theorem on the uniqueness of solutions for a system of differential equations.

Now suppose the body to be rotating this way and then slightly disturbed, so that now the body has acquired the very small angular velocities ω_y, ω_z. We can neglect $\omega_y\omega_z$ as compared to $\omega_y\omega_0$ and $\omega_z\omega_0$. Euler's equations now become

$$\begin{aligned} B\frac{d\omega_y}{dt} + (A - C)\omega_z\omega_0 &= 0 \\ C\frac{d\omega_z}{dt} + (B - A)\omega_y\omega_0 &= 0 \\ \omega_0 &= \text{constant} \end{aligned} \tag{375}$$

Differentiating the first equation of (375) with respect to time and eliminating $\frac{d\omega_z}{dt}$, we obtain

$$B\frac{d^2\omega_y}{dt^2} + \frac{(A - C)(A - B)}{C}\omega_0{}^2\omega_y = 0 \tag{376}$$

If A is greater than B and C or smaller than B and C, then $a^2 = \frac{(A - C)(A - B)}{C} > 0$, and the solution to (376) is

$$\omega_y = L\cos(at + \alpha)$$

Also $\omega_z = \frac{aBL\sin(at + \alpha)}{\omega_0(A - C)}$ by replacing ω_y in (375).

Problems

1. Solve the free body with $A = B$ for ω_x, ω_y, ω_z.

2. A disk ($B = C$) rotates about its x axis (perpendicular to the plane of the disk) with constant angular speed ω_0. A constant torque $\mathbf{L}_0$ is applied constantly in the y direction. Find ω_y and ω_z.

3. Show that a necessary and sufficient condition that a rigid body be in static equilibrium is that the sum of the external forces and external torques vanish.

4. A sphere rotates about its fixed center. If the only forces acting on the sphere are applied at the center, show that the initial motion continues.

5. In Prob. 2 a constant torque $\mathbf{L}_0$ is also applied in the z direction. Find ω_y and ω_z.

104. Euler's Angular Coordinates. More complicated problems can be solved by use of Euler's angular coordinates. Let O-x'-y'-z' be a cartesian coordinate system fixed in space, and let O-x-y-z be fixed in the moving body (Fig. 92).

The x-y plane will intersect the x'-y' plane in a line, called the nodal line N. Let θ be the angle between the z and z' axes, ψ the angle between the x' and N axes, and φ the angle between the nodal line and the x axis. The positive directions of these angles are indicated in the figure.

The three angles ψ, θ, φ completely specify the configuration of the body. Now $\frac{d\psi}{dt}$ represents the rotation of the O-z'-N-T' frame relative to the O-x'-y'-z' frame; $\frac{d\theta}{dt}$ represents the rotation of the O-z-N-T frame relative to the O-z'-N-T' frame, and finally, $\frac{d\varphi}{dt}$ represents the rotation of the O-x-y-z frame relative to the O-z-N-T frame. Therefore $\frac{d\boldsymbol{\psi}}{dt} + \frac{d\boldsymbol{\theta}}{dt} + \frac{d\boldsymbol{\varphi}}{dt}$ gives us the angular velocity of the O-x-y-z frame relative to the fixed O-x'-y'-z' frame, and

$$\boldsymbol{\omega} = \frac{d\boldsymbol{\psi}}{dt} + \frac{d\boldsymbol{\theta}}{dt} + \frac{d\boldsymbol{\varphi}}{dt} \tag{377}$$

The three angular velocities are not mutually perpendicular.

We now define **i**, **j**, **k**, **i**′, **j**′, **k**′, **N**, **T**′, **T** as unit vectors along the x, y, z, x', y', z', N, T', T axes, respectively. Thus

$$\begin{aligned}\boldsymbol{\omega} &= \frac{d\psi}{dt}\mathbf{k}' + \frac{d\theta}{dt}\mathbf{N} + \frac{d\varphi}{dt}\mathbf{k} \\ &= \omega_x\mathbf{i} + \omega_y\mathbf{j} + \omega_z\mathbf{k} \\ &= \omega_{x'}\mathbf{i}' + \omega_{y'}\mathbf{j}' + \omega_{z'}\mathbf{k}' \end{aligned} \tag{378}$$

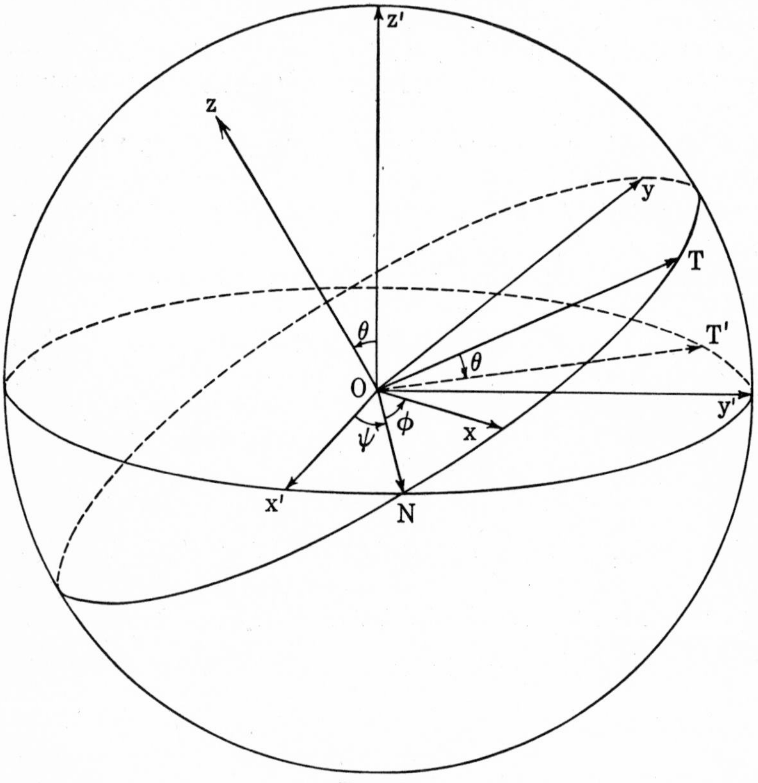

FIG. 92.

Now it is easy to verify that

$$\begin{aligned}\mathbf{i} &= \cos\varphi\,\mathbf{N} + \sin\varphi\,\mathbf{T} \\ \mathbf{j} &= -\sin\varphi\,\mathbf{N} + \cos\varphi\,\mathbf{T} \\ \mathbf{i}' &= \cos\psi\,\mathbf{N} - \sin\psi\,\mathbf{T}'\end{aligned}$$

$$\begin{aligned} \mathbf{j}' &= \sin\psi\,\mathbf{N} + \cos\psi\,\mathbf{T}' \\ \mathbf{i}\cdot\mathbf{k}' &= \sin\varphi\,\mathbf{T}\cdot\mathbf{k}' = \sin\varphi\sin\theta \\ \mathbf{j}\cdot\mathbf{k}' &= \cos\varphi\,\mathbf{T}\cdot\mathbf{k}' = \cos\varphi\sin\theta \end{aligned}$$

so that

$$\begin{aligned} \omega_x &= \boldsymbol{\omega}\cdot\mathbf{i} = \frac{d\psi}{dt}\mathbf{k}'\cdot\mathbf{i} + \frac{d\theta}{dt}\mathbf{N}\cdot\mathbf{i} + \frac{d\varphi}{dt}\mathbf{k}\cdot\mathbf{i} \\ &= \sin\varphi\sin\theta\frac{d\psi}{dt} + \cos\varphi\frac{d\theta}{dt} \\ \omega_y &= \cos\varphi\sin\theta\frac{d\psi}{dt} - \sin\varphi\frac{d\theta}{dt} \\ \omega_z &= \cos\theta\frac{d\psi}{dt} + \frac{d\varphi}{dt} \end{aligned}$$

Rewriting this, we have

$$\begin{aligned} \omega_x &= \frac{d\psi}{dt}\sin\theta\sin\varphi + \frac{d\theta}{dt}\cos\varphi \\ \omega_y &= \frac{d\psi}{dt}\sin\theta\cos\varphi - \frac{d\theta}{dt}\sin\varphi \\ \omega_z &= \frac{d\psi}{dt}\cos\theta + \frac{d\varphi}{dt} \end{aligned} \tag{379}$$

Also

$$\omega^2 = \omega_x^2 + \omega_y^2 + \omega_z^2 = \left(\frac{d\psi}{dt}\right)^2 + \left(\frac{d\theta}{dt}\right)^2 + \left(\frac{d\varphi}{dt}\right)^2 + 2\cos\theta\frac{d\varphi}{dt}\frac{d\psi}{dt} \tag{380}$$

For the fixed frame

$$\begin{aligned} \omega_{x'} &= \boldsymbol{\omega}\cdot\mathbf{i}' = \cos\psi\frac{d\theta}{dt} + \sin\psi\sin\theta\frac{d\varphi}{dt} \\ \omega_{y'} &= \boldsymbol{\omega}\cdot\mathbf{j}' = \sin\psi\frac{d\theta}{dt} - \cos\psi\sin\theta\frac{d\varphi}{dt} \\ \omega_{z'} &= \boldsymbol{\omega}\cdot\mathbf{k}' = \frac{d\psi}{dt} + \cos\theta\frac{d\varphi}{dt} \end{aligned} \tag{381}$$

105. Motion of a Free Top about a Fixed Point. Let us assume that no torques exist and that the top is symmetric ($A = B$). Euler's equations become

$$
\begin{aligned}
&\text{(i)} && A\frac{d\omega_x}{dt} + (C - A)\omega_y\omega_z = 0 \\
&\text{(ii)} && A\frac{d\omega_y}{dt} + (A - C)\omega_x\omega_z = 0 \qquad (382) \\
&\text{(iii)} && C\frac{d\omega_z}{dt} = 0
\end{aligned}
$$

Integrating (iii), we obtain $\omega_z = \omega_0 =$ constant. Multiply (ii) by $i = \sqrt{-1}$ and add to (i). We obtain

$$A\frac{d}{dt}(\omega_x + i\omega_y) + (C - A)\omega_0(\omega_y - i\omega_x) = 0$$

or

$$A\frac{d}{dt}(\omega_x + i\omega_y) = i\omega_0(C - A)(\omega_x + i\omega_y)$$

Integrating,

$$\omega_x + i\omega_y = \alpha e^{[i(C-A)/A]\omega_0 t}$$

so that

$$
\begin{aligned}
\omega_x &= \alpha \cos \sigma t \\
\omega_y &= \alpha \sin \sigma t
\end{aligned}
\qquad (383)
$$

where $\sigma = [(C - A)/A]\omega_0$ and α is a constant of integration.

Now $\omega^2 = \omega_x{}^2 + \omega_y{}^2 + \omega_z{}^2 = \alpha^2 + \omega_0{}^2 =$ constant, so that the magnitude of the angular velocity remains constant during the motion. Moreover, $\dfrac{d\mathbf{H}}{dt} = 0$, so that $\mathbf{H}$ is a constant vector in fixed space. We choose the z' axis for the direction of $\mathbf{H}$. Now

$$
\begin{aligned}
\mathbf{H} &= A\omega_x\mathbf{i} + B\omega_y\mathbf{j} + C\omega_z\mathbf{k} \\
&= A\alpha \cos \sigma t\, \mathbf{i} + A\alpha \sin \sigma t\, \mathbf{j} + C\omega_0\mathbf{k}
\end{aligned}
\qquad (384)
$$

This shows that $\mathbf{H}$ rotates around the z axis (of the body) with constant angular speed $\sigma = [(C - A)/A]\omega_0$, and since $\mathbf{H}$ is fixed in space, it is the z axis of the body which is rotating about the fixed z' axis with constant angular speed $-\sigma = [(A - C)/A]\omega_0$. Also $\mathbf{H} \cdot \mathbf{k} = |H| \cos \theta = C\omega_0$, so that θ is a constant since $|H| =$ constant. We say that the top precesses about the z' axis.

106. The Top (*Continued*). We have assumed above that the weight of the top or gyroscope was negligible, or that the gyroscope was balanced, that is, suspended with its center of mass at the point of support, so that no torques were produced. We shall now assume that the center of mass, while still located on the axis of symmetry, is not at the point of support. We now have the following situation (Fig. 93):

$$\begin{aligned} \mathbf{L} &= l\mathbf{k} \times (-W\mathbf{k}') \\ &= Wl \sin\theta\, \mathbf{N} \end{aligned}$$

The three components of the torque are

$$\begin{aligned} L_x &= Wl \sin\theta \cos\varphi \\ L_y &= -Wl \sin\theta \sin\varphi \\ L_z &= 0 \end{aligned}$$

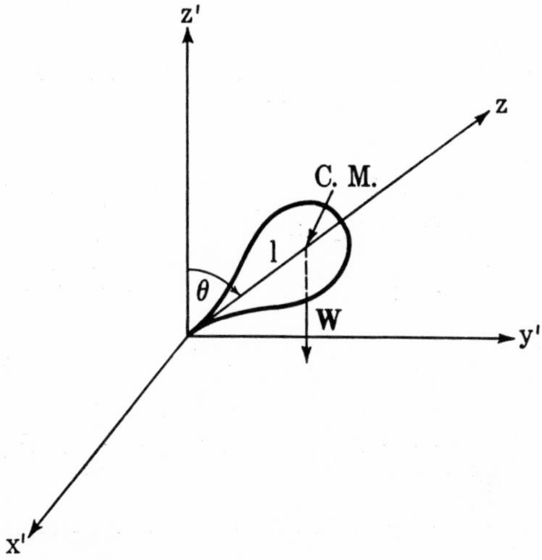

FIG. 93.

Euler's equations become

$$\begin{aligned} Wl \sin\theta \cos\varphi &= A\frac{d\omega_x}{dt} + (C - A)\omega_y\omega_z \\ -Wl \sin\theta \sin\varphi &= A\frac{d\omega_y}{dt} + (A - C)\omega_z\omega_x \\ 0 &= C\frac{d\omega_z}{dt} \text{ for } A = B \end{aligned} \tag{385}$$

Hence $\omega_z \equiv \omega_0$. Multiplying Eqs. (385) by ω_x, ω_y, ω_z, respectively, and adding, we obtain

$$\frac{1}{2}\frac{d}{dt}(A\omega_x^2 + B\omega_y^2 + C\omega_z^2) = Wl \sin\theta(\omega_x \cos\varphi - \omega_y \sin\varphi) \tag{386}$$

From (379) we have $\omega_x \cos \varphi - \omega_y \sin \varphi = \dfrac{d\theta}{dt}$, so that (386) becomes

$$\frac{1}{2}\frac{d}{dt}(A\omega_x^2 + B\omega_y^2 + C\omega_z^2) = Wl \sin \theta \frac{d\theta}{dt}$$

and integrating

$$A\omega_x^2 + B\omega_y^2 + C\omega_z^2 = -2Wl \cos \theta + k$$

or, again using (379),

$$\left(\frac{d\psi}{dt}\right)^2 \sin^2 \theta + \left(\frac{d\theta}{dt}\right)^2 = \alpha - a \cos \theta \tag{387}$$

α and a are constants.

Now since $L_{z'} = \mathbf{L} \cdot \mathbf{k}' = 0$, we have $H_{z'} =$ constant. Also $\mathbf{H} = A\omega_x\mathbf{i} + B\omega_y\mathbf{j} + C\omega_z\mathbf{k}$, so that

$$H_{z'} = \mathbf{H} \cdot \mathbf{k}' = A\omega_x \sin \theta \sin \varphi + A\omega_y \cos \varphi \sin \theta + C\omega_0 \cos \theta = \text{constant}$$

Replacing ω_x and ω_y by their equals from (379), we have

$$A\left(\frac{d\psi}{dt} \sin^2 \theta \sin^2 \varphi + \frac{d\theta}{dt} \sin \varphi \sin \theta \cos \varphi + \frac{d\psi}{dt} \cos^2 \varphi \sin^2 \theta - \frac{d\theta}{dt} \cos \varphi \sin \varphi \sin \theta\right) + C\omega_0 \cos \theta = \text{constant} \tag{388}$$

or $A \dfrac{d\psi}{dt} \sin^2 \theta + C\omega_0 \cos \theta = \text{constant} = H_{z'}$.

Let $\beta = H_{z'}/A$, $b = C\omega_0/A$, so that (388) becomes

$$\frac{d\psi}{dt} = \frac{\beta - b \cos \theta}{\sin^2 \theta} \tag{389}$$

From (379)

$$\omega_z = \omega_0 = \frac{d\psi}{dt} \cos \theta + \frac{d\varphi}{dt} \tag{390}$$

Using (389), (387) becomes

$$\left(\frac{\beta - b \cos \theta}{\sin \theta}\right)^2 + \left(\frac{d\theta}{dt}\right)^2 = \alpha - a \cos \theta \tag{391}$$

Let $z = \cos\theta$, so that $\dfrac{dz}{dt} = -\sin\theta\,\dfrac{d\theta}{dt}$, and

$$(\beta - bz)^2 + \left(\frac{dz}{dt}\right)^2 = (\alpha - az)(1 - z^2)$$

Hence

$$t = \int_0^z [(\alpha - az)(1 - z^2) - (\beta - bz)^2]^{-\frac{1}{2}}\,dz \tag{392}$$

This integral belongs to the class of elliptic integrals. If we can integrate and find z, then we shall know

$$\frac{d\psi}{dt} = \frac{\beta - bz}{1 - z^2}, \qquad \frac{d\varphi}{dt} = \omega_0 - \frac{d\psi}{dt}\,z$$

The reader should look up a complete discussion of elliptic integrals in the literature.

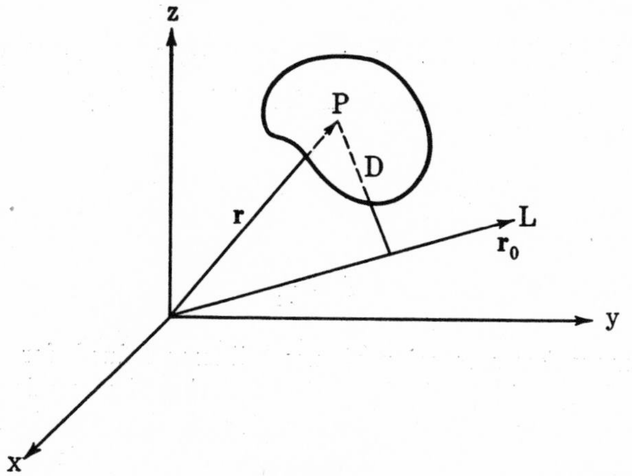

FIG. 94.

107. Inertia Tensor. The moment of inertia of a rigid body about a line through the origin may be computed as follows. Let the line L be given by the unit vector $\mathbf{r}_0 = l\mathbf{i} + m\mathbf{j} + n\mathbf{k}$, and let $\mathbf{r}$ be the vector from O to any point P in the body,

$$\mathbf{r} = x\mathbf{i} + y\mathbf{j} + z\mathbf{k}$$

(see Fig. 94). The shortest distance from P to L is given by

$$\begin{aligned}
D^2 &= \mathbf{r}^2 - (\mathbf{r} \cdot \mathbf{r}_0)^2 \\
&= (x^2 + y^2 + z^2) - (lx + my + nz)^2 \\
&= (l^2 + m^2 + n^2)(x^2 + y^2 + z^2) - (lx + my + nz)^2 \\
&= l^2(y^2 + z^2) + m^2(z^2 + x^2) + n^2(x^2 + y^2) - 2mnyz \\
&\qquad - 2lnzx - 2lmxy
\end{aligned}$$

Thus

$$\begin{aligned}
I &= \iiint \rho D^2 \, dz \, dy \, dx \\
&= Al^2 + Bm^2 + Cn^2 - 2mnD - 2nlE - 2lmF
\end{aligned}$$

Let us replace l, m, n by the variables x, y, z, and let us consider the surface

$$\varphi(x, y, z) = Ax^2 + By^2 + Cz^2 - 2Dyz - 2Ezx - 2Fxy = 1 \tag{393}$$

A line L through the origin is given by the equation $x = lt$, $y = mt$, $z = nt$. This line intersects the ellipsoid $\varphi(x, y, z) = 1$ for t satisfying

$$(Al^2 + Bm^2 + Cn^2 - 2Dmn - 2nlE - 2lmF)t^2 = 1$$

or $t^2 = 1/I$. The distance from the origin to this point of intersection is given by

$$d = (l^2t^2 + m^2t^2 + n^2t^2)^{\frac{1}{2}} = t = I^{-\frac{1}{2}}$$

so that

$$I = \frac{1}{d^2} \tag{394}$$

We know that a rotation of axes will keep I fixed, for the line and the body will be similarly situated after the rotation. We now attempt to simplify the equation of the quadric surface $\varphi(x, y, z) = 1$. First, let us find a point P on this surface at which the normal will be parallel to the radius vector to this point. The normal to the surface is given by $\nabla\varphi$, so that we desire $\mathbf{r}$ parallel to $\nabla\varphi$, which yields the equations

$$\frac{Ax - Ez - Fy}{x} = \frac{By - Dz - Fx}{y} = \frac{Cz - Dy - Ex}{z} \tag{395}$$

Any orthogonal transformation (Example 8) will preserve the form of (393) and (395) with x, y, z replaced by x', y', z' and $A, B, \ldots, F$ replaced by $A', B', \ldots, F'$. Now choose the

z' axis through P so that $x' = 0$, $y' = 0$, $z' = \zeta$ satisfy (395). This yields $-E'/0 = -D'/0 = C'$, which means that

$$E' = D' = 0,$$

and (393) reduces to

$$A'x'^2 + B'y'^2 + C'z'^2 - 2F'x'y' = 1 \tag{396}$$

The rotation

$$\begin{aligned} x'' &= x' \cos\theta - y' \sin\theta \\ y'' &= x' \sin\theta + y' \cos\theta \\ z'' &= z' \end{aligned}$$

with $\tan 2\theta = F'/(B' - A')$ reduces (396) to

$$\boxed{A''x''^2 + B''y''^2 + C''z''^2 = 1} \tag{397}$$

This is the canonical form desired. We have thus proved the important theorem that a quadratic form of the type (393) can always be reduced to a sum of squares of the form (397) by a rotation of axes. In the proof we made the assumption that there was a point P such that $\mathbf{r}$ is parallel to $\nabla\varphi$, which yielded (395). We could have arrived at Eqs. (395) by asking at what point on the sphere $x^2 + y^2 + z^2 = 1$ is $\varphi(x, y, z)$ a maximum. Since $\varphi(x, y, z)$ is continuous on the compact set

$$x^2 + y^2 + z^2 = 1$$

such a point always exists. Equations (395) are then easily deduced by Lagrange's method of multipliers.

We can arrange the constants of inertia into a square matrix

$$I = \begin{pmatrix} A & -F & -E \\ -F & B & -D \\ -E & -D & C \end{pmatrix} \tag{398}$$

The elements of the matrix (an array of elements) are called the components of I. Under a proper rotation we have shown that we can write

$$I = \begin{pmatrix} A'' & 0 & 0 \\ 0 & B'' & 0 \\ 0 & 0 & C'' \end{pmatrix} \tag{399}$$

In general, under an orthogonal transformation, I will become

$$I = \begin{pmatrix} A' & -F' & -E' \\ -F' & B' & -D' \\ -E' & -D' & C' \end{pmatrix} \tag{400}$$

and the components of I in (400) will be related to the components of I in (398) according to a certain law. We shall see in Chap. 8 that I is a tensor and so is called the inertia tensor.

Referring back to (367), we may write

$$\begin{pmatrix} H_x \\ H_y \\ H_z \end{pmatrix} = \begin{pmatrix} A & -F & -E \\ -F & B & -D \\ -E & -D & C \end{pmatrix} \begin{pmatrix} \omega_x \\ \omega_y \\ \omega_z \end{pmatrix} \tag{401}$$

from the definition of multiplication of matrices, where

$$\mathbf{H}_O{}^r = H_x\mathbf{i} + H_y\mathbf{j} + H_z\mathbf{k}$$

$\boldsymbol{\omega} = \omega_x\mathbf{i} + \omega_y\mathbf{j} + \omega_z\mathbf{k}$.

If we write (398) as $\begin{pmatrix} I_1{}^1 & I_2{}^1 & I_3{}^1 \\ I_1{}^2 & I_2{}^2 & I_3{}^2 \\ I_1{}^3 & I_2{}^3 & I_3{}^3 \end{pmatrix}$ and

$$\mathbf{H}_O{}^r = H_1\mathbf{i} + H_2\mathbf{j} + H_3\mathbf{k}$$

$\boldsymbol{\omega} = \omega_1\mathbf{i} + \omega_2\mathbf{j} + \omega_3\mathbf{k}$, then (367) may be written

$$H_j = \sum_{\alpha=1}^{3} I_j{}^\alpha \omega_\alpha, \qquad j = 1, 2, 3 \tag{402}$$

which is equivalent to the matrix form (401).

Problems

1. Find the moments and products of inertia for a uniform cube, taking the cube edges as axes.

2. Show that the moment of inertia of a body about any line is equal to its moment of inertia about a parallel line through the center of mass, plus the product of the total mass and the square of the distance from the line to the center of mass.

3. Find the angular-momentum vector of a thin rectangular sheet rotating about one of its diagonals with constant angular speed ω_0.

4. If

$$H_\beta = \sum_{\alpha=1}^{3} I_\beta{}^\alpha \omega_\alpha, \qquad \bar{H}_\beta = \sum_{\alpha=1}^{3} \bar{I}_\beta{}^\alpha \bar{\omega}_\alpha$$

$$\bar{H}_\beta = \sum_{\alpha=1}^{3} a_\beta{}^\alpha H_\alpha, \qquad \bar{\omega}_\beta = \sum_{\alpha=1}^{3} a_\beta{}^\alpha \omega_\alpha, \quad \beta = 1, 2, 3$$

for arbitrary ω_α, show that

$$\sum_{\alpha=1}^{3} \bar{I}_\beta{}^\alpha a_\alpha{}^\sigma = \sum_{\alpha=1}^{3} I_\alpha{}^\sigma a_\beta{}^\alpha, \qquad \beta, \sigma = 1, 2, 3$$

5. Let us consider the form

$$I = x^2 + 9y^2 + 18z^2 - 2xy - 2xz + 18yz$$

We may write

$$\begin{aligned} I &= (x^2 - 2xy - 2xz) + 9y^2 + 18yz + 18z^2 \\ &\equiv (x - y - z)^2 + 8y^2 + 16yz + 17z^2 \\ &\equiv (x - y - z)^2 + 8(y + z)^2 + 9z^2 \\ &\equiv X^2 + Y^2 + Z^2 \end{aligned}$$

where $X = x - y - z$, $Y = \sqrt{8}\,(y + z)$, $Z = 3z$, a set of linear transformations from x, y, z to X, Y, Z.

This method may be employed to reduce any quadratic form to normal form. However, the linear transformations may not be a rotation of axes. Reduce I to normal form by a rotation of axes.

CHAPTER 7

HYDRODYNAMICS AND ELASTICITY

108. Pressure. The science of hydrodynamics deals with the motion of fluids. We shall be interested in liquids and gases, a liquid or gas being defined as a collection of molecules, which, when studied macroscopically, appear to be continuous in structure. A liquid differs from a solid in that the liquid will yield to any shearing stress, however small, if the stress is continued long enough. All liquids are compressible to a slight extent, but for many purposes it is simpler to consider the liquid as being incompressible. We shall also be highly interested in perfect fluids. These are liquids which possess no shearing stresses.

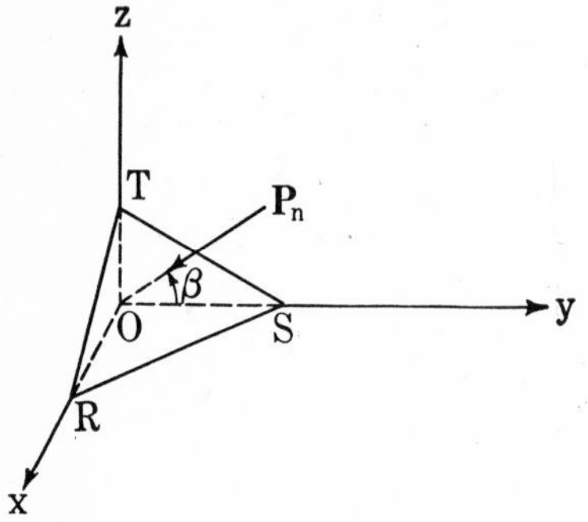

FIG. 95.

We now show that the pressure is the same in all directions for a perfect fluid. Let us consider the motion of the tetrahedron $ORST$ (see Fig. 95). The face ORT has a force acting on it, since it is in contact with other parts of the liquid. Under the above assumption, this force acts normal to the face. Call it Δf_y. If we divide Δf_y by the area of the face ORT, ΔA_y, we obtain the pressure on this face, $P_y = \dfrac{\Delta f_y}{\Delta A_y}\cdot$ The limit of this quotient is called the pressure in the direction normal to the face ORT. The y component of the pressure on the face RST is $P_n \cos \beta$. Let f_y be the y component of the external force per unit volume, and let ρ be the density of the fluid. The equation of motion in the y direction is given by

$$P_y \, \Delta A_y - P_n \cos \beta \, \Delta A_n + f_y \, \Delta \tau = \frac{d}{dt}\left(\rho \, \Delta \tau \frac{dy}{dt}\right)$$

$$= \rho \, \Delta \tau \frac{d^2 y}{dt^2} \qquad (403)$$

since $\dfrac{d}{dt}(\rho\,\Delta\tau) = \dfrac{dm}{dt} = 0$. Now $\Delta A_y = \Delta A_n \cos \beta$, so that (403) becomes

$$(P_y - P_n) + f_y \frac{\Delta\tau}{\Delta A_y} = \frac{1}{\Delta A_y}\frac{d}{dt}\left(\rho\,\Delta\tau\,\frac{dy}{dt}\right) \tag{404}$$

As $\Delta A_y \to 0$, we have $\dfrac{\Delta\tau}{\Delta A_y} \to 0$, so that if we assume f_y, $\dfrac{d^2y}{dt^2}$, ρ finite, we must have $P_y = P_n$. Similarly, $P_y = P_x = P_z = p$. Since the normal $\mathbf{n}$ for the tetrahedron can be chosen arbitrarily, the pressure is the same in all directions and p is a point function, $p = p(x, y, z, t)$. We leave it to the student to prove that at the boundary of two perfect fluids the pressure is continuous.

109. The Equation of Continuity. Consider a surface S bounding a simply connected region lying entirely inside the liquid. Let ρ be the density of the fluid, so that the total mass of the fluid inside S is given by

$$M = \iiint_R \rho(x, y, z, t)\,d\tau$$

Differentiating with respect to time and remembering that x, y, z are variables of integration, we obtain

$$\frac{dM}{dt} = \iiint_R \frac{\partial\rho}{\partial t}\,d\tau \tag{405}$$

Now there are only three ways in which the mass of the fluid inside S can change: (1) fluid may be entering or leaving the surface. The contribution due to this effect is $\iint_S \mathbf{v}\rho \cdot d\boldsymbol{\sigma}$. (2) matter may be created (source), or (3) matter may be destroyed (sink). Let $\psi(x, y, z, t)$ be the amount of matter created or destroyed per unit volume. For a source, $\psi > 0$, and for a sink, $\psi < 0$. The net gain of fluid is therefore

$$\iiint_R \psi\,d\tau - \iint_S \rho\mathbf{v} \cdot d\boldsymbol{\sigma} \tag{406}$$

Equating (405) and (406) and applying the divergence theorem, we obtain

$$\frac{\partial \rho}{\partial t} + \nabla \cdot (\rho \mathbf{v}) = \psi(x, y, z, t) \tag{407}$$

This is the equation of continuity. For no source and sink, (407) reduces to

$$\frac{\partial \rho}{\partial t} + \nabla \cdot (\rho \mathbf{v}) = 0 \tag{408}$$

If furthermore the liquid is incompressible, $\rho =$ constant, $\frac{\partial \rho}{\partial t} = 0$, and (408) becomes

$$\nabla \cdot \mathbf{v} = 0 \tag{409}$$

If the motion is irrotational, that is, if $\oint \mathbf{v} \cdot d\mathbf{r} = 0$, then $\mathbf{v} = \nabla \varphi$, so that the equation of continuity for an incompressible fluid possessing no sources and sinks and having irrotational motion is given by

$$\nabla^2 \varphi = 0 \tag{410}$$

We call φ the velocity potential. We solve Laplace's equation for φ, then compute the velocity from $\mathbf{v} = \nabla \varphi$.

Problems

1. If the velocity of a fluid is radial, $u = u(r, t)$, show that the equation of continuity is

$$\frac{\partial \rho}{\partial t} + u \frac{\partial \rho}{\partial r} + \frac{\rho}{r^2} \frac{\partial}{\partial r} (r^2 u) = \psi(r, t)$$

Solve this equation for an incompressible fluid, if $\psi(r, t) = 1/r^2$.

2. Show that $\mathbf{v} = \dfrac{-2xyz}{(x^2 + y^2)^2} \mathbf{i} + \dfrac{(x^2 - y^2)z}{(x^2 + y^2)^2} \mathbf{j} + \dfrac{y}{x^2 + y^2} \mathbf{k}$ is a possible motion for an incompressible perfect fluid. Is this motion irrotational?

3. Prove that, if the normal velocity is zero at every point of the boundary of a liquid occupying a simply connected region, and moving irrotationally, φ is constant throughout the interior of that region.

4. Prove that if φ is constant over the boundary of any simply connected region, then φ has the same constant value throughout the interior.

5. Express (407) in cylindrical coordinates, spherical coordinates, rectangular coordinates.

110. Equations of Motion for a Perfect Fluid. Let us consider the motion of a fluid inside a simply connected region of volume V and boundary S. The forces acting on this volume are

(1) external forces (gravity, etc.), say, $\mathbf{f}$ per unit mass; (2) pressure thrust on the surface, $-p\,d\boldsymbol{\sigma}$, since $d\boldsymbol{\sigma}$ points outward. The total force acting on V is

$$\mathbf{F} = \iiint_V \rho\mathbf{f}\,d\tau - \iint_S p\,d\boldsymbol{\sigma} = \iiint_V (\rho\mathbf{f} - \nabla p)\,d\tau$$

The linear momentum of V is

$$\mathbf{M} = \iiint_V \rho\mathbf{v}\,d\tau$$

and the time rate of change of linear momentum is

$$\begin{aligned}\frac{d\mathbf{M}}{dt} &= \frac{d}{dt}\iiint_V \rho\mathbf{v}\,d\tau \\ &= \iiint_V \rho\frac{d\mathbf{v}}{dt}\,d\tau + \iiint_V \mathbf{v}\frac{d}{dt}(\rho\,d\tau)\end{aligned}$$

since the volume V changes with time. However, $\rho\,d\tau$ is the mass of the volume $d\tau$, and this remains constant throughout the motion, so that $\frac{d}{dt}(\rho\,d\tau) = 0$. Since $\mathbf{F} = \frac{d\mathbf{M}}{dt}$, we obtain

$$\iiint_V (\rho\mathbf{f} - \nabla p)\,d\tau = \iiint_V \rho\frac{d\mathbf{v}}{dt}\,d\tau$$

This equation is true for all V, so that

$$\rho\mathbf{f} - \nabla p = \rho\frac{d\mathbf{v}}{dt}$$

or

$$\boxed{\frac{d\mathbf{v}}{dt} = \mathbf{f} - \frac{1}{\rho}\nabla p} \qquad (411)$$

This is Euler's equation of motion.

From (76) we have that $\frac{d\mathbf{v}}{dt} = \frac{\partial \mathbf{v}}{\partial t} + (\mathbf{v} \cdot \nabla)\mathbf{v}$, so that an alternative form of (411) is

$$\frac{\partial \mathbf{v}}{\partial t} + (\mathbf{v} \cdot \nabla)\mathbf{v} = \mathbf{f} - \frac{1}{\rho} \nabla p \tag{412}$$

Also from Eq. (9) of Sec. 22, $\nabla \mathbf{v}^2 = 2\mathbf{v} \times (\nabla \times \mathbf{v}) + 2(\mathbf{v} \cdot \nabla)\mathbf{v}$, so that (412) becomes

$$\frac{\partial \mathbf{v}}{\partial t} + \frac{1}{2} \nabla \mathbf{v}^2 - \mathbf{v} \times (\nabla \times \mathbf{v}) = \mathbf{f} - \frac{1}{\rho} \nabla p \tag{413}$$

111. Equations of Motion for an Incompressible Fluid under the Action of a Conservative Field. If the external field is conservative, $\mathbf{f} = -\nabla \chi$, so that $\mathbf{f} - (1/\rho) \nabla p = -\nabla[\chi + (p/\rho)]$ if $\rho = $ constant. Hence (413) becomes

$$\frac{\partial \mathbf{v}}{\partial t} - \mathbf{v} \times (\nabla \times \mathbf{v}) = -\nabla \left(\chi + \frac{p}{\rho} + \frac{1}{2} \mathbf{v}^2 \right) \tag{414}$$

We consider two special cases:

(*a*) *Irrotational motion.* $\mathbf{v} = \nabla \varphi$ and $\nabla \times \mathbf{v} = 0$, so that (414) becomes $\frac{\partial \mathbf{v}}{\partial t} = -\nabla \left(\chi + \frac{p}{\rho} + \frac{1}{2} \mathbf{v}^2 \right)$.

(*b*) *Steady motion.* $\frac{\partial \mathbf{v}}{\partial t} = 0$, so that (414) becomes

$$\mathbf{v} \times (\nabla \times \mathbf{v}) = \nabla \left(\chi + \frac{p}{\rho} + \frac{1}{2} \mathbf{v}^2 \right)$$

For this case we immediately have that

$$\mathbf{v} \cdot \left[\nabla \left(\chi + \frac{p}{\rho} + \frac{1}{2} \mathbf{v}^2 \right) \right] = 0$$

Hence $\nabla[\chi + (p/\rho) + \frac{1}{2}\mathbf{v}^2]$ is normal everywhere to the velocity field $\mathbf{v}$. Thus $\mathbf{v}$ is parallel to the surface $\chi + (p/\rho) + \frac{1}{2}\mathbf{v}^2 = $ constant. The curve drawn in the fluid so that its tangents are parallel to the velocity vectors at corresponding points is called a streamline. We have proved that for an incompressible perfect fluid, which moves moving under the action of conservative forces and whose motion is steady, the expression $\chi + (p/\rho) + \frac{1}{2}\mathbf{v}^2$ remains constant along a streamline. This is the general form of Bernoulli's theorem. If χ remains essentially constant, then an increase of velocity demands a decrease of pressure, and conversely.

Problems

1. If the motion of a perfect incompressible fluid is both steady and irrotational, show that $\chi + (p/\rho) + \frac{1}{2}\mathbf{v}^2 = \text{constant}$.

2. If the fluid is at rest, $\dfrac{d\mathbf{v}}{dt} = 0$. Show that $\nabla \times (\rho\mathbf{f}) = 0$, and hence that $\mathbf{f} \cdot \nabla \times \mathbf{f} = 0$. This is a necessary condition for equilibrium of a fluid. Why must $\rho\mathbf{f}$ be the gradient of a scalar if equilibrium is to be possible?

3. If a liquid rotates like a rigid body with constant angular velocity $\boldsymbol{\omega} = \omega\mathbf{k}$ and if gravity is the only external force, prove that $p/\rho = \frac{1}{2}\omega^2 r^2 - gz + \text{constant}$, where r is the distance from the z axis.

4. Write (411) in rectangular, cylindrical, and spherical coordinates.

5. A liquid is in equilibrium under the action of an external force $\mathbf{f} = (y + z)\mathbf{i} + (z + x)\mathbf{j} + (x + y)\mathbf{k}$. Find the surfaces of equal pressure.

6. If the motion of the fluid is referred to a moving frame of reference which rotates with angular velocity $\boldsymbol{\omega}$ and has translational velocity $\mathbf{u}$, show that the equation of motion is

$$\mathbf{f} - \frac{1}{\rho}\nabla p = \frac{d\mathbf{u}}{dt} + \frac{d\boldsymbol{\omega}}{dt} \times \mathbf{r} + \boldsymbol{\omega} \times (\boldsymbol{\omega} \times \mathbf{r}) + 2\boldsymbol{\omega} \times \frac{D\mathbf{r}}{dt} + \frac{D^2\mathbf{r}}{dt^2}$$

and that the equation of continuity is

$$\frac{\partial \rho}{\partial t} + \nabla \cdot \left(\rho \frac{D\mathbf{r}}{dt}\right) = 0$$

7. *The energy equation.* For a simply connected region R with boundary S, the kinetic energy of R is

$$T = \frac{1}{2} \iiint_R \rho \mathbf{v}^2 \, d\tau$$

Let the surface S move so that it always contains all the original mass of R. Show that

$$\frac{dT}{dt} = \frac{1}{2}\iiint\limits_R \rho \frac{d\mathbf{v}^2}{dt}\, d\tau$$
$$= \iiint\limits_R \mathbf{v}\cdot\left(\mathbf{f} - \frac{1}{\rho}\nabla p\right) d\tau$$
$$= \iiint\limits_R \mathbf{v}\cdot\mathbf{f}\, d\tau - \iint\limits_S \rho\mathbf{v}\cdot d\boldsymbol{\sigma} + \iiint\limits_R p\frac{d}{dt}(d\tau) \quad (415)$$

Analyze each term of (415).

8. For irrotational flow show that

$$\frac{\partial\varphi}{\partial t} = -\left(\chi + \frac{p}{\rho} + \frac{1}{2}\mathbf{v}^2\right) + C(t),$$

and if $\rho = \rho(p)$, $\dfrac{\partial\varphi}{\partial t} + \dfrac{\mathbf{v}^2}{2} + \chi + \displaystyle\int \frac{dp}{\rho} = D(t)$.

112. The General Motion of a Fluid. Let us consider the velocities of the particles occupying an element of volume of a fluid. Let P be a point of the volume or region, and let $\mathbf{v}_P$ represent the velocity of the fluid at P (Fig. 96). The velocity at a nearby point Q is

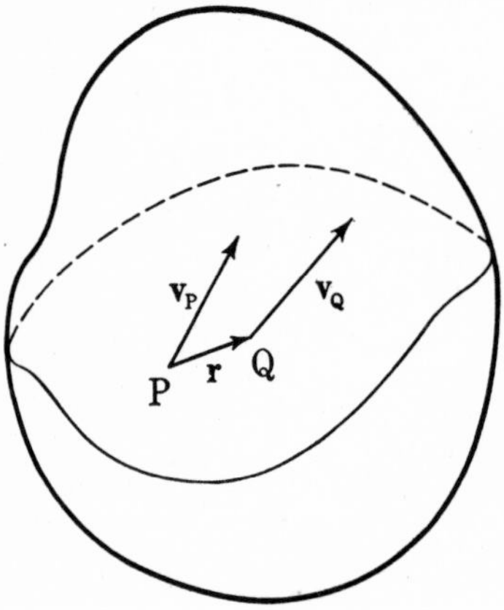

FIG. 96.

$$\mathbf{v}_Q = \mathbf{v}_P + d\mathbf{v}_P$$
$$= \mathbf{v}_P + (d\mathbf{r}\cdot\nabla)\mathbf{v}_P \quad (416)$$

from (75). By $(d\mathbf{r}\cdot\nabla)\mathbf{v}_P$ we mean that after differentiation, the partial derivatives of $\mathbf{v}$ are calculated at P. We now replace $d\mathbf{r}$ by $\mathbf{r}$ for convenience, so that $\mathbf{r} = x\mathbf{i} + y\mathbf{j} + z\mathbf{k}$ if we consider P as the origin and x, y, z large in comparison with x^2, y^2, z^2, zy, etc. Equation (416) now becomes $\mathbf{v}_Q = \mathbf{v}_P + (\mathbf{r}\cdot\nabla)\mathbf{v}_P$. Now

$$\nabla(\mathbf{r}\cdot\mathbf{w}) = \mathbf{r}\times(\nabla\times\mathbf{w}) + (\mathbf{r}\cdot\nabla)\mathbf{w} + \mathbf{w} \quad (417)$$

from (9), (10), (12) of Sec. 22.

Now let

$$\mathbf{w} \equiv (\mathbf{r}\cdot\nabla)\mathbf{v}_P = x\left.\frac{\partial\mathbf{v}}{\partial x}\right|_P + y\left.\frac{\partial\mathbf{v}}{\partial y}\right|_P + z\left.\frac{\partial\mathbf{v}}{\partial z}\right|_P \quad (418)$$

and hence

$$(\mathbf{r}\cdot\nabla)\mathbf{w} = x\frac{\partial x}{\partial x}\frac{\partial \mathbf{v}}{\partial x}\bigg|_P + y\frac{\partial y}{\partial y}\frac{\partial \mathbf{v}}{\partial y}\bigg|_P + z\frac{\partial z}{\partial z}\frac{\partial \mathbf{v}}{\partial z}\bigg|_P$$
$$= \mathbf{w}$$

We did not differentiate the $\left.\frac{\partial \mathbf{v}}{\partial x}\right|_P, \left.\frac{\partial \mathbf{v}}{\partial y}\right|_P, \left.\frac{\partial \mathbf{v}}{\partial z}\right|_P$, since they have been evaluated at P and so are constants for the moment. Thus, using (417), we obtain

$$\mathbf{w} = \tfrac{1}{2}\nabla(\mathbf{r}\cdot\mathbf{w}) + \tfrac{1}{2}(\nabla\times\mathbf{w})\times\mathbf{r} \tag{419}$$

Moreover, $\mathbf{v} = \mathbf{v}_P + \mathbf{w}$, so that $\nabla\times\mathbf{v} = \nabla\times\mathbf{w} = (\nabla\times\mathbf{v})_P$ [see (418)], and

$$\mathbf{v}_Q = \mathbf{v}_P + \tfrac{1}{2}(\nabla\times\mathbf{v})_P\times\mathbf{r} + \tfrac{1}{2}\nabla(\mathbf{r}\cdot\mathbf{w}) \tag{420}$$

It is easy to verify that $\mathbf{r}\cdot\mathbf{w}$ is a quadratic form, that is,

$$\mathbf{r}\cdot\mathbf{w} = Ax^2 + By^2 + Cz^2 + 2Dyz + 2Ezx + 2Fxy$$

and so by a rotation (Sec. 107), we can write

$$\mathbf{r}\cdot\mathbf{w} = ax^2 + by^2 + cz^2$$

and

$$\tfrac{1}{2}\nabla(\mathbf{r}\cdot\mathbf{w}) = ax\mathbf{i} + by\mathbf{j} + cz\mathbf{k}$$

We may now write (420) as

$$\mathbf{v}_Q = \mathbf{v}_P + \boldsymbol{\omega}\times\mathbf{r} + (ax\mathbf{i} + by\mathbf{j} + cz\mathbf{k}) \tag{421}$$

where $\boldsymbol{\omega} = \tfrac{1}{2}(\nabla\times\mathbf{v})_P$.

Let us analyze (421), which states that the velocity of Q is the sum of three parts:

1. The velocity $\mathbf{v}_P$ of P, which corresponds to a translation of the element.

2. $\boldsymbol{\omega}\times\mathbf{r}$ represents the velocity due to a rotation about a line through P with angular velocity $\tfrac{1}{2}(\nabla\times\mathbf{v})_P$.

3. $ax\mathbf{i} + by\mathbf{j} + cz\mathbf{k}$ represents a velocity relative to P with components ax, by, cz, respectively, along the x, y, z axes.

The first two are rigid-body motions; they could still take place if the fluid were a solid. The third term shows that particles at different distances from P move at different rates relative to P. If we consider a sphere surrounding P, the spherical element is translated, rotated, and stretched in the directions of the principal axes by amounts proportional to a, b, c. Hence the sphere is deformed into an ellipsoid. This third motion is called a pure strain and takes place only when a substance is deformable. Each point of the fluid will have the three principal directions associated with it. Unfortunately, these directions are not the same at all points, so that no single coordinate system will suffice for the complete fluid.

The most general motion of a fluid is that described above and is independent of the coordinate system used to describe the motion. It is therefore an intrinsic property of the fluid.

113. Vortex Motion. If at each point of a curve the tangent vector is parallel to the vector $\boldsymbol{\omega} = \frac{1}{2}(\nabla \times \mathbf{v})$, we say that the curve is a vortex line. This implies that $\dfrac{dx}{\omega_x} = \dfrac{dy}{\omega_y} = \dfrac{dz}{\omega_z}$ where dx, dy, dz are the components of the tangent vector and

$$\boldsymbol{\omega} = \omega_x\mathbf{i} + \omega_y\mathbf{j} + \omega_z\mathbf{k}$$

The integration of this system of differential equations yields the vortex lines. The vortex lines may change as time goes on, since, in general, $\boldsymbol{\omega}$ will depend on the time.

Let us now calculate the *circulation* around any closed curve in the fluid.

$$C = \oint_{\Gamma} \mathbf{v} \cdot d\mathbf{r} = \iint_S (\nabla \times \mathbf{v}) \cdot d\boldsymbol{\sigma} \tag{422}$$

If $\nabla \times \mathbf{v} \equiv 0$, then $C = 0$. This is true while we keep the curve Γ fixed in space. Let us now find out how the circulation changes with time if we let the particles which comprise Γ move according to the motion of the fluid. As time goes on, assuming continuity of flow, the closed curve will remain closed.

Now

$$C = \oint_{\Gamma'} \mathbf{v} \cdot d\mathbf{r} = \oint_{\Gamma'} \mathbf{v} \cdot \frac{d\mathbf{r}}{ds}\, ds \tag{423}$$

where s is arc length along the particular curve Γ', at some time t. At an instant later the curve Γ' has moved to a new position given by the curve Γ''. The velocity of the particles over this path is slightly different from that over Γ', and, moreover, the unit tangents $\frac{d\mathbf{r}}{ds}$ have changed. The parameter s is still a variable of integration and has nothing to do with the time. Therefore

$$\frac{dC}{dt} = \oint_{\Gamma'} \frac{d\mathbf{v}}{dt} \cdot \frac{d\mathbf{r}}{ds}\, ds + \oint_{\Gamma'} \mathbf{v} \cdot \frac{d}{dt}\left(\frac{d\mathbf{r}}{ds}\right) ds$$

$$= \oint_{\Gamma'} \frac{d\mathbf{v}}{dt} \cdot \frac{d\mathbf{r}}{ds}\, ds + \oint_{\Gamma'} \mathbf{v} \cdot \frac{d}{ds}\left(\frac{d\mathbf{r}}{dt}\right) ds \tag{424}$$

Euler's equation of motion (411) for a conservative field,

$$\mathbf{f} = -\nabla\chi$$

is $\frac{d\mathbf{v}}{dt} = -\nabla\chi - \frac{1}{\rho}\nabla p = -\nabla V$, where $V = \chi + \int dp/\rho$. Therefore

$$\frac{dC}{dt} = -\oint \nabla V \cdot d\mathbf{r} + \oint \frac{1}{2}\frac{dv^2}{ds}\, ds$$

$$= -\oint d(V - \tfrac{1}{2}v^2) \equiv 0 \tag{425}$$

We have arrived at a theorem by Lord Kelvin that the circulation around a closed curve composed of a given set of particles remains constant if the field is conservative, provided that the density ρ is a function only of the pressure p.

If we now consider a closed curve lying on a tube made up of vortex lines, but not encircling the tube (see Fig. 97), then

$$C = \oint \mathbf{v} \cdot d\mathbf{r}$$

$$= \iint_S \nabla \times \mathbf{v} \cdot d\boldsymbol{\sigma} = 0$$

since $d\boldsymbol{\sigma}$ is normal to $\nabla \times \mathbf{v}$. From Kelvin's theorem, $C = 0$ for all time, so that the curve Γ always lies on the vortex tube.

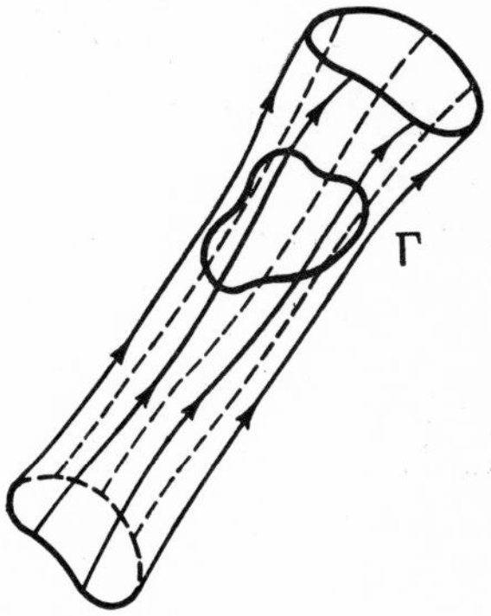

FIG. 97.

114. Applications

Example 109. Let us consider the steady irrotational motion of an incompressible fluid when a sphere moves through the fluid with constant velocity. Let the center of the sphere travel along the z axis with velocity $\mathbf{v}_0$. We choose the center of the sphere as the origin of our coordinate system. From Sec. 110, Prob. 6, we have

$$\mathbf{f} - \frac{1}{\rho}\nabla p = \frac{d\mathbf{v}}{dt}$$

and

$$\nabla \cdot \left(\frac{D\mathbf{r}}{dt}\right) = 0$$

Hence $\dfrac{D\mathbf{r}}{dt} = \nabla\varphi$, so that $\nabla^2\varphi = 0$. Now at points on the surface of the sphere we must have $\left(\dfrac{D\mathbf{r}}{dt}\right)_{\text{radially}} = 0$, so that $\left(\dfrac{\partial\varphi}{\partial r}\right)_{r=a} = 0$.

We look for a solution of Laplace's equation satisfying this boundary condition, so that we try

$$\varphi = \left(Ar + \frac{B}{r^2}\right)\cos\theta \tag{426}$$

(see Sec. 67). We need

$$\left(\frac{\partial\varphi}{\partial r}\right)_{r=a} = \left(A - \frac{2B}{a^3}\right)\cos\theta = 0$$

so that $B = a^3A/2$. Moreover, at infinity we expect the velocity of the fluid to be zero, so that the velocity relative to the sphere should be $-\mathbf{v}_0$. Hence

$$v_z = \left(\frac{\partial\varphi}{\partial z}\right)_{z=\infty} = A = -v_0$$

and

$$\varphi = -v_0\left(r + \frac{a^3}{2r^2}\right)\cos\theta \tag{427}$$

The velocity of the fluid relative to the sphere is given by $\mathbf{v} = \nabla\varphi$ and the velocity of the fluid is $\mathbf{v} = \nabla\varphi + v_0\mathbf{k}$.

Example 110. Let us consider a fluid resting on a horizontal surface (x-y plane) and take z vertical. Let us assume a transverse wave traveling in the x direction. For an incompressible fluid

$$\nabla^2\varphi = \frac{\partial^2\varphi}{\partial x^2} + \frac{\partial^2\varphi}{\partial z^2} = 0 \tag{428}$$

We assume a solution of the form $\varphi = A(z)e^{\frac{2\pi}{\lambda}(x-vt)i}$. Substituting into (428), we obtain

$$e^{\frac{2\pi}{\lambda}(x-vt)i}\left[-\frac{4\pi^2}{\lambda^2}A(z) + \frac{d^2A}{dz^2}\right] = 0$$

so that

$$\frac{d^2A}{dz^2} = \frac{4\pi^2}{\lambda^2}A \tag{429}$$

The solution to (429) is $A = A_0e^{(2\pi/\lambda)z} + B_0e^{-(2\pi/\lambda)z}$, and a real solution to (428) is

$$\varphi = (A_0e^{(2\pi/\lambda)z} + B_0e^{-(2\pi/\lambda)z}) \cos\left[\frac{2\pi}{\lambda}(x - vt)\right] \tag{430}$$

The fluid has no vertical velocity at the bottom of the plane on which it rests, so that $v_z = \frac{\partial\varphi}{\partial z} = 0$ at $z = 0$. This yields $A_0 = B_0$, so that

$$\begin{aligned}\varphi &= A_0(e^{(2\pi/\lambda)z} + e^{-(2\pi/\lambda)z}) \cos\left[\frac{2\pi}{\lambda}(x - vt)\right]\\ &= \frac{A_0}{2}\cosh\left(\frac{2\pi}{\lambda}z\right)\cos\left[\frac{2\pi}{\lambda}(x - vt)\right]\end{aligned} \tag{431}$$

From Prob. 8, Sec. 110, we have

$$\frac{\partial\varphi}{\partial t} = -\left(\chi + \frac{p}{\rho} + \frac{1}{2}v^2\right) + C(t)$$

and for a gravitational potential, $\chi = gz$, so that

$$\frac{\partial\varphi}{\partial t} = -\left(gz + \frac{p}{\rho} + \frac{1}{2}v^2\right) + C(t) \tag{432}$$

We now assume that the waves are restricted to small amplitudes and velocities, so that we neglect $\frac{1}{2}v^2$. Moreover, at the surface, p, the atmospheric pressure, is essentially constant, so that $\frac{dp}{dt} = 0$. Differentiating (432), we obtain

$$\frac{\partial^2 \varphi}{\partial t^2} = -g \frac{\partial z}{\partial t} + \frac{dC}{dt} \tag{433}$$

and again at the surface $\frac{\partial z}{\partial t} = v_z = \frac{\partial \varphi}{\partial z}$, so that (433) becomes

$$\frac{\partial^2 \varphi}{\partial t^2} = -g \frac{\partial \varphi}{\partial z} + \frac{dC}{dt} \tag{434}$$

Substituting (431) into (434), we obtain

$$-v^2 \frac{2\pi^2}{\lambda^2} A_0 \cosh \frac{2\pi}{\lambda} z \cos \left[\frac{2\pi}{\lambda}(x - vt)\right] = -\frac{g\pi A_0}{\lambda} \sinh \frac{2\pi}{\lambda} z \cos \left[\frac{2\pi}{\lambda}(x - vt)\right] + \frac{dC}{dt} \tag{435}$$

In order for C to be dependent only on t, we must have the coefficient of $\cos [(2\pi/\lambda)(x - vt)]$ identically zero in (435). Hence

$$A_0 \left(-\frac{v^2 2\pi^2}{\lambda^2} \cosh \frac{2\pi}{\lambda} z + \frac{\pi g}{\lambda} \sinh \frac{2\pi}{\lambda} z \right) = 0 \tag{435a}$$

or

$$v^2 = \frac{\lambda g}{2\pi} \tanh \frac{2\pi}{\lambda} z$$

In deep water z/λ is large so that $\tanh \frac{2\pi}{\lambda} z \approx 1$, and the velocity of the wave is $v = (\lambda g / 2\pi)^{\frac{1}{2}}$.

Problems

1. Show that for steady motion of an incompressible fluid under the action of conservative forces, $(\mathbf{v} \cdot \nabla)\boldsymbol{\omega} - (\boldsymbol{\omega} \cdot \nabla)\mathbf{v} = 0$, where $\boldsymbol{\omega} = \frac{1}{2} \nabla \times \mathbf{v}$.

2. Show that $\frac{d}{dt}\left(\frac{\boldsymbol{\omega}}{\rho}\right) = \left(\frac{\boldsymbol{\omega}}{\rho} \cdot \nabla\right) \mathbf{v}$ for a conservative system.

3. If C is the circulation around any closed circuit moving with the fluid, prove that $\frac{dC}{dt} = \oint p\, d\left(\frac{1}{\rho}\right)$ if the field is conservative and if the pressure depends only on the density.

4. Show that $\mathbf{v} = 2axy\mathbf{i} + a(x^2 - y^2)\mathbf{j}$ is a possible velocity of an incompressible fluid.

5. Verify that the velocity potential $\varphi = A[r + (a^2/r)] \cos \theta$ represents a stream motion past a fixed circular cylinder.

115. Small Displacements. Strain Tensor. In the absence of external forces, a solid body remains in equilibrium and the forces between the various particles of the solid are in equilibrium because of the configuration of the particles. If external forces are added, the particles (atoms, molecules) tend to redistribute themselves so that equilibrium will occur again. Here we are interested in the kinematic relationship between the old positions of equilibrium and the new. We shall assume that the deformations are small and continuous. We expect, from Sec. 112, that in the neighborhood of a given point P_0, the remaining points will be rotated about P_0 and will suffer a pure strain relative to P_0. Let $\mathbf{r}$ be the position vector of P relative to P_0, and let $\mathbf{s}$ be the displacement vector suffered by P, and $\mathbf{s}_0$ the displacement suffered by P_0 (Fig. 98). Then

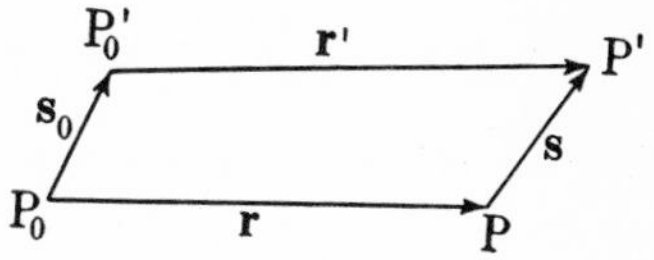

FIG. 98.

$$\mathbf{s} = \mathbf{s}_0 + d\mathbf{s} = \mathbf{s}_0 + (\mathbf{r} \cdot \nabla)\mathbf{s}_0 \tag{436}$$

Let $\mathbf{s} = u(x, y, z, t)\mathbf{i} + v(x, y, z, t)\mathbf{j} + w(x, y, z, t)\mathbf{k}$. Since we will be dealing with static conditions,

$$\mathbf{s} = u(x, y, z)\mathbf{i} + v(x, y, z)\mathbf{j} + w(x, y, z)\mathbf{k}$$

From (420),

$$\mathbf{s} = \mathbf{s}_0 + \tfrac{1}{2}(\nabla \times \mathbf{s})_{P_0} \times \mathbf{r} + \tfrac{1}{2}\nabla(\mathbf{r} \cdot \mathbf{w}) \tag{437}$$

where

$$\mathbf{w} = x \left.\frac{\partial \mathbf{s}}{\partial x}\right|_{P_0} + y \left.\frac{\partial \mathbf{s}}{\partial y}\right|_{P_0} + z \left.\frac{\partial \mathbf{s}}{\partial z}\right|_{P_0}$$

since $\mathbf{s} = \mathbf{v}\,\Delta t$.

We are interested in the position of P after the deformation (now P') relative to the new position of P_0 (now P_0'). This is the vector $\mathbf{r}' = \mathbf{r} + \mathbf{s} - \mathbf{s}_0$, or

$$\mathbf{r}' = \mathbf{r} + \tfrac{1}{2}(\nabla \times \mathbf{s})_{P_0} \times \mathbf{r} + \tfrac{1}{2}\nabla(\mathbf{r} \cdot \mathbf{w}) \tag{438}$$

Since $\frac{1}{2}(\nabla \times \mathbf{s})_{P_0} \times \mathbf{r}$ represents a rigid-body rotation about P_0, we ignore this nondeformation term and so are interested in $\mathbf{r} + \frac{1}{2}\nabla(\mathbf{r} \cdot \mathbf{w})$. Now

$$\begin{aligned}\mathbf{r} + \frac{1}{2}\nabla(\mathbf{r} \cdot \mathbf{w}) = x\mathbf{i} + y\mathbf{j} + z\mathbf{k} \\ + \frac{1}{2}\nabla\left(x^2 \frac{\partial u}{\partial x}\bigg|_{P_0} + xy \frac{\partial v}{\partial x}\bigg|_{P_0} + xz \frac{\partial w}{\partial x}\bigg|_{P_0}\right) \\ + \frac{1}{2}\nabla\left(xy \frac{\partial u}{\partial y}\bigg|_{P_0} + y^2 \frac{\partial v}{\partial y}\bigg|_{P_0} + yz \frac{\partial w}{\partial y}\bigg|_{P_0}\right) \\ + \frac{1}{2}\nabla\left(zx \frac{\partial u}{\partial z}\bigg|_{P_0} + zy \frac{\partial v}{\partial z}\bigg|_{P_0} + z^2 \frac{\partial w}{\partial z}\bigg|_{P_0}\right)\end{aligned}$$

and

$$\begin{aligned}\mathbf{r} + \frac{1}{2}\nabla(\mathbf{r} \cdot \mathbf{w}) \\ = \left[x\left(1 + \frac{\partial u}{\partial x}\right) + \frac{y}{2}\left(\frac{\partial u}{\partial y} + \frac{\partial v}{\partial x}\right) + \frac{z}{2}\left(\frac{\partial w}{\partial x} + \frac{\partial u}{\partial z}\right)\right]\mathbf{i} \\ + \left[\frac{x}{2}\left(\frac{\partial u}{\partial y} + \frac{\partial v}{\partial x}\right) + y\left(1 + \frac{\partial v}{\partial y}\right) + \frac{z}{2}\left(\frac{\partial w}{\partial y} + \frac{\partial v}{\partial z}\right)\right]\mathbf{j} \\ + \left[\frac{x}{2}\left(\frac{\partial w}{\partial x} + \frac{\partial u}{\partial z}\right) + \frac{y}{2}\left(\frac{\partial w}{\partial y} + \frac{\partial v}{\partial z}\right) + z\left(1 + \frac{\partial w}{\partial z}\right)\right]\mathbf{k}\end{aligned} \tag{439}$$

The partial derivatives are evaluated at the point P_0.

Let us now consider the matrix

$$\|s_j{}^i\| = \left\| \begin{matrix} 1 + \dfrac{\partial u}{\partial x} & \dfrac{1}{2}\left(\dfrac{\partial u}{\partial y} + \dfrac{\partial v}{\partial x}\right) & \dfrac{1}{2}\left(\dfrac{\partial w}{\partial x} + \dfrac{\partial u}{\partial z}\right) \\ \dfrac{1}{2}\left(\dfrac{\partial u}{\partial y} + \dfrac{\partial v}{\partial x}\right) & 1 + \dfrac{\partial v}{\partial y} & \dfrac{1}{2}\left(\dfrac{\partial w}{\partial y} + \dfrac{\partial v}{\partial z}\right) \\ \dfrac{1}{2}\left(\dfrac{\partial w}{\partial x} + \dfrac{\partial u}{\partial z}\right) & \dfrac{1}{2}\left(\dfrac{\partial w}{\partial y} + \dfrac{\partial v}{\partial z}\right) & 1 + \dfrac{\partial w}{\partial z} \end{matrix} \right\| \tag{440}$$

The nine components of this matrix form the *strain tensor*. If we write $\mathbf{r} = x^1\mathbf{i} + x^2\mathbf{j} + x^3\mathbf{k}$ and $\mathbf{r}' = y^1\mathbf{i} + y^2\mathbf{j} + y^3\mathbf{k}$ (see

Example 8), then $\mathbf{r}' = \mathbf{r} + \frac{1}{2}\nabla(\mathbf{r}\cdot\mathbf{w})$ may be written

$$y^i = s_1{}^i x^1 + s_2{}^i x^2 + s_3{}^i x^3, \qquad i = 1, 2, 3$$

or

$$y^i = \sum_{\alpha=1}^{3} s_\alpha{}^i x^\alpha \tag{441}$$

We shall see in Chap. 8 that since $\mathbf{r}$ and $\mathbf{r}'$ are vectors, then, of necessity, the $s_j{}^i$ are the components of a tensor. Notice that $s_j{}^i = s_i{}^j$, so that the tensor is symmetric.

The ellipsoid which has the equation

$$\left(1+\frac{\partial u}{\partial x}\right)x^2 + \left(1+\frac{\partial v}{\partial y}\right)y^2 + \left(1+\frac{\partial w}{\partial z}\right)z^2 + \left(\frac{\partial u}{\partial y}+\frac{\partial v}{\partial x}\right)xy + \left(\frac{\partial w}{\partial y}+\frac{\partial v}{\partial z}\right)yz + \left(\frac{\partial w}{\partial x}+\frac{\partial u}{\partial z}\right)zx = 1 \tag{442}$$

is called the *strain ellipsoid*. From Sec. 107 we know that we can reduce the ellipsoid to the form

$$Ax'^2 + By'^2 + Cz'^2 = 1$$

by a proper rotation. The strain tensor becomes entirely diagonal,

$$\|\bar{s}_j{}^i\| = \begin{Vmatrix} A & 0 & 0 \\ 0 & B & 0 \\ 0 & 0 & C \end{Vmatrix}$$

In the directions of the new x', y', z' axes, the deformation is a pure translation, and these directions are called the principal directions of the strain ellipsoid.

Let us now compute the change in the unit vectors, neglecting the rotation term. The unit vector $\mathbf{i}$ has the components (1, 0, 0), so that from (438) and (439)

$$\mathbf{i} \to \mathbf{r}_1 = \left(1+\frac{\partial u}{\partial x}\right)\mathbf{i} + \frac{1}{2}\left(\frac{\partial u}{\partial y}+\frac{\partial v}{\partial x}\right)\mathbf{j} + \frac{1}{2}\left(\frac{\partial w}{\partial x}+\frac{\partial u}{\partial z}\right)\mathbf{k}$$

By neglecting higher terms such as $\left(\frac{\partial u}{\partial x}\right)^2, \frac{\partial u}{\partial y}\frac{\partial v}{\partial x}, \ldots$, we have $|\mathbf{r}_1| = \left|1+\frac{\partial u}{\partial x}\right|$. Similarly $\mathbf{j}\to\mathbf{r}_2$, and $|\mathbf{r}_2| = \left|1+\frac{\partial v}{\partial y}\right|$; $\mathbf{k}\to\mathbf{r}_3$, and $|\mathbf{r}_3| = \left|1 + \frac{\partial w}{\partial z}\right|$. The angle between $\mathbf{r}_1$ and $\mathbf{r}_2$ is given by

$$\cos\theta = \frac{\mathbf{r}_1 \cdot \mathbf{r}_2}{|\mathbf{r}_1||\mathbf{r}_2|} \approx \frac{\partial u}{\partial y} + \frac{\partial v}{\partial x}, \quad \cdots$$

The terms of the strain tensor are now fully understood. The volume of the parallelepiped formed by $\mathbf{r}_1, \mathbf{r}_2, \mathbf{r}_3$ is

$$V' = \mathbf{r}_1 \cdot \mathbf{r}_2 \times \mathbf{r}_3 \approx 1 + \frac{\partial u}{\partial x} + \frac{\partial v}{\partial y} + \frac{\partial w}{\partial z}$$

so that

$$\frac{V' - V}{V} = \frac{\partial u}{\partial x} + \frac{\partial v}{\partial y} + \frac{\partial w}{\partial z} = \nabla \cdot \mathbf{s} \tag{443}$$

The left-hand side of (443) is independent of the coordinate system, so that $\nabla \cdot \mathbf{s}$ is an invariant.

Finally, we see that the deformation tensor due to the tensor $\frac{1}{2}\nabla(\mathbf{r} \cdot \mathbf{w})$ has the components

$$\|e_j^{\,i}\| = \left\| \begin{matrix} \dfrac{\partial u}{\partial x} & \dfrac{1}{2}\left(\dfrac{\partial u}{\partial y} + \dfrac{\partial v}{\partial x}\right) & \dfrac{1}{2}\left(\dfrac{\partial w}{\partial x} + \dfrac{\partial u}{\partial z}\right) \\ \dfrac{1}{2}\left(\dfrac{\partial u}{\partial y} + \dfrac{\partial v}{\partial x}\right) & \dfrac{\partial v}{\partial y} & \dfrac{1}{2}\left(\dfrac{\partial u}{\partial y} + \dfrac{\partial v}{\partial z}\right) \\ \dfrac{1}{2}\left(\dfrac{\partial w}{\partial x} + \dfrac{\partial u}{\partial z}\right) & \dfrac{1}{2}\left(\dfrac{\partial u}{\partial y} + \dfrac{\partial v}{\partial z}\right) & \dfrac{\partial w}{\partial z} \end{matrix} \right\|$$

$$= \left\| \frac{1}{2}\left(\frac{\partial u^i}{\partial x^j} + \frac{\partial u^j}{\partial x^i}\right) \right\| \tag{444}$$

where

$$\begin{aligned} u^1 &= u, & x^1 &= x \\ u^2 &= v, & x^2 &= y \\ u^3 &= w, & x^3 &= z \end{aligned}$$

116. The Stress Tensor. Corresponding to any strain in the body must be an impressed force which produces this strain. Let us consider a cube with faces perpendicular to the coordinate axes. In Sec. 108 we assumed no shearing stresses, but now we consider all forces possible between two neighboring surfaces.

Let us consider the face $ABCD$ (Fig. 99). It is in immediate contact with other particles of the body. As a consequence, the resultant force t_x on the face $ABCD$ can be decomposed into three forces: t_{xx}, t_{yx}, t_{zx}, where t_{xx} is the component of t_x in the x direction, t_{yx} is the component of t_x in the y direction, and t_{zx} is the component of t_x in the z direction. We have similar results for the other two faces and so obtain the matrix

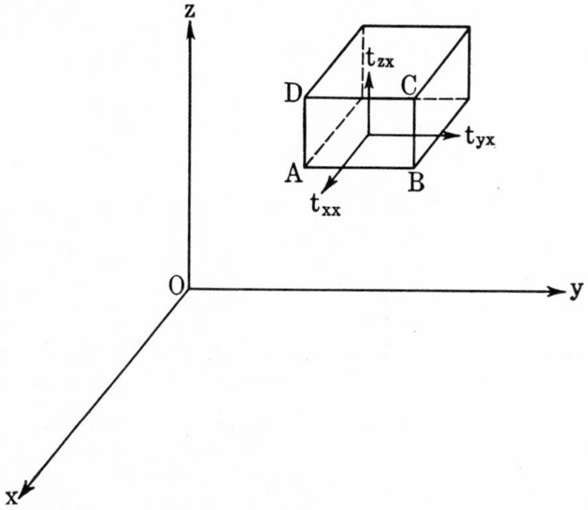

FIG. 99.

$$\|t\| = \begin{Vmatrix} t_{xx} & t_{xy} & t_{xz} \\ t_{yx} & t_{yy} & t_{yz} \\ t_{zx} & t_{zy} & t_{zz} \end{Vmatrix} \tag{445}$$

These are the components of the stress tensor.

By considering a tetrahedron as in Sec. 108, we immediately see that if $d\mathbf{s}$ is the vectoral area of the slant face, then the components of the force $\mathbf{f}$ on this face are

$$\begin{aligned} f_x &= t_{xx}\, ds_x + t_{xy}\, ds_y + t_{xz}\, ds_z \\ f_y &= t_{yx}\, ds_x + t_{yy}\, ds_y + t_{yz}\, ds_z \\ f_z &= t_{zx}\, ds_x + t_{zy}\, ds_y + t_{zz}\, ds_z \end{aligned} \tag{446}$$

where $ds_x = \mathbf{i} \cdot d\mathbf{s}$, $ds_y = \mathbf{j} \cdot d\mathbf{s}$, $ds_z = \mathbf{k} \cdot d\mathbf{s}$.

We immediately see that

$$t_{xx} = \frac{\partial f_x}{\partial s_x}, \qquad t_{xy} = \frac{\partial f_x}{\partial s_y}, \quad \cdots$$

and that $f_i = \sum_{\alpha=1}^{3} t_{i\alpha}\, ds_\alpha$, where $f_1 = f_x$, $f_2 = f_y$, $f_3 = f_z$, $t_{12} = t_{xy}$, ... We shall see later that this explains why the t_{ij} are called the components of a tensor.

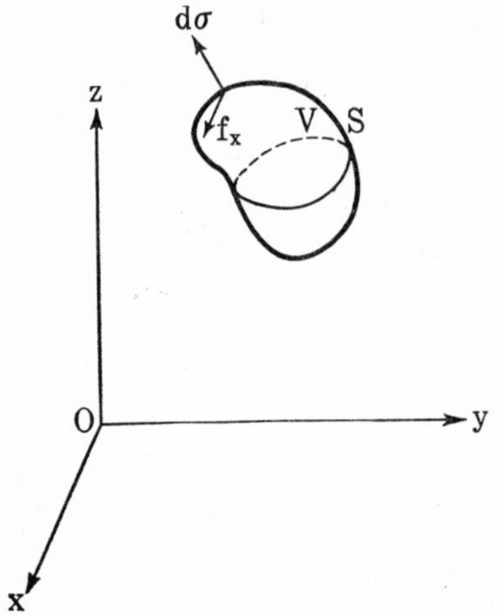

FIG. 100.

Let us now consider the resultant force acting on a volume V with boundary S (see Fig. 100). We have from (446)

$$f_x = t_{xx}\, ds_x + t_{xy}\, ds_y + t_{xz}\, ds_z$$

so that

$$F_x = \sum f_x = \iint_S t_{xx}\, ds_x + t_{xy}\, ds_y + t_{xz}\, ds_z = \iint_S \mathbf{t} \cdot d\boldsymbol{\sigma}$$

where $\mathbf{t} = t_{xx}\mathbf{i} + t_{xy}\mathbf{j} + t_{xz}\mathbf{k}$.

Applying the divergence theorem, we obtain

$$F_x = \iiint_V \left(\frac{\partial t_{xx}}{\partial x} + \frac{\partial t_{xy}}{\partial y} + \frac{\partial t_{xz}}{\partial z} \right) d\tau \tag{447}$$

with similar expressions for F_y, F_z.

By letting $V \to 0$, we have that the x component of the force per unit volume must be $\left(\frac{\partial t_{xx}}{\partial x} + \frac{\partial t_{xy}}{\partial y} + \frac{\partial t_{xz}}{\partial z} \right)$.

117. Relationship between the Strain and Stress Tensors. In the neighborhood of a point P in our region, let us choose the three principal directions of the stress tensor for the axes of our cartesian coordinate system. If we assume that the region is isotropic (only contractions and extensions exist), a cube with faces normal to the principal directions will suffer distortions only along the principal axes. Hence the principal directions of the strain ellipsoid will coincide with those of the stress ellipsoid. In this coordinate system

$$\|e_{ij}\| = \begin{Vmatrix} e_1 & 0 & 0 \\ 0 & e_2 & 0 \\ 0 & 0 & e_3 \end{Vmatrix}, \qquad \|t_{ij}\| = \begin{Vmatrix} t_1 & 0 & 0 \\ 0 & t_2 & 0 \\ 0 & 0 & t_3 \end{Vmatrix} \tag{448}$$

Our fundamental postulate relating the shear components with those of the stress will be Hooke's law, which states that every tension produces an extension in the direction of the tension and is proportional to it. We let E (Young's modulus) be the factor of proportionality. Experiments also show that extensions in fibers produce transverse contractions. The constant for this phenomenon is called Poisson's ratio σ. We thus obtain for the relative elongations of the cube in the three principal directions the following:

$$\begin{aligned} e_1 &= \frac{1}{E} t_1 - \frac{\sigma}{E}(t_2 + t_3) = \frac{1+\sigma}{E} t_1 - \frac{\sigma}{E}(t_1 + t_2 + t_3) \\ e_2 &= \frac{1}{E} t_2 - \frac{\sigma}{E}(t_3 + t_1) = \frac{1+\sigma}{E} t_2 - \frac{\sigma}{E}(t_1 + t_2 + t_3) \\ e_3 &= \frac{1}{E} t_3 - \frac{\sigma}{E}(t_1 + t_2) = \frac{1+\sigma}{E} t_3 - \frac{\sigma}{E}(t_1 + t_2 + t_3) \end{aligned} \tag{449}$$

The formulas for e_1, e_2, e_3 apply only in the immediate neighborhood of a point P. Since points far removed from P will have different stress ellipsoids, the principal directions will vary from point to point. Hence no single coordinate system will exist that would enable the stress and strain components to be related by the simple law of (449). Let us therefore transform the components of the stress and strain tensors so that they may be referred to a single coordinate system. The reader should read Chap. 8 to understand what follows. If he desires not to break the continuity of the present paragraph, he may take formula (456) with a grain of salt, at least for the present. Example 8, Probs. 21 and 22 of Sec. 11, and Prob. 21 of Sec. 15 will aid the reader in what follows.

If x^1, x^2, x^3 are the coordinates above, and if we change to a new coordinate system $\bar{x}^1$, $\bar{x}^2$, $\bar{x}^3$ where

$$\bar{x}^i = \sum_{\alpha=1}^{3} a_\alpha{}^i x^\alpha, \qquad i = 1, 2, 3 \tag{450}$$

then the transformation (450) is said to be linear. Notice that the origin (0, 0, 0) remains invariant. If, furthermore, we desire distance to be preserved, we must have

$$\sum_{i=1}^{3} (\bar{x}^i)^2 = \sum_{i=1}^{3} (x^i)^2$$

In Chap. 8 we shall easily show that this requires

$$\sum_{\alpha=1}^{3} a_i^{\alpha} a_j^{\alpha} = \delta_{ij} \begin{matrix} = 0 \text{ if } i \neq j \\ = 1 \text{ if } i = j \end{matrix} \tag{451}$$

Equation (451) is the requirement that (450) be a rotation of axes. Moreover, since we are dealing with tensors, we shall see that the components of the strain tensor in the x^1-x^2-x^3 coordinate system are related to the components in the $\bar{x}^1$-$\bar{x}^2$-$\bar{x}^3$ system by the following rule:

$$\bar{e}_{ij} = \sum_{\beta=1}^{3} \sum_{\alpha=1}^{3} a_i^{\alpha} a_j^{\beta} e_{\alpha\beta} \tag{452}$$

If we now let $i = j$ and sum on i, we obtain

$$\begin{aligned} \sum_{i=1}^{3} \bar{e}_{ii} &= \sum_{i=1}^{3} \sum_{\beta=1}^{3} \sum_{\alpha=1}^{3} a_i^{\alpha} a_i^{\beta} e_{\alpha\beta} \\ &= \sum_{\beta=1}^{3} \sum_{\alpha=1}^{3} \delta^{\alpha\beta} e_{\alpha\beta} = \sum_{\alpha=1}^{3} e_{\alpha\alpha} \end{aligned}$$

so that

$$\bar{e}_{11} + \bar{e}_{22} + \bar{e}_{33} = e_{11} + e_{22} + e_{33} = e_1 + e_2 + e_3 \tag{453}$$

This is an invariant obtained from the strain tensor [see (443)]. A similar expression is obtained for the stress tensor; namely, that

$$\bar{t}_{11} + \bar{t}_{22} + \bar{t}_{33} = t_{11} + t_{22} + t_{33} = t_1 + t_2 + t_3$$

Equations (449) may now be written as

$$\begin{aligned} e_1 &= \frac{1+\sigma}{E} t_1 + \psi \\ e_2 &= \frac{1+\sigma}{E} t_2 + \psi \\ e_3 &= \frac{1+\sigma}{E} t_3 + \psi \end{aligned} \tag{454}$$

where ψ is the invariant

$$-\frac{\sigma}{E}(t_1 + t_2 + t_3) = -\frac{\sigma}{E}(\bar{t}_{11} + \bar{t}_{22} + \bar{t}_{33})$$

From Eq. (452) we have

$$\bar{e}_{ij} = \sum_{\beta=1}^{3} \sum_{\alpha=1}^{3} a_i^{\alpha} a_j^{\beta} e_{\alpha\beta}$$

and since $e_{\alpha\beta} = 0$ unless $\alpha = \beta$ [see (448)], we obtain

$$\begin{aligned} \bar{e}_{ij} &= \sum_{\alpha=1}^{3} a_i^{\alpha} a_j^{\alpha} e_{\alpha\alpha} = \sum_{\alpha=1}^{3} a_i^{\alpha} a_j^{\alpha} e_{\alpha} \\ &= \sum_{\alpha=1}^{3} a_i^{\alpha} a_j^{\alpha} \left(\frac{1+\sigma}{E} t_{\alpha} + \psi \right) \\ &= \frac{1+\sigma}{E} \sum_{\alpha=1}^{3} a_i^{\alpha} a_j^{\alpha} t_{\alpha} + \psi \sum_{\alpha=1}^{3} a_i^{\alpha} a_j^{\alpha} \end{aligned}$$

and

$$\bar{e}_{ij} = \frac{1+\sigma}{E} \bar{t}_{ij} + \psi \delta_{ij} \tag{455}$$

since $\bar{t}_{ij} = \sum_{\beta=1}^{3} \sum_{\alpha=1}^{3} a_i^{\alpha} a_j^{\beta} t_{\alpha\beta} = \sum_{\alpha=1}^{3} a_i^{\alpha} a_j^{\alpha} t_{\alpha}$.

Equation (455) is the relationship between the components of the strain and stress tensors when referred to a single coordinate system. We have

$$\begin{aligned} \bar{e}_{11} &= \frac{1+\sigma}{E} \bar{t}_{11} - \frac{\sigma}{E}(\bar{t}_{11} + \bar{t}_{22} + \bar{t}_{33}) \\ \bar{e}_{22} &= \frac{1+\sigma}{E} \bar{t}_{22} - \frac{\sigma}{E}(\bar{t}_{11} + \bar{t}_{22} + \bar{t}_{33}) \\ \bar{e}_{33} &= \frac{1+\sigma}{E} \bar{t}_{33} - \frac{\sigma}{E}(\bar{t}_{11} + \bar{t}_{22} + \bar{t}_{33}) \\ \bar{e}_{12} &= \frac{1+\sigma}{E} \bar{t}_{12} \\ \bar{e}_{23} &= \frac{1+\sigma}{E} \bar{t}_{23} \\ \bar{e}_{31} &= \frac{1+\sigma}{E} \bar{t}_{31} \end{aligned} \tag{456}$$

Solving Eqs. (456) for the $\bar{t}_{ij}$ and removing the bars, we obtain

$$
\begin{aligned}
t_{11} &= \frac{E}{1+\sigma}\left[e_{11} + \frac{\sigma}{1-2\sigma}(e_{11}+e_{22}+e_{33})\right] \\
&= \frac{E}{1+\sigma}\left[\frac{\partial u}{\partial x} + \frac{\sigma}{1-2\sigma}\left(\frac{\partial u}{\partial x} + \frac{\partial v}{\partial y} + \frac{\partial w}{\partial z}\right)\right] \\
t_{22} &= \frac{E}{1+\sigma}\left[\frac{\partial v}{\partial y} + \frac{\sigma}{1-2\sigma}\left(\frac{\partial u}{\partial x} + \frac{\partial v}{\partial y} + \frac{\partial w}{\partial z}\right)\right] \\
t_{33} &= \frac{E}{1+\sigma}\left(\frac{\partial w}{\partial z} + \frac{\sigma}{1-2\sigma}\nabla\cdot\mathbf{s}\right) \\
t_{12} &= t_{21} = \frac{E}{1+\sigma}e_{12} = \frac{E}{2(1+\sigma)}\left(\frac{\partial u}{\partial y} + \frac{\partial v}{\partial x}\right) \\
t_{23} &= t_{32} = \frac{E}{2(1+\sigma)}\left(\frac{\partial v}{\partial z} + \frac{\partial w}{\partial y}\right) \\
t_{31} &= t_{13} = \frac{E}{2(1+\sigma)}\left(\frac{\partial w}{\partial x} + \frac{\partial u}{\partial z}\right)
\end{aligned}
\tag{457}
$$

Equation (447) now becomes

$$
\begin{aligned}
F_x &= \left\{\frac{E}{1+\sigma}\left[\frac{\partial^2 u}{\partial x^2} + \frac{\sigma}{1-2\sigma}\left(\frac{\partial^2 u}{\partial x^2} + \frac{\partial^2 v}{\partial x\,\partial y} + \frac{\partial^2 w}{\partial x\,\partial z}\right)\right]\right. \\
&\qquad \left. + \frac{E}{2(1+\sigma)}\left(\frac{\partial^2 u}{\partial y^2} + \frac{\partial^2 v}{\partial y\,\partial x} + \frac{\partial^2 w}{\partial z\,\partial x} + \frac{\partial^2 u}{\partial z^2}\right)\right\} \\
&= \frac{E}{2(1+\sigma)}\left[\nabla^2 u + \frac{1}{1-2\sigma}\frac{\partial}{\partial x}\left(\frac{\partial u}{\partial x} + \frac{\partial v}{\partial y} + \frac{\partial w}{\partial z}\right)\right] \\
&= \frac{E}{2(1+\sigma)}\left[\nabla^2 u + \frac{1}{1-2\sigma}\frac{\partial}{\partial x}(\nabla\cdot\mathbf{s})\right]
\end{aligned}
$$

The forces per unit volume in the y and z directions are

$$
\begin{aligned}
F_y &= \frac{E}{2(1+\sigma)}\left[\nabla^2 v + \frac{1}{1-2\sigma}\frac{\partial}{\partial y}(\nabla\cdot\mathbf{s})\right] \\
F_z &= \frac{E}{2(1+\sigma)}\left[\nabla^2 w + \frac{1}{1-2\sigma}\frac{\partial}{\partial z}(\nabla\cdot\mathbf{s})\right]
\end{aligned}
$$

so that

$$
\mathbf{f} = \frac{E}{2(1+\sigma)}\left[\nabla^2\mathbf{s} + \frac{1}{1-2\sigma}\nabla(\nabla\cdot\mathbf{s})\right] \tag{458}
$$

If we let $\mathbf{R} = R_x\mathbf{i} + R_y\mathbf{j} + \mathbf{R}_z\mathbf{k}$ be the external body force per unit volume, ρ the density of the medium, then Newton's second law of motion yields

$$\mathbf{R} + \frac{E}{2(1+\sigma)}\left[\nabla^2\mathbf{s} + \frac{1}{1-2\sigma}\nabla(\nabla\cdot\mathbf{s})\right] = \rho\,\frac{\partial^2\mathbf{s}}{\partial t^2} \tag{459}$$

For the case $\mathbf{R} = 0$, (459) reduces to

$$\frac{E}{2(1+\sigma)}\left[\nabla^2\mathbf{s} + \frac{1}{1-2\sigma}\nabla(\nabla\cdot\mathbf{s})\right] = \rho\,\frac{\partial^2\mathbf{s}}{\partial t^2} \tag{460}$$

In Sec. 70 we saw that a vector could be written as the sum of a solenoidal and an irrotational vector. Let $\mathbf{s} = \mathbf{s}_1 + \mathbf{s}_2$, where $\nabla\cdot\mathbf{s}_1 = 0$ and $\nabla\times\mathbf{s}_2 = 0$. Since (460) is linear in $\mathbf{s}$, we can consider it as satisfied by $\mathbf{s}_1$ and $\mathbf{s}_2$. This yields

$$\frac{E}{2(1+\sigma)}\nabla^2\mathbf{s}_1 = \rho\,\frac{\partial^2\mathbf{s}_1}{\partial t^2}$$

and

$$\frac{E}{2(1+\sigma)}\left[\nabla^2\mathbf{s}_2 + \frac{1}{1-2\sigma}\nabla(\nabla\cdot\mathbf{s}_2)\right] = \rho\,\frac{\partial^2\mathbf{s}_2}{\partial t^2} \tag{461}$$

However,

$$\nabla\times(\nabla\times\mathbf{s}_2) = \nabla(\nabla\cdot\mathbf{s}_2) - \nabla^2\mathbf{s}_2 = 0$$

so that

$$\frac{E(1-\sigma)}{(1+\sigma)(1-2\sigma)}\nabla^2\mathbf{s}_2 = \rho\,\frac{\partial^2\mathbf{s}_2}{\partial t^2} \tag{462}$$

In Sec. 80, we saw that (461) leads to a transverse wave moving with speed $V_t = \sqrt{E/2(1+\sigma)\rho}$.

Equation (462) is also a wave equation, but the wave is not transverse. Let us assume that the wave is traveling along the x axis. Then

$$\begin{aligned}
\mathbf{s}_2 &= \mathbf{s}_2(x - Vt) \\
&= u(x - Vt)\mathbf{i} + v(x - Vt)\mathbf{j} + w(x - Vt)\mathbf{k} \\
\nabla\times\mathbf{s}_2 &= \begin{vmatrix} \mathbf{i} & \mathbf{j} & \mathbf{k} \\ \dfrac{\partial}{\partial x} & \dfrac{\partial}{\partial y} & \dfrac{\partial}{\partial z} \\ u & v & w \end{vmatrix} = 0 \\
&= -\frac{\partial w}{\partial x}\mathbf{j} + \frac{\partial v}{\partial x}\mathbf{k} = 0
\end{aligned}$$

so that w and v are independent of x and therefore are independent of $x - Vt$. We are not interested in constant displacements, so that $\mathbf{s}_2 = u(x - Vt)\mathbf{i}$, and the displacement of $\mathbf{s}_2$ is parallel to the direction of propagation of the wave. The wave is therefore longitudinal. The speed of the wave is

$$V_l = \sqrt{\frac{E(1 - \sigma)}{\rho(1 + \sigma)(1 - 2\sigma)}}.$$

In general, both types of waves are produced, this result being useful in the study of earthquakes.

Problems

1. Derive (451).

2. If f^1, f^2, f^3 are the components of a vector for a cartesian coordinate system, prove that the components $\bar{f}^1, \bar{f}^2, \bar{f}^3$ of this vector in a new cartesian coordinate system are related to the old components by the rule $\bar{f}^i = \sum_{\alpha=1}^{3} a_\alpha{}^i f^\alpha$, $i = 1, 2, 3$, using the coordinate transformation (450).

3. If the body forces are negligible and if the medium is in a state of equilibrium, show that $\nabla^2\mathbf{s} + \dfrac{1}{1 - 2\sigma}\nabla(\nabla \cdot \mathbf{s}) = 0$.

4. If the strain of Prob. 3 is radial, that is, if $\mathbf{s} = s(r)\mathbf{r}$, find the differential equation satisfied by $s(r)$.

5. Assuming $\sigma = 0$ for a long thin bar, find the velocity of propagation of the longitudinal waves.

6. If $\mu = E/2(1 + \sigma)$ (modulus of rigidity) and

$$\lambda = \frac{E\sigma}{(1 + \sigma)(1 - 2\sigma)}$$

show that Eq. (459) becomes

$$\mathbf{R} + \mu\,\nabla^2\mathbf{s} + (\lambda + \mu)\nabla(\nabla \cdot \mathbf{s}) = \rho\frac{\partial^2\mathbf{s}}{\partial t^2}$$

7. Why do we use $\dfrac{\partial^2\mathbf{s}}{\partial t^2}$ instead of $\dfrac{d^2\mathbf{s}}{dt^2}$ in (459)?

8. A coaxial cable is made by filling the space between a solid core of radius a and a concentric cylindrical shell of internal radius b with rubber. If the core is displaced a small distance

axially, find the displacement in the rubber. Assume that end effects, gravity, and the distortion of the metal can be neglected.

118. Navier-Stokes Equation. We are now in a position to derive the equations of motion of a viscous fluid. In the case of nonviscous fluids, we assumed no friction between adjacent layers of fluid. As a result of friction (viscosity), rapidly moving layers tend to drag along the slower layers of fluid, and, conversely, the slower layers tend to retard the motion of the faster layers. It is found by experiment that the force of viscosity is directly proportional to the common area A of the two layers and to the gradient of the velocity normal to the flow. If the fluid is moving in the x-y plane with speed v, then the viscous force is

$$F = \eta A \frac{\partial v}{\partial z}$$

since $\frac{\partial v}{\partial z}$ is the gradient of the speed normal to the direction of flow. η is called the coefficient of viscosity.

We shall let P_{ij} be the stress tensor and σ_{ij} the strain tensor for the fluid analogous to t_{ij} and e_{ij} of the previous paragraphs. We have

$$P_{12} = \frac{E}{2(1+\sigma)}\left(\frac{\partial u}{\partial y} + \frac{\partial v}{\partial x}\right)$$

where u, v, w are the components of the velocity vector $\mathbf{v}$ (see Sec. 112). For a fluid moving in the y direction with a gradient in the x direction, we have $u = 0$ and $\frac{\partial u}{\partial y} = 0$, so that

$$P_{12} = \frac{E}{2(1+\sigma)} \frac{\partial v}{\partial x} = \eta \frac{\partial v}{\partial x}$$

Hence the term $\frac{E}{2(1+\sigma)}$ must be replaced by η.

In addition to the stress components due to viscosity, we must add the stress components due to the pressure field, which we assume to be

$$\begin{Vmatrix} -p & 0 & 0 \\ 0 & -p & 0 \\ 0 & 0 & -p \end{Vmatrix}$$

The equations of (457) become

$$P_{ij} = 2\eta\sigma_{ij} + \lambda(\sigma_{11} + \sigma_{22} + \sigma_{33})\delta_{ij} - p\delta_{ij} \tag{463}$$

where λ is undetermined as yet. Now let $i = j$ and sum on j. We see that

$$P_{11} + P_{22} + P_{33} = (2\eta + 3\lambda)(\sigma_{11} + \sigma_{22} + \sigma_{33}) - 3p$$

We know that $P_{11} + P_{22} + P_{33}$ is an invariant and that for the static case $P_{11} + P_{22} + P_{33} = -3p$. Consequently we choose $2\eta + 3\lambda = 0$, so that (463) becomes

$$p_{ij} \equiv P_{ij} = 2\eta\sigma_{ij} - \tfrac{2}{3}\eta(\sigma_{11} + \sigma_{22} + \sigma_{33}) - p\delta_{ij} \tag{464}$$

for small velocities. Moreover, the velocity vector is given by

$$\mathbf{v} = u_1\mathbf{i} + u_2\mathbf{j} + u_3\mathbf{k}$$

and $\sigma_{ij} = \dfrac{1}{2}\left(\dfrac{\partial u_i}{\partial x^j} + \dfrac{\partial u_j}{\partial x^i}\right)$, so that div $\mathbf{v} = \nabla \cdot \mathbf{v} = \sigma_{11} + \sigma_{22} + \sigma_{33}$.

To obtain the equations of motion, we note that from (447)

$$f_1 = \frac{\partial p_{11}}{\partial x^1} + \frac{\partial p_{12}}{\partial x^2} + \frac{\partial p_{13}}{\partial x^3} \qquad \text{where } \begin{aligned} x^1 &= x \\ x^2 &= y \\ x^3 &= z \end{aligned}$$

$$= \sum_{j=1}^{3} \frac{\partial p_{1j}}{\partial x^j}$$

and in general $f_i = \displaystyle\sum_{j=1}^{3} \frac{\partial p_{ij}}{\partial x^j}$. Hence

$$\rho F_i + f_i = \rho \frac{du_i}{dt}$$

becomes

$$\rho F_i + \sum_{j=1}^{3} \frac{\partial p_{ij}}{\partial x^j} = \rho \frac{du_i}{dt} \tag{465}$$

where F_i is the external force per unit mass. From (464)

$$\begin{aligned} \frac{\partial p_{ij}}{\partial x^j} &= 2\eta \frac{\partial \sigma_{ij}}{\partial x^j} - \frac{2}{3}\eta \frac{\partial(\text{div } \mathbf{v})}{\partial x^j}\delta_{ij} - \frac{\partial p}{\partial x^j}\delta_{ij} \\ &= \eta \frac{\partial}{\partial x^j}\left(\frac{\partial u_i}{\partial x^j} + \frac{\partial u_j}{\partial x^i}\right) - \frac{2}{3}\eta \frac{\partial(\nabla \cdot \mathbf{v})}{\partial x^j}\delta_{ij} - \frac{\partial p}{\partial x^j}\delta_{ij} \end{aligned}$$

and

$$\sum_{j=1}^{3} \frac{\partial p_{ij}}{\partial x^j} = \eta \sum_{j=1}^{3} \frac{\partial}{\partial x^j}\left(\frac{\partial u_i}{\partial x^j} + \frac{\partial u_j}{\partial x^i}\right) - \frac{2}{3}\eta \frac{\partial(\nabla \cdot \mathbf{v})}{\partial x^i} - \frac{\partial p}{\partial x^i}$$

The equations of motion (465) are

$$\rho \frac{du_i}{dt} = \rho F_i + \eta \sum_{j=1}^{3} \frac{\partial}{\partial x^j}\left(\frac{\partial u_i}{\partial x^j} + \frac{\partial u_j}{\partial x^i}\right) - \frac{2}{3}\eta \frac{\partial(\nabla \cdot \mathbf{v})}{\partial x^i} - \frac{\partial p}{\partial x^i} \tag{466}$$

or

$$\rho \frac{d\mathbf{v}}{dt} = \rho\mathbf{f} + \eta \nabla^2\mathbf{v} - \frac{1}{3}\nabla(\nabla \cdot \mathbf{v}) - \nabla p \tag{467}$$

For an incompressible fluid $\nabla \cdot \mathbf{v} = 0$, and

$$\rho \frac{d\mathbf{v}}{dt} = \rho\mathbf{f} + \eta \nabla^2\mathbf{v} - \nabla p$$

Along with (467) we have the equation of continuity

$$\frac{\partial \rho}{\partial t} + \nabla \cdot (\rho\mathbf{v}) = 0$$

Problems

1. Derive (467) from (466).

2. Consider the steady flow of an incompressible fluid through a small cylindrical tube of radius a in a nonexternal field. Let $\mathbf{v} = v\mathbf{k}$ and show that $p = p(z)$ and $\eta \nabla^2 v = \frac{\partial p}{\partial z}$. Show that the boundary conditions are $v = 0$ when $r = a$, and $v = v(r)$, $r^2 = x^2 + y^2$, and that $\frac{dp}{dz} = \frac{\eta}{r}\frac{d}{dr}\left(r\frac{dv}{dr}\right)$. Hence show that $v = (A/4\eta)(r^2 - a^2)$, where A is a constant and $\frac{dp}{dz} = A$.

3. Consider a sphere moving with constant velocity $v_0\mathbf{k}$ (along the z axis) in an infinite mass of incompressible fluid. Choose the center of the sphere as the origin of our coordinate system. Show that the equation of motion is $\rho \frac{d\mathbf{v}}{dt} = \eta \nabla^2\mathbf{v} - \nabla p$ and that the boundary conditions are $\mathbf{v} = 0$ for $r = a$, $\mathbf{v} = -v_0\mathbf{k}$ at $r = \infty$,

and that for steady motion $\dfrac{\partial \psi}{\partial t} = 0$ for any quantity ψ associated with the motion. Moreover $\nabla \cdot \mathbf{v} = 0$. We shall assume that $\mathbf{v}$ and the partial derivatives of $\mathbf{v}$ are small. Show that this implies

$$\text{(i)} \qquad \nabla p = \eta \, \nabla^2 \mathbf{v}$$

Hence prove that $\nabla^2 p = 0$. Now let $\mathbf{v} = -\nabla \varphi + \omega_1 \mathbf{k}$ and show that

$$\text{(ii)} \qquad \frac{\partial \omega_1}{\partial z} = \nabla^2 \varphi$$

Assuming $p = -\eta \, \nabla^2 \varphi$ and $\nabla^2 \omega_1 = 0$, show that (i) and (ii) will be satisfied.

Let $\omega_1 = 3v_0 a/2r$, $\varphi = c_0 z - (v_0 a^3 z/4r^3) + (3v_0 az/4r)$,

$$p = \frac{3\eta v_0 az}{2r^3}$$

and show that

$$\nabla^2 \omega_1 = 0, \qquad \nabla^2 p = 0, \qquad p = -\eta \, \nabla^2 \varphi$$
$$\mathbf{v} = -\nabla \varphi + \omega_1 \mathbf{k} = 0 \text{ for } r = a$$
$$\mathbf{v} = -v_0 \mathbf{k} \text{ for } r = \infty$$

4. Solve for the steady motion of an incompressible viscous fluid between two parallel plates, one of the plates fixed, the other moving at a constant velocity, the distance between the plates remaining constant.

5. Find the steady motion of an incompressible, viscous fluid surrounding a sphere rotating about a diameter with constant angular velocity. No external forces exist.

CHAPTER 8

TENSOR ANALYSIS AND RIEMANNIAN GEOMETRY

119. Summation Notation. We shall be interested in sums of the type

$$S = a_1x_1 + a_2x_2 + \cdots + a_nx_n \tag{468}$$

We can shorten the writing of (468) and write

$$S = \sum_{i=1}^{n} a_ix_i \tag{469}$$

Now it will be much more convenient to replace the subscripts of the quantities $x_1, x_2, \ldots, x_n$ by superscripts, $x^1, x^2, \ldots, x^n$. The superscripts do not stand for powers but are labels that allow us to distinguish between the various x's. Our sum S now becomes

$$S = \sum_{i=1}^{n} a_ix^i \tag{470}$$

We can get rid of the summation sign and write

$$\boxed{S = a_ix^i} \tag{471}$$

where the repeated index i is to be summed from 1 to n. This notation is due to Einstein.

Whenever a letter appears once as a subscript and once as a superscript, we shall mean that a summation is to occur on this letter. If we are dealing with n dimensions, we shall sum from 1 to n. The index of summation is a dummy index since the final result is independent of the letter used. We can write $S = a_ix^i = a_jx^j = a_\alpha x^\alpha = a_\beta x^\beta$.

Example 111. If $f = f(x^1, x^2, \ldots, x^n)$, we have from the calculus

$$df = \frac{\partial f}{\partial x^1}\,dx^1 + \frac{\partial f}{\partial x^2}\,dx^2 + \cdots + \frac{\partial f}{\partial x^n}\,dx^n$$
$$= \sum_{i=1}^{n} \frac{\partial f}{\partial x^i}\,dx^i$$
$$= \frac{\partial f}{\partial x^\alpha}\,dx^\alpha$$

and

$$\frac{df}{dt} = \frac{\partial f}{\partial x^\alpha}\frac{dx^\alpha}{dt}$$

Example 112. Let $S = g_{\alpha\beta}x^\alpha x^\beta$. The index α occurs both as a subscript and superscript. Hence we first sum on α, say from 1 to 3. This yields

$$S = g_{1\beta}x^1x^\beta + g_{2\beta}x^2x^\beta + g_{3\beta}x^3x^\beta$$

Now each term of S has the repeated index β summed, say, from 1 to 3. Hence

$$S = g_{11}x^1x^1 + g_{12}x^1x^2 + g_{13}x^1x^3 + g_{21}x^2x^1 + g_{22}x^2x^2 + g_{23}x^2x^3 + g_{31}x^3x^1 + g_{32}x^3x^2 + g_{33}x^3x^3$$

and $S = g_{\alpha\beta}x^\alpha x^\beta$ represents the double sum

$$S = \sum_{\beta=1}^{3}\sum_{\alpha=1}^{3} g_{\alpha\beta}x^\alpha x^\beta$$

We also notice that the $g_{\alpha\beta}$ can be thought of as elements of a square matrix

$$\begin{Vmatrix} g_{11} & g_{12} & g_{13} \\ g_{21} & g_{22} & g_{23} \\ g_{31} & g_{32} & g_{33} \end{Vmatrix}$$

120. The Kronecker Deltas. We define the Kronecker δ^i_j to be equal to zero if $i \neq j$ and to equal one if $i = j$:

$$\delta^i_j = \begin{matrix} 0, & i \neq j \\ 1, & i = j \end{matrix} \qquad (472)$$

We notice that $\delta^1_1 = \delta^2_2 = \cdots = \delta^n_n = 1$,

$$\delta^1_2 = \delta^1_3 = \cdots = \delta^r_{r+1} = 0$$
$$\delta^\alpha_\alpha = \delta^1_1 + \delta^2_2 + \cdots + \delta^n_n = n$$

If $x^1, x^2, \ldots, x^n$ are n independent variables, then $\dfrac{\partial x^i}{\partial x^j} = \delta^i_j$, for if $i = j$, $\dfrac{\partial x^i}{\partial x^j} = 1$, and if $i \neq j$, there is no change in the variable x^i if we change x^j since they are independent variables, so that $\dfrac{\partial x^i}{\partial x^j} = 0$.

Example 113. Let $S = a_\alpha x^\alpha$. Then $\dfrac{\partial S}{\partial x^\mu} = a_\alpha \dfrac{\partial x^\alpha}{\partial x^\mu}$, and

$$\frac{\partial S}{\partial x^\mu} = a_\alpha \delta^\alpha_\mu$$

Now $\delta^\alpha_\mu = 0$ except when $\alpha = \mu$, so that on summing on α we obtain $\dfrac{\partial(a_\alpha x^\alpha)}{\partial x^\mu} = a_\mu$.

Example 114. Let $S = a_{\alpha\beta}x^\alpha x^\beta \equiv 0$ for all values of the variables $x^1, x^2, \ldots, x^n$. We show that $a_{ij} + a_{ji} = 0$. First differentiate S with respect to x^i and obtain

$$\begin{aligned}\frac{\partial S}{\partial x^i} &= a_{\alpha\beta}x^\alpha \frac{\partial x^\beta}{\partial x^i} + a_{\alpha\beta} \frac{\partial x^\alpha}{\partial x^i} x^\beta = 0 \\ &= a_{\alpha\beta}x^\alpha \delta^\beta_i + a_{\alpha\beta}\delta^\alpha_i x^\beta = 0 \\ &= a_{\alpha i}x^\alpha + a_{i\beta}x^\beta = 0\end{aligned}$$

Now differentiate with respect to x^j so that

$$\frac{\partial^2 S}{\partial x^j\, \partial x^i} = a_{\alpha i}\delta^\alpha_j + a_{i\beta}\delta^\beta_j = 0$$

and

$$a_{ji} + a_{ij} \equiv 0$$

We define the generalized Kronecker delta $\delta^{i_1 i_2 \cdots i_m}_{j_1 j_2 \cdots j_m}$ as follows: The superscripts and subscripts can have any value from 1 to n. If at least two superscripts or at least two subscripts have the same value, or if the subscripts are not the same set of numbers as the superscripts, then we define the generalized Kronecker delta to be zero. If all the superscripts and subscripts are separately distinct, and the subscripts are the same set of numbers as the superscripts, the delta has the value of $+1$, or -1, according to whether it requires an even or odd number of permutations to arrange the superscripts in the same order as the subscripts. For example, $\delta^{123}_{123} = 1$, $\delta^{123}_{213} = -1$, $\delta^{123}_{231} = 1$, $\delta^{173}_{163} = 0$,

$$\delta^{1348}_{1438} = -1,\ \delta^{122}_{123} = \delta^{123}_{221} = \delta^{312}_{323} = 0,\ \delta^{123}_{iij} = 0$$

It is convenient to define

$$\left.\begin{aligned} \epsilon^{i_1 i_2 \cdots i_n} &= \delta^{i_1 i_2 \cdots i_n}_{12 \cdots n} \\ \text{and} \qquad \epsilon_{i_1 i_2 \cdots i_n} &= \delta^{12 \cdots n}_{i_1 i_2 \cdots i_n} \end{aligned}\right\} \tag{473}$$

Problems

1. Write in full $a^i_\alpha x^\alpha = b^i$, $\alpha, i = 1, 2, 3$.
2. If $a_{\alpha\beta\gamma}x^\alpha x^\beta x^\gamma \equiv 0$, show that $a_{ijk} + a_{kij} + a_{jki} = 0$.
3. Show that $\delta^{i_1 i_2 \cdots i_n}_{12 \cdots n} \delta^{12 \cdots n}_{j_1 j_2 \cdots j_n} = \delta^{i_1 i_2 \cdots i_n}_{j_1 j_2 \cdots j_n}$.
4. If $y^i = a^i_\alpha x^\alpha$, $z^i = b^i_\alpha y^\alpha$, show that $z^i = b^i_\alpha a^\alpha_\beta x^\beta$.
5. Prove that

$$\delta^{rs}_{\alpha\beta} a^{\alpha\beta} = a^{rs} - a^{sr}$$
$$\delta^{rst}_{\alpha\beta\gamma} a^{\alpha\beta\gamma} = a^{rst} + a^{trs} + a^{str} - a^{srt} - a^{tsr} - a^{rts}$$

6. Show that the determinant $\begin{vmatrix} a^1_1 & a^1_2 \\ a^2_1 & a^2_2 \end{vmatrix}$ can be written

$$\begin{vmatrix} a^1_1 & a^1_2 \\ a^2_1 & a^2_2 \end{vmatrix} = \epsilon^{ij} a^1_i a^2_j = \epsilon_{ij} a^i_1 a^j_2$$

and that

$$\begin{vmatrix} a^1_1 & a^1_2 & a^1_3 \\ a^2_1 & a^2_2 & a^2_3 \\ a^3_1 & a^3_2 & a^3_3 \end{vmatrix} = \epsilon^{ijk} a^1_i a^2_j a^3_k = \epsilon_{ijk} a^i_1 a^j_2 a^k_3$$

7. Prove that $\epsilon_{i_1 i_2 \cdots i_n} = \epsilon^{i_1 i_2 \cdots i_n}$.
8. Show that $\delta^{\alpha\beta}_{ij} \dfrac{\partial \varphi_\alpha}{\partial x^\beta} = \dfrac{\partial \varphi_i}{\partial x^j} - \dfrac{\partial \varphi_j}{\partial x^i}$.
9. If $y^i = y^i(x^1, x^2, \ldots, x^n)$, $i = 1, 2, \ldots, n$, show that $\delta^i_j = \dfrac{\partial y^i}{\partial x^\alpha}\dfrac{\partial x^\alpha}{\partial y^j}$ assuming the existence of the derivatives. Also show that $\dfrac{\partial y^i}{\partial x^\alpha}\dfrac{\partial x^\beta}{\partial y^j}\delta^\alpha_\beta = \delta^i_j$. Show that

$$\frac{\partial^2 y^i}{\partial x^\alpha\, \partial x^\beta}\frac{\partial x^\alpha}{\partial y^j}\frac{\partial x^\beta}{\partial y^k} + \frac{\partial y^i}{\partial x^\alpha}\frac{\partial^2 x^\alpha}{\partial y^j\, \partial y^k} = 0$$

10. If $\varphi = \varphi(x^1, x^2, \ldots, x^n)$,

$$x^i = x^i(y^1, y^2, \ldots, y^n) = x^i(y)$$

$i = 1, 2, \ldots, n$, and if

$$\bar{\varphi} = \bar{\varphi}(y^1, y^2, \ldots, y^n) \equiv \varphi[x^1(y), x^2(y), \ldots, x^n(y)]$$

show that $\dfrac{\partial \bar{\varphi}}{\partial y^\alpha} = \dfrac{\partial \varphi}{\partial x^\beta} \dfrac{\partial x^\beta}{\partial y^\alpha}$. Given $\varphi_\alpha \equiv \dfrac{\partial \varphi}{\partial x^\alpha}$, show that

$$\frac{\partial \bar{\varphi}_i}{\partial y^j} - \frac{\partial \bar{\varphi}_j}{\partial y^i} = \left(\frac{\partial \varphi_\alpha}{\partial x^\beta} - \frac{\partial \varphi_\beta}{\partial x^\alpha}\right) \frac{\partial x^\alpha}{\partial y^i} \frac{\partial x^\beta}{\partial y^j}$$

121. Determinants. We define the determinant $|a^i_j|$ by the equation

$$|a^i_j| = \begin{vmatrix} a^1_1 & a^1_2 & \cdots & a^1_n \\ a^2_1 & a^2_2 & \cdots & a^2_n \\ \cdots & \cdots & \cdots & \cdots \\ \cdots & \cdots & \cdots & \cdots \\ \cdots & \cdots & \cdots & \cdots \\ a^n_1 & a^n_2 & \cdots & a^n_n \end{vmatrix} \equiv \epsilon_{i_1 i_2 \ldots i_n} a^{i_1}_1 a^{i_2}_2 \cdots a^{i_n}_n$$

$$|a^i_j| \equiv \epsilon^{i_1 i_2 \cdots i_n} a^1_{i_1} a^2_{i_2} \cdots a^n_{i_n} \tag{474}$$

The reader should note that this definition agrees with the definition for the special case of second- and third-order determinants which he has encountered in elementary algebra. The definition of a determinant as given by (474) shows that it consists of a sum of terms, n^n in number. Of these, $n!$ are, in general, different from zero. Each term consists of a product of elements, one element from each row and column. The sign of the term $a^1_{i_1} a^2_{i_2} a^3_{i_3} \cdots a^n_{i_n}$ depends on whether it takes an even or odd permutation to regroup $i_1 i_2 \cdots i_n$ into $12 \cdots n$.

Since i_1 and i_2 are dummy indices, we can interchange them so that

$$|a^i_j| = \epsilon_{i_1 i_2 \ldots i_n} a^{i_1}_1 a^{i_2}_2 \cdots a^{i_n}_n \equiv \epsilon_{i_2 i_1 \ldots i_n} a^{i_2}_1 a^{i_1}_2 \cdots a^{i_n}_n$$

An interchange of the subscripts 1 and 2, however, will mean that an extra permutation will be needed on $i_2 i_1 i_3 \cdots i_n$. This changes the sign of the determinant. Hence interchanging two columns (or rows) changes the sign of the determinant. As an immediate corollary, we have $|a_j^i| = 0$ if two rows (or columns) are identical.

Let us now examine the sum

$$b^{j_1 j_2 \cdots j_n} = \epsilon^{i_1 i_2 \cdots i_n} a_{i_1}^{j_1} a_{i_2}^{j_2} \cdots a_{i_n}^{j_n} \tag{475}$$

If $j_1, j_2, \ldots, j_n$ take on the values $1, 2, \ldots, n$, respectively, we know that (475) reduces to (474). If the $j_1, j_2, \ldots, j_n$ take on the values $1, 2, \ldots, n$, but not respectively, then we have interchanged the rows. An even permutation reduces (475) to $|a_j^i|$, and an odd permutation of the j's reduces (475) to $-|a_j^i|$. If two of the j's have the same value, (475) is zero, since if two rows of a determinant are alike it has zero value. Hence

$$\epsilon^{i_1 i_2 \cdots i_n} a_{i_1}^{j_1} a_{i_2}^{j_2} \cdots a_{i_n}^{j_n} = |a_j^i| \epsilon^{j_1 j_2 \cdots j_n} \tag{475a}$$

and

$$\epsilon_{i_1 i_2 \cdots i_n} a_{j_1}^{i_1} a_{j_2}^{i_2} \cdots a_{j_n}^{i_n} = |a_j^i| \epsilon_{j_1 j_2 \cdots j_n} \tag{475b}$$

Example 115. We now derive the law of multiplication for two determinants of the same order. We have

$$\begin{aligned}
|a_j^i|\,|b_j^i| &= |a_j^i| \epsilon_{i_1 i_2 \cdots i_n} b_1^{i_1} b_2^{i_2} \cdots b_n^{i_n} \\
&= \epsilon_{j_1 j_2 \cdots j_n} a_{i_1}^{j_1} a_{i_2}^{j_2} \cdots a_{i_n}^{j_n} b_1^{i_1} b_2^{i_2} \cdots b_n^{i_n} \\
&= \epsilon_{j_1 j_2 \cdots j_n} (a_{i_1}^{j_1} b_1^{i_1})(a_{i_2}^{j_2} b_2^{i_2}) \cdots (a_{i_n}^{j_n} b_n^{i_n}) \\
&= \epsilon_{j_1 j_2 \cdots j_n} c_1^{j_1} c_2^{j_2} \cdots c_n^{j_n} = |c_j^i|
\end{aligned}$$

where

$$\boxed{c_j^i = a_\alpha^i b_j^\alpha} \tag{476}$$

Example 116. We now derive an expansion of a determinant in terms of the cofactors of the elements. We have

$$\begin{aligned}
|a_j^i| &= \epsilon_{i_1 i_2 \cdots i_n} a_1^{i_1} a_2^{i_2} \cdots a_n^{i_n} \\
&= a_1^{i_1} (\epsilon_{i_1 i_2 \cdots i_n} a_2^{i_2} \cdots a_n^{i_n}) \\
&= a_1^\alpha A_\alpha^1
\end{aligned}$$

where $A_\alpha^1 = \epsilon_{\alpha i_2 \cdots i_n} a_2^{i_2} \cdots a_n^{i_n}$. A_α^1 is called the cofactor of a_1^α. In general,

$$\begin{aligned}
|a_j^i| &= a_\beta^\alpha (\epsilon_{i_1 i_2 \cdots i_n} a_1^{i_1} \cdots a_{\beta-1}^{i_{\beta-1}} \delta_\alpha^{i_\beta} a_{\beta+1}^{i_{\beta+1}} \cdots a_n^{i_n}) \\
&= a_\beta^\alpha A_\alpha^\beta \qquad (\beta \text{ not summed})
\end{aligned}$$

Hence

$$\boxed{a_j^\alpha A_\alpha^i = |a|\,\delta_j^i} \tag{477}$$

Also

$$a^i_\alpha A^\alpha_j = |a|\,\delta^i_j \tag{477a}$$

Example 117. Let us consider the n linear equations

$$y^i = a^i_\alpha x^\alpha, \quad \alpha, i = 1, 2, \ldots, n \tag{478}$$

Multiplying by A^β_i, we obtain

$$A^\beta_i y^i = a^i_\alpha A^\beta_i x^\alpha$$

so that summing on i, we have from (477)

$$A^\beta_i y^i = |a|\,\delta^\beta_\alpha x^\alpha = |a|\,x^\beta$$

If $|a| \neq 0$, then

$$x^\beta = \frac{A^\beta_i y^i}{|a|} = \frac{y^i(\text{cofactor of } a^i_\beta \text{ in } |a|)}{|a|} \tag{479}$$

Example 118. Let $y^i = y^i(x^1, x^2, \ldots, x^n)$, $i = 1, 2, \ldots, n$. In the calculus it is shown that if $\left|\dfrac{\partial y^i}{\partial x^j}\right| \neq 0$ at a point P and if the partial derivatives are continuous, we can solve for the x^i in terms of the y's, that is, $x^i = x^i(y^1, y^2, \ldots, y^n)$ in a neighborhood of P.

Now we have identically

$$\delta^i_j = \frac{\partial y^i}{\partial y^j} \equiv \frac{\partial y^i}{\partial x^\alpha}\frac{\partial x^\alpha}{\partial y^j}$$

Forming the determinant of both sides,

$$|\delta^i_j| = \left|\frac{\partial y^i}{\partial x^\alpha}\frac{\partial x^\alpha}{\partial y^j}\right| = \left|\frac{\partial y}{\partial x}\right|\left|\frac{\partial x}{\partial y}\right|$$

from (476). Hence

$$1 \equiv \left|\frac{\partial y}{\partial x}\right|\left|\frac{\partial x}{\partial y}\right|$$

The determinant $\left|\dfrac{\partial y^i}{\partial x^j}\right|$ is called the Jacobian of the y's with respect to the x's. We have shown that

$$J\left(\frac{y^1, y^2, \ldots, y^n}{x^1, x^2, \ldots, x^n}\right) J\left(\frac{x^1, x^2, \ldots, x^n}{y^1, y^2, \ldots, y^n}\right) = 1 \tag{480}$$

Example 119. If the elements of the determinant $|a|$ are functions of the variables $x^1, x^2, \ldots, x^n$, we leave it to the student to prove that

$$\boxed{\frac{\partial |a|}{\partial x^\mu} = A^\alpha_\beta \frac{\partial a^\beta_\alpha}{\partial x^\mu}} \tag{481}$$

As a special case, suppose $a^i_j = \dfrac{\partial y^i}{\partial x^j}$, where $y^i = y^i(x^1, x^2, \ldots, x^n)$.

Now $\delta^i_j = \dfrac{\partial y^i}{\partial x^\alpha} \cdot \dfrac{\partial x^\alpha}{\partial y^j}$. Let us consider j to be fixed for the moment.

Let $Y^i = \delta^i_j$, $\dfrac{\partial y^i}{\partial x^j} = a^i_j$, $X^i = \dfrac{\partial x^i}{\partial y^j}$, so that $Y^i = a^i_\alpha X^\alpha$. If

$$|a| = \left|\frac{\partial y^i}{\partial x^j}\right| \neq 0$$

then

$$X^\beta = \frac{Y^\alpha(\text{cofactor of } a^\alpha_\beta \text{ in } |a|)}{|a|}$$

from (479), or

$$\frac{\partial x^\beta}{\partial y^j} = \frac{\delta^\alpha_j \left(\text{cofactor of } \dfrac{\partial y^\alpha}{\partial x^\beta} \text{ in } \left|\dfrac{\partial y}{\partial x}\right|\right)}{\dfrac{\partial y}{\partial x}}$$

$$= \frac{\left(\text{cofactor of } \dfrac{\partial y^j}{\partial x^\beta} \text{ in } \left|\dfrac{\partial y}{\partial x}\right|\right)}{\left|\dfrac{\partial y}{\partial x}\right|}$$

and so

$$\left(\text{cofactor of } \frac{\partial y^j}{\partial x^\beta} \text{ in } \left|\frac{\partial y}{\partial x}\right|\right) \equiv \left|\frac{\partial y}{\partial x}\right| \frac{\partial x^\beta}{\partial y^j}$$

Applying this to (481), we have

or

$$\left.\begin{aligned}\frac{\partial \left|\frac{\partial y}{\partial x}\right|}{\partial x^\mu} &= \left|\frac{\partial y}{\partial x}\right| \frac{\partial x^\alpha}{\partial y^\beta} \frac{\partial^2 y^\beta}{\partial x^\mu \, \partial x^\alpha} \\ \frac{\partial \log \left|\frac{\partial y}{\partial x}\right|}{\partial x^\mu} &= \frac{\partial x^\alpha}{\partial y^\beta} \frac{\partial^2 y^\beta}{\partial x^\mu \, \partial x^\alpha}\end{aligned}\right\} \tag{482}$$

We shall make use of this result later.

Example 120. Let $y^i = y^i(x^1, x^2, \ldots, x^n)$, $\left|\frac{\partial y}{\partial x}\right| \neq 0$. We wish to prove that

$$\frac{\partial^2 x^\mu}{\partial y^k \, \partial y^j} = -\frac{\partial x^\alpha}{\partial y^j} \frac{\partial x^\beta}{\partial y^k} \frac{\partial x^\mu}{\partial y^i} \frac{\partial^2 y^i}{\partial x^\beta \, \partial x^\alpha}$$

Now $\delta^i_j = \frac{\partial y^i}{\partial x^\alpha} \frac{\partial x^\alpha}{\partial y^j}$, so that upon differentiating with respect to y^k, we obtain

$$0 = \frac{\partial y^i}{\partial x^\alpha} \frac{\partial^2 x^\alpha}{\partial y^k \, \partial y^j} + \frac{\partial x^\alpha}{\partial y^j} \frac{\partial^2 y^i}{\partial x^\beta \, \partial x^\alpha} \frac{\partial x^\beta}{\partial y^k} \tag{483}$$

Multiplying both sides of (483) by $\frac{\partial x^\mu}{\partial y^i}$ and summing on i, we obtain

$$0 = \frac{\partial^2 x^\mu}{\partial y^k \, \partial y^j} + \frac{\partial x^\alpha}{\partial y^j} \frac{\partial x^\beta}{\partial y^k} \frac{\partial x^\mu}{\partial y^i} \frac{\partial^2 y^i}{\partial x^\beta \, \partial x^\alpha} \qquad \text{Q.E.D.}$$

As a special case, if $y = f(x)$, $\frac{d^2 x}{dy^2} = -\left(\frac{dx}{dy}\right)^3 \frac{d^2 y}{dx^2}$.

Problems

1. What is the cofactor of each term of $\begin{vmatrix} a^1_1 & a^1_2 & a^1_3 \\ a^2_1 & a^2_2 & a^2_3 \\ a^3_1 & a^3_2 & a^3_3 \end{vmatrix}$?

2. Prove (481), (477).

3. If $|A|$ is the derminant of the cofactors of the determinant $|a|$, show that $|A| = |a|^{n-1}$.

4. If $a^i_j = a^j_i$, show that $A^i_j = A^j_i$. Is $|a^i_j| = |a^j_i|$ in all cases?

5. Find an expression for $\dfrac{\partial^3 x^\alpha}{\partial y^i\,\partial y^j\,\partial y^k}$.

6. If $z^i = z^i(y^1, y^2, \ldots, y^n)$, $y^i = y^i(x^1, x^2, \ldots, x^n)$, show that $J(z/y)J(y/x) = J(z/x)$.

7. If $\bar g_{ij} = g_{\alpha\beta} \dfrac{\partial x^\alpha}{\partial \bar x^i} \dfrac{\partial x^\beta}{\partial \bar x^j}$, show that $|\bar g| = |g| \left|\dfrac{\partial x}{\partial \bar x}\right|^2$.

8. If $\bar u^i = u^\alpha \dfrac{\partial \bar x^i}{\partial x^\alpha}$, $\bar v_i = v_\alpha \dfrac{\partial x^\alpha}{\partial \bar x^i}$, show that $\bar u^\alpha \bar v_\alpha = u^\alpha v_\alpha$, where $\bar x^i = \bar x^i(x^1, x^2, \ldots, x^n)$.

9. If $\bar u^i = u^\alpha \dfrac{\partial \bar x^i}{\partial x^\alpha}$, show that $u^i = \bar u^\alpha \dfrac{\partial x^i}{\partial \bar x^\alpha}$.

10. If $\bar g_{ij} = g_{\alpha\beta} \dfrac{\partial x^\alpha}{\partial \bar x^i} \dfrac{\partial x^\beta}{\partial \bar x^j}$, show that $g_{ij} = \bar g_{\alpha\beta} \dfrac{\partial \bar x^\alpha}{\partial x^i} \dfrac{\partial \bar x^\beta}{\partial x^j}$.

11. If $\bar u^i = u^\alpha \dfrac{\partial \bar x^i}{\partial x^\alpha}$, $\bar{\bar u}^i = \bar u^\alpha \dfrac{\partial \bar{\bar x}^i}{\partial \bar x^\alpha}$, show that $\bar{\bar u}^i = u^\alpha \dfrac{\partial \bar{\bar x}^i}{\partial x^\alpha}$.

12. Apply (476) for the product of two third-order determinants.

13. If $A^i_j = B^\alpha_\beta \dfrac{\partial x^\beta}{\partial y^j} \dfrac{\partial y^i}{\partial x^\alpha}$, show that $A^i_i = B^i_i$, and that $|A| = |B|$, $A^i_j A^j_i = B^i_j B^j_i$.

14. If λ is a root of the equation $|a_{ij} - \lambda b_{ij}| = 0$, show that λ is also a root of $|\bar a_{ij} - \lambda \bar b_{ij}| = 0$ provided that $\bar a_{ij} = a_{\alpha\beta} \dfrac{\partial x^\alpha}{\partial \bar x^i} \dfrac{\partial x^\beta}{\partial \bar x^j}$, $\bar b_{ij} = b_{\alpha\beta} \dfrac{\partial x^\alpha}{\partial \bar x^i} \dfrac{\partial x^\beta}{\partial \bar x^j}$, $\left|\dfrac{\partial x}{\partial \bar x}\right| \neq 0$.

122. Arithmetic, or Vector, n-space. In the vector analysis studied in the previous chapters, we set up a coordinate system with three independent variables x, y, z. We chose three mutually perpendicular vectors **i, j, k,** and all other vectors could be written as a linear combination of these three vectors. Any vector could have been represented by the number triple (x, y, z), where we imply that $(x, y, z) \equiv x\mathbf{i} + y\mathbf{j} + z\mathbf{k}$. The unit vectors could have been represented by $(1, 0, 0)$, $(0, 1, 0)$, and $(0, 0, 1)$. A system of mathematics could have been derived solely by defining relationships and operations for these triplets, and we need never have introduced a geometrical picture of a vector. For example, two triplets (a, b, c), (α, β, γ) are defined to be equal if and only if $a = \alpha$, $b = \beta$, $c = \gamma$. We define the scalar product as $(a, b, c)(\alpha, \beta, \gamma) \equiv a\alpha + b\beta + c\gamma$. The vector product, differentiation, etc., can easily be defined. Addition of triples is defined by $(a, b, c) + (\alpha, \beta, \gamma) \equiv (a + \alpha, b + \beta, c + \gamma)$. If A is a real number, then $A(a, b, c)$ is defined as (Aa, Ab, Ac). The set of all triples obeying the rules

$$\begin{aligned} &\text{(i)} \qquad (a, b, c) + (\alpha, \beta, \gamma) = (a + \alpha, b + \beta, c + \gamma) \\ &\text{(ii)} \qquad A(a, b, c) = (Aa, Ab, Ac) \end{aligned} \tag{484}$$

is called a three-dimensional vector space, or the arithmetic space of three dimensions.

It is easy to generalize all this to obtain the arithmetic n-space. Elements of this space are of the form $(x^1, x^2, \ldots, x^n)$, the x^i taken as real. In particular, the unit or basic vectors are $(1, 0, 0, \ldots, 0)$, $(0, 1, 0, \ldots, 0)$, $\ldots$, $(0, 0, \ldots, 0, 1)$. We shall designate V_n as the arithmetic n-space.

By a space of n dimensions we mean any set of objects which can be put in one-to-one reciprocal correspondence with the arithmetic n-space. We call the correspondence a coordinate system. The one-to-one correspondence between the elements or points of the n-space and the arithmetic n-space can be chosen in many ways, and, in general, the choice depends on the nature of the physical problem which determines or sets up the desired coordinate system.

Let the point P correspond to the n-tuple $(x^1, x^2, \ldots, x^n)$. We now consider the n equations

$$\boxed{y^i = y^i(x^1, x^2, \ldots, x^n)}, \quad i = 1, 2, \ldots, n \tag{485}$$

and assume that we can solve for the x^i, so that

$$x^i = x^i(y^1, y^2, \ldots, y^n), \quad i = 1, 2, \ldots, n \tag{486}$$

We assume (485) and (486) are single-valued. It is at once obvious that the point P can be put into correspondence with the n-tuple $(y^1, y^2, \ldots, y^n)$. The n-space of which P is an element is also in one-to-one correspondence with the set of $(y^1, y^2, \ldots, y^n)$, so that we have a new coordinate system. The point P has not changed, but we have a new method for attaching numbers to the points. We call (485) a transformation of coordinates.

123. Contravariant Vectors. We consider the arithmetic n-space and define a space curve in this V_n by

$$x^i = x^i(t), \quad i = 1, 2, \ldots, n \qquad \alpha \leqq t \leqq \beta \tag{487}$$

Note the immediate generalization from the space curve $x = x(t)$, $y = y(t)$, $z = z(t)$: In our new notation $x = x^1$, $y = x^2$, $z = x^3$.

We remember that $\frac{dx}{dt}, \frac{dy}{dt}, \frac{dz}{dt}$ are the components of a tangent vector to this curve. Generalizing, we define a tangent vector to the space curve (487) as having the components

$$\frac{dx^i}{dt}, \quad i = 1, 2, \ldots, n \tag{488}$$

Now let us consider an allowable (one-to-one and single-valued) coordinate transformation, of the type (485). We immediately have that

$$y^i = y^i(x^1, x^2, \ldots, x^n) = y^i[x^1(t), x^2(t), \ldots, x^n(t)] = y^i(t)$$

as the equation of our space curve for observers using the y coordinate system. The components of a tangent vector to the same space curve (remember the points of the curve have not changed; only the labels attached to these points have changed) are given by

$$\frac{dy^i}{dt}, \quad i = 1, 2, \ldots, n \tag{489}$$

Certainly the x coordinate system is no more important than the y coordinate system. We cannot say that $\frac{dx^i}{dt}$ is the tangent vector any more than we can say $\frac{dy^i}{dt}$ is the tangent vector. If we considered all allowable coordinate transformations, we would obtain the whole class of tangent elements, each element claiming to be the tangent vector for that particular coordinate system. It is the abstract collection of all these elements that is said to be the tangent vector. We now ask what relationship exists between the components of the tangent vector in the x coordinate system and the components of the tangent vector in the y coordinate system. We can easily answer this question, for

$$\frac{dy^i}{dt} = \frac{\partial y^i}{\partial x^\alpha}\frac{dx^\alpha}{dt} \tag{490}$$

We also notice that $\dfrac{dx^i}{dt} = \dfrac{\partial x^i}{\partial y^\alpha}\dfrac{dy^\alpha}{dt}$. We leave it as an exercise that this result follows from (490) as well as from (486).

We now make the following generalization: Any set of numbers $A^i(x^1, x^2, \ldots, x^n)$, $i = 1, 2, \ldots, n$, which transform according to the law

$$\bar{A}^i(\bar{x}^1, \bar{x}^2, \ldots, \bar{x}^n) = A^\alpha(x^1, x^2, \ldots, x^n)\frac{\partial \bar{x}^i}{\partial x^\alpha} \tag{491}$$

under the coordinate transformation $\bar{x}^i = \bar{x}^i(x^1, x^2, \ldots, x^n)$, are said to be the components of a contravariant vector. The vector is not just the set of components in one coordinate system but is rather the abstract quantity which is represented in each coordinate system x by the set of components $A^i(x)$.

We immediately see that the law of transformation for a contravariant vector is transitive. Let

$$\bar{A}^i = A^\alpha \frac{\partial \bar{x}^i}{\partial x^\alpha}, \qquad \bar{\bar{A}}^i = \bar{A}^\alpha \frac{\partial \bar{\bar{x}}^i}{\partial \bar{x}^\alpha}$$

Then

$$\bar{\bar{A}}^i = \bar{A}^\beta \frac{\partial \bar{\bar{x}}^i}{\partial \bar{x}^\beta} = A^\alpha \frac{\partial \bar{x}^\beta}{\partial x^\alpha}\frac{\partial \bar{\bar{x}}^i}{\partial \bar{x}^\beta} = A^\alpha \frac{\partial \bar{\bar{x}}^i}{\partial x^\alpha}$$

which proves our statement.

If the components of a contravariant vector are known in one coordinate system, then the components are known in all other allowable coordinate systems by (491). A coordinate transformation does not give a new vector; it merely changes the components of the same vector. We thus say that a contravariant vector is an invariant under a coordinate transformation. An object of any sort which is not changed by transformations of coordinates is called an invariant.

Example 121. Let X, Y, Z be the components of a contravariant vector in a Euclidean space, for an orthogonal coordinate system, and let $ds^2 = dx^2 + dy^2 + dz^2$. The components of this vector in a polar coordinate system are

$$R = X\frac{\partial r}{\partial x} + Y\frac{\partial r}{\partial y} + Z\frac{\partial r}{\partial z} = \cos\theta\, X + \sin\theta\, Y$$

$$\Theta = X\frac{\partial \theta}{\partial x} + Y\frac{\partial \theta}{\partial y} + Z\frac{\partial \theta}{\partial z} = \frac{-\sin\theta}{r} X + \frac{\cos\theta}{r} Y$$

$$Z = X\frac{\partial z}{\partial x} + Y\frac{\partial z}{\partial y} + Z\frac{\partial z}{\partial z} = Z$$

where $r = (x^2 + y^2)^{\frac{1}{2}}$, $\theta = \tan^{-1}(y/x)$, $z = z$.

The components R, Θ, Z are not the projections of the vector $\mathbf{A} = X\mathbf{i} + Y\mathbf{j} + Z\mathbf{k}$ on the r, θ, z directions. However, if the θ-component is given the dimensions of a length by multiplying by r, we obtain $\Theta r = -\sin\theta\, X + \cos\theta\, Y$, which is the projection of $\mathbf{A}$ in the θ-direction. We multiplied by r because $r\,d\theta$ is arc length along the θ-curve. R, Θ, Z are the vector components of the vector $\mathbf{A}$ in the r-θ-z coordinate system, whereas R, $r\Theta$, Z are the physical components of the same vector.

Problems

1. If $A^i(x)$, $B^i(x)$ are components of two contravariant vectors, show that $C^{ij}(x) = A^i(x)B^j(x)$ transforms according to the law $\bar{C}^{ij} = C^{\alpha\beta}\dfrac{\partial \bar{x}^i}{\partial x^\alpha}\dfrac{\partial \bar{x}^j}{\partial x^\beta}$, where $\bar{C}^{ij} = \bar{A}^i\bar{B}^j$.

2. Show that if the components of a contravariant vector vanish in one coordinate system, they vanish in all coordinate systems. What can be said of two contravariant vectors whose components are equal in one coordinate system?

3. Show that the sum and difference of two contravariant vectors of order n is another contravariant vector.

4. If X, Y, Z are the components of a contravariant vector in an orthogonal coordinate system, find the components in a spherical coordinate system. By what must the θ and φ components be multiplied so that we can obtain the projections of the vector on the θ- and φ-directions?

5. If $\bar{A}^i = A^\alpha \dfrac{\partial \bar{x}^i}{\partial x^\alpha}$, show that $A^i = \bar{A}^\alpha \dfrac{\partial x^i}{\partial \bar{x}^\alpha}$.

6. Referring to Prob. 1, show that $C^{ij} = \bar{C}^{\alpha\beta} \dfrac{\partial x^i}{\partial \bar{x}^\alpha} \dfrac{\partial x^j}{\partial \bar{x}^\beta}$.

7. If $\bar{A}^i = \left|\dfrac{\partial x}{\partial \bar{x}}\right|^N A^\alpha \dfrac{\partial \bar{x}^i}{\partial x^\alpha}$, show that $A^i = \left|\dfrac{\partial \bar{x}}{\partial x}\right|^N \bar{A}^\alpha \dfrac{\partial x^i}{\partial \bar{x}^\alpha}$.

124. Covariant Vectors. We consider the scalar point function $\varphi = \varphi(x^1, x^2, \ldots, x^n)$, and form the n-tuple

$$\left(\frac{\partial \varphi}{\partial x^1}, \frac{\partial \varphi}{\partial x^2}, \ldots, \frac{\partial \varphi}{\partial x^n}\right) \tag{492}$$

Now under a coordinate transformation

$$\frac{\partial \varphi}{\partial y^i} = \frac{\partial \varphi}{\partial x^\alpha} \frac{\partial x^\alpha}{\partial y^i} \tag{493}$$

so that the elements of the n-tuple $\left(\dfrac{\partial \varphi}{\partial y^1}, \dfrac{\partial \varphi}{\partial y^2}, \ldots, \dfrac{\partial \varphi}{\partial y^n}\right)$ are related to the elements of (492) by (493). We say that the $\dfrac{\partial \varphi}{\partial x^i}$ are the components of a covariant vector, called the gradient of φ. More generally, if

$$\bar{A}_i = A_\alpha \frac{\partial x^\alpha}{\partial \bar{x}^i} \tag{494}$$

the A_i are said to be the components of a covariant vector. The remarks of Sec. 123 apply here. What is the difference between a contravariant and a covariant vector? It is the law of transformation! The reader is asked to compare (494) with (491). We might ask why it was that no such distinction was made in the elementary vector analysis. We shall answer this question in a later paragraph.

125. Scalar Product of Two Vectors. Let $A^i(x)$ and $B_i(x)$ be the components of a contravariant and a covariant vector. We form the scalar $A^\alpha B_\alpha$. What is the form of $A^\alpha B_\alpha$ if we make a coordinate transformation?

Now

$$A^\alpha = \bar{A}^\beta \frac{\partial x^\alpha}{\partial \bar{x}^\beta}, \qquad B_\alpha = \bar{B}_\sigma \frac{\partial \bar{x}^\sigma}{\partial x^\alpha}$$

so that

$$A^\alpha B_\alpha = \bar{A}^\beta \bar{B}_\sigma \frac{\partial x^\alpha}{\partial \bar{x}^\beta} \frac{\partial \bar{x}^\sigma}{\partial x^\alpha}$$

$$\begin{aligned} A^\alpha B_\alpha &= \bar{A}^\beta \bar{B}_\sigma \frac{\partial \bar{x}^\sigma}{\partial \bar{x}^\beta} \\ &= \bar{A}^\beta \bar{B}_\sigma \delta^\sigma_\beta \\ &= \bar{A}^\beta \bar{B}_\beta = \bar{A}^\alpha \bar{B}_\alpha \end{aligned}$$

Hence $A^\alpha B_\alpha$ is a scalar invariant under a coordinate transformation. The product $(A^\alpha B_\alpha)$ is called the scalar, or dot, product, or inner product, of the two vectors.

Problems

1. If A_i and B_i are components of two covariant vectors, show that $C_{ij} = A_i B_j$ transforms according to the law $\bar{C}_{ij} = C_{\alpha\beta} \dfrac{\partial x^\alpha}{\partial \bar{x}^i} \dfrac{\partial x^\beta}{\partial \bar{x}^j}$.

2. Show that $C^i_j = A^i B_j$ transforms according to the law $\bar{C}^i_j = C^\alpha_\beta \dfrac{\partial x^\beta}{\partial \bar{x}^j} \dfrac{\partial \bar{x}^i}{\partial x^\alpha}$.

3. If $\bar{A}_i = A_\alpha \dfrac{\partial x^\alpha}{\partial \bar{x}^i}$ show that $A_i = \bar{A}_\alpha \dfrac{\partial \bar{x}^\alpha}{\partial x^i}$.

4. If φ and ψ are scalar invariants, show that

$$\begin{aligned} \text{grad } (\varphi\psi) &= \varphi \text{ grad } \psi + \psi \text{ grad } \varphi \\ \text{grad } [F(\varphi)] &= F'(\varphi) \text{ grad } \varphi \end{aligned}$$

5. If $A^i B_i$ is a scalar invariant for all contravariant vectors A^i, show that B_i is a covariant vector.

126. Tensors. The contravariant and covariant vectors defined above are special cases of differential invariants called tensors. The components of the tensor are of the form $T^{a_1 a_2 \cdots a_r}_{b_1 b_2 \cdots b_s}$, where the indices $a_1, a_2, \ldots, a_r, b_1, b_2, \ldots, b_s$ run through the integers $1, 2, \ldots, n$, and the components transform according to the rule

$$\bar{T}^{a_1 a_2 \cdots a_r}_{b_1 b_2 \cdots b_s} = \left| \frac{\partial x}{\partial \bar{x}} \right|^N T^{\alpha_1 \alpha_2 \cdots \alpha_r}_{\beta_1 \beta_2 \cdots \beta_s} \frac{\partial \bar{x}^{a_1}}{\partial x^{\alpha_1}} \cdots \frac{\partial \bar{x}^{a_r}}{\partial x^{\alpha_r}} \frac{\partial x^{\beta_1}}{\partial \bar{x}^{b_1}} \cdots \frac{\partial x^{\beta_s}}{\partial \bar{x}^{b_s}} \tag{495}$$

We call the exponent N of the Jacobian $\left|\frac{\partial x}{\partial \bar{x}}\right|$ the weight of the tensor field. If $N = 0$, we say that the tensor field is absolute; otherwise the tensor field is relative of weight N. A tensor density occurs for $N = 1$. The vectors of Secs. 123 and 124 are absolute tensors of order 1. The tensor of (495) is said to be contravariant of order r and covariant of order s. If $s = 0$, the tensor is purely contravariant, and if $r = 0$, purely covariant; otherwise it is called a mixed tensor.

Two tensors are said to be of the same kind if the tensors have the same number of covariant indices and the same number of contravariant indices and are of the same weight. We can construct further tensors as follows:

(*a*) The sum of two tensors of the same kind is a tensor of this kind. The proof is obvious, for if

$$\bar{T}^{a \cdots b}_{c \cdots d} = \left|\frac{\partial x}{\partial \bar{x}}\right|^N T^{\alpha \cdots \beta}_{\sigma \cdots \tau} \frac{\partial x^\sigma}{\partial \bar{x}^c} \cdots \frac{\partial x^\tau}{\partial \bar{x}^d} \frac{\partial \bar{x}^a}{\partial x^\alpha} \cdots \frac{\partial \bar{x}^b}{\partial x^\beta}$$

$$\bar{S}^{a \cdots b}_{c \cdots d} = \left|\frac{\partial x}{\partial \bar{x}}\right|^N S^{\alpha \cdots \beta}_{\sigma \cdots \tau} \frac{\partial x^\sigma}{\partial \bar{x}^c} \cdots \frac{\partial x^\tau}{\partial \bar{x}^d} \frac{\partial \bar{x}^a}{\partial x^\alpha} \cdots \frac{\partial \bar{x}^b}{\partial x^\beta}$$

then

$$\bar{U}^{a \cdots b}_{c \cdots d} = (\bar{T}^{a \cdots b}_{c \cdots d} + \bar{S}^{a \cdots b}_{c \cdots d}) = U^{\alpha \cdots \beta}_{\sigma \cdots \tau} \left|\frac{\partial x}{\partial \bar{x}}\right|^N \frac{\partial x^\sigma}{\partial \bar{x}^c} \cdots \frac{\partial \bar{x}^b}{\partial x^\beta}$$

(*b*) The product of two tensors is a tensor. We show this for a special case. Let

$$\bar{T}^a_b = \left|\frac{\partial x}{\partial \bar{x}}\right|^2 T^\alpha_\beta \frac{\partial x^\beta}{\partial \bar{x}^b} \frac{\partial \bar{x}^a}{\partial x^\alpha}$$

$$\bar{S}^i = \left|\frac{\partial x}{\partial \bar{x}}\right|^3 S^\sigma \frac{\partial \bar{x}^i}{\partial x^\sigma}$$

so that

$$(\bar{T}^a_b \bar{S}^i) = \left|\frac{\partial x}{\partial \bar{x}}\right|^5 (T^\alpha_\beta S^\sigma) \frac{\partial x^\beta}{\partial \bar{x}^b} \frac{\partial \bar{x}^a}{\partial x^\alpha} \frac{\partial \bar{x}^i}{\partial x^\sigma}$$

The new tensor is of weight $N + N' = 3 + 2 = 5$.

(*c*) *Contraction.* Consider the absolute tensor

$$\bar{A}^i_{jk} = A^\alpha_{\beta\gamma} \frac{\partial x^\beta}{\partial \bar{x}^j} \frac{\partial x^\gamma}{\partial \bar{x}^k} \frac{\partial \bar{x}^i}{\partial x^\alpha}$$

Replace k by i and sum. We obtain

$$\begin{aligned}\bar{A}^i_{ji} &= A^\alpha_{\beta\gamma} \frac{\partial x^\beta}{\partial \bar{x}^j} \frac{\partial x^\gamma}{\partial \bar{x}^i} \frac{\partial \bar{x}^i}{\partial x^\alpha} \\ &= A^\alpha_{\beta\gamma} \frac{\partial x^\beta}{\partial \bar{x}^j} \frac{\partial x^\gamma}{\partial x^\alpha} = A^\alpha_{\beta\gamma} \frac{\partial x^\beta}{\partial \bar{x}^j} \delta^\gamma_\alpha \\ &= A^\alpha_{\beta\alpha} \frac{\partial x^\beta}{\partial \bar{x}^j}\end{aligned}$$

so that A^i_{ji} are the components of an absolute contravariant vector. In general, we equate a certain covariant index to a contravariant index, sum on the repeated indices, and obtain a new tensor. We call this process a contraction.

(*d*) *Quotient law.* We illustrate the quotient law as follows: Assume that A^iB_{jk} is a tensor for all contravariant vectors A^i. We prove that B_{jk} is a tensor, for

$$\begin{aligned}\bar{A}^i\bar{B}_{jk} &= A^\alpha B_{\beta\gamma} \frac{\partial x^\beta}{\partial \bar{x}^j} \frac{\partial x^\gamma}{\partial \bar{x}^k} \frac{\partial \bar{x}^i}{\partial x^\alpha} \\ &= \bar{A}^i B_{\beta\gamma} \frac{\partial x^\beta}{\partial \bar{x}^j} \frac{\partial x^\gamma}{\partial \bar{x}^k}\end{aligned}$$

or

$$\bar{A}^i \left(\bar{B}_{jk} - B_{\beta\gamma} \frac{\partial x^\beta}{\partial \bar{x}^j} \frac{\partial x^\gamma}{\partial \bar{x}^k} \right) = 0$$

Since $\bar{A}^i$ is arbitrary, we must have $\bar{B}_{jk} = B_{\beta\gamma} \dfrac{\partial x^\beta}{\partial \bar{x}^j} \dfrac{\partial x^\gamma}{\partial \bar{x}^k}$, the desired result.

Example 122. The Kronecker delta, δ^i_j, is a mixed absolute tensor, for

$$\delta^i_j \frac{\partial x^j}{\partial \bar{x}^\alpha} \frac{\partial \bar{x}^\beta}{\partial x^i} = \frac{\partial x^i}{\partial \bar{x}^\alpha} \frac{\partial \bar{x}^\beta}{\partial x^i} = \delta^\beta_\alpha = \bar{\delta}^\beta_\alpha$$

Example 123. If A^i and B_i are the components of a contravariant and a covariant vector, then $C^i_j = A^iB_j$ are the components of a mixed tensor, for

$$\bar{A}^i = A^\alpha \frac{\partial \bar{x}^i}{\partial x^\alpha}, \qquad \bar{B}_j = B_\beta \frac{\partial x^\beta}{\partial \bar{x}^j}$$

so that

$$\bar{C}^i_j = \bar{A}^i \bar{B}_j = A^\alpha B_\beta \frac{\partial \bar{x}^i}{\partial x^\alpha} \frac{\partial x^\beta}{\partial \bar{x}^j} = C^\alpha_\beta \frac{\partial \bar{x}^i}{\partial x^\alpha} \frac{\partial x^\beta}{\partial \bar{x}^j}$$

Example 124. Let g_{ij} be the components of a covariant tensor so that $\bar{g}_{ij} = g_{\alpha\beta} \frac{\partial x^\alpha}{\partial \bar{x}^i} \frac{\partial x^\beta}{\partial \bar{x}^j}$. Taking determinants and applying Example 115 twice, we obtain

$$|\bar{g}| = |g| \left|\frac{\partial x}{\partial \bar{x}}\right|^2 \qquad \text{or} \qquad |\bar{g}|^{\frac{1}{2}} = |g|^{\frac{1}{2}} \left|\frac{\partial x}{\partial \bar{x}}\right|$$

Now if A^i are the components of an absolute contravariant vector, then $\bar{A}^i = A^\alpha \frac{\partial \bar{x}^i}{\partial x^\alpha}$, so that

$$\begin{aligned} \bar{B}^i \equiv |\bar{g}|^{\frac{1}{2}} \bar{A}^i &= \left|\frac{\partial x}{\partial \bar{x}}\right| |g|^{\frac{1}{2}} A^\alpha \frac{\partial \bar{x}^i}{\partial x^\alpha} \\ &= \left|\frac{\partial x}{\partial \bar{x}}\right| B^\alpha \frac{\partial \bar{x}^i}{\partial x^\alpha} \end{aligned}$$

so that $B^i \equiv |g|^{\frac{1}{2}} A^i$ are the components of a vector density. This method affords a means of changing absolute tensors into relative tensors.

Example 125. Assume $g_{\alpha\beta}\, dx^\alpha\, dx^\beta$ an invariant, that is,

$$\bar{g}_{\alpha\beta}\, d\bar{x}^\alpha\, d\bar{x}^\beta = g_{\alpha\beta}\, dx^\alpha\, dx^\beta$$

Now $d\bar{x}^\alpha = \frac{\partial \bar{x}^\alpha}{\partial x^\mu} dx^\mu$, so that

$$\bar{g}_{\alpha\beta} \frac{\partial \bar{x}^\alpha}{\partial x^\mu} \frac{\partial \bar{x}^\beta}{\partial x^\nu} dx^\mu\, dx^\nu = g_{\mu\nu}\, dx^\mu\, dx^\nu$$

or

$$\left(\bar{g}_{\alpha\beta} \frac{\partial \bar{x}^\alpha}{\partial x^\mu} \frac{\partial \bar{x}^\beta}{\partial x^\nu} - g_{\mu\nu}\right) dx^\mu\, dx^\nu = 0 \tag{496}$$

If we assume $g_{\alpha\beta} = g_{\beta\alpha}$, then since (496) is identically zero for arbitrary dx^i, we must have (see Example 114)

$$g_{\mu\nu} = \bar{g}_{\alpha\beta} \frac{\partial \bar{x}^\alpha}{\partial x^\mu} \frac{\partial \bar{x}^\beta}{\partial x^\nu} \tag{496a}$$

or the $g_{\mu\nu}$ are components of a covariant tensor of rank 2.

Example 126. If the components of a tensor are zero in one coordinate system, it follows from the law of transformation (495) that the components are zero in all coordinate systems. This is an important result.

Example 127. *Outer product of two vectors.* Let A_i and B_i be the components of two covariant vectors, so that

$$C_{ij} \equiv A_i B_j = \bar{A}_\alpha \bar{B}_\beta \frac{\partial \bar{x}^\alpha}{\partial x^i} \frac{\partial \bar{x}^\beta}{\partial x^j} = \bar{C}_{\alpha\beta} \frac{\partial \bar{x}^\alpha}{\partial x^i} \frac{\partial \bar{x}^\beta}{\partial x^j}$$

The C_{ij} are the components of a covariant tensor of second order, the outer product of A_i and B_i.

Example 128. By the same reasoning as in Example 127, we have that $C_{ij} = A_i B_j - A_j B_i$ are the components of a covariant tensor of the second order. Notice that C_{ij} is skew-symmetric, for $C_{ij} = -C_{ji}$. For a three-dimensional space

$$\|C_{ij}\| = \left\| \begin{matrix} 0 & A_1B_2 - A_2B_1 & A_1B_3 - A_3B_1 \\ -(A_1B_2 - A_2B_1) & 0 & A_2B_3 - A_3B_2 \\ -(A_1B_3 - A_3B_1) & -(A_2B_3 - A_3B_2) & 0 \end{matrix} \right\|$$

The nonvanishing terms are similar to the components of the vector cross product.

Problems

1. If $\bar{A}_{ij} = A_{\alpha\beta} \dfrac{\partial x^\alpha}{\partial \bar{x}^i} \dfrac{\partial x^\beta}{\partial \bar{x}^j}$, show that $A_{ij} = \bar{A}_{\alpha\beta} \dfrac{\partial \bar{x}^\alpha}{\partial x^i} \dfrac{\partial \bar{x}^\beta}{\partial x^j}$.

2. Show that A_{ij} can be written as the sum of a symmetric and a skew-symmetric component.

3. If A^i_j are the components of an absolute mixed tensor, show that A^i_i is a scalar invariant.

4. If $A_{\alpha\beta}$ are the components of an absolute covariant tensor, and if $A^{\alpha\beta} A_{\beta\gamma} = \delta^\alpha_\gamma$, show that the $A^{\alpha\beta}$ are the components of an absolute contravariant tensor. The two tensors are said to be reciprocal.

5. If A^{ij} and A_{ij} are reciprocal symmetric tensors, and if u_i are components of a covariant vector, show that $A_{ij}u^iu^j = A^{ij}u_iu_j$, where $u^i = A^{i\alpha}u_\alpha$.

6. Let A_{ij} and B_{ij} be symmetric tensors and let u^i, v^i be components of contravariant vectors satisfying

$$\begin{aligned} (A_{ij} - \kappa B_{ij})u^i &= 0 \quad i, j = 1, 2, \ldots, n \\ (A_{ij} - \kappa' B_{ij})v^i &= 0, \qquad \kappa \neq \kappa' \end{aligned}$$

Prove that $A_{ij}u^iv^j = B_{ij}u^iv^j = 0$, and that $\kappa = \dfrac{A_{ij}u^iu^j}{B_{ij}u^iu^j}\cdot$ Why is κ an invariant?

7. From the relative tensor A^i_j of weight N, derive a relative scalar of weight N.

8. If A^{ij}_{mn} is a mixed tensor of weight N, show that A^{mj}_{mn} is a mixed tensor of weight N.

9. Show that the cofactors of the determinant $|a_{ij}|$ are the components of a relative tensor of weight 2 if a_{ij} is an absolute covariant tensor.

10. If A^i are the components of an absolute contravariant vector, show that $\dfrac{\partial A^i}{\partial x^j}$ are *not* the components of a mixed tensor.

127. The Line Element. In the Euclidean space of three dimensions we have assumed that

$$ds^2 = dx^2 + dy^2 + dz^2$$

In the Euclidean n-space we have

$$\begin{aligned} ds^2 &= (dx^1)^2 + (dx^2)^2 + \cdots + (dx^n)^2 \\ &= \delta_{\alpha\beta}\, dx^\alpha\, dx^\beta \end{aligned} \tag{497}$$

If we apply a transformation of coordinates

$$x^i = x^i(\bar{x}^1, \bar{x}^2, \ldots, \bar{x}^n)$$

we have that $dx^i = \dfrac{\partial x^i}{\partial \bar{x}^\alpha}\, d\bar{x}^\alpha$, so that (497) takes the form

$$ds^2 = \delta_{\alpha\beta} \frac{\partial x^\alpha}{\partial \bar{x}^\mu} \frac{\partial x^\beta}{\partial \bar{x}^\nu}\, d\bar{x}^\mu\, d\bar{x}^\nu$$

We may write $ds^2 = \bar{g}_{\mu\nu}\, d\bar{x}^\mu\, d\bar{x}^\nu$, where

$$\bar{g}_{\mu\nu} = \delta_{\alpha\beta} \frac{\partial x^\alpha}{\partial \bar{x}^\mu} \frac{\partial x^\beta}{\partial \bar{x}^\nu} = \sum_{\alpha=1}^{n} \frac{\partial x^\alpha}{\partial \bar{x}^\mu} \frac{\partial x^\alpha}{\partial \bar{x}^\nu}$$

Thus the most general form for the line element $(ds)^2$ for a Euclidean space is the quadratic form

$$ds^2 = g_{\alpha\beta}\, dx^\alpha\, dx^\beta \tag{498}$$

The $g_{\alpha\beta}$ are the components of the metric tensor (see Example 125). The quadratic differential form (498) is called a Riemannian metric. Any space characterized by such a metric is called a Riemannian space. It does not follow that there exists a coordinate transformation which reduces (498) to a sum of squares. If there is a coordinate transformation

$$x^i = x^i(y^1, y^2, \ldots, y^n)$$

such that $ds^2 = \delta_{\alpha\beta}\, dy^\alpha\, dy^\beta$, we say that the Riemannian space is Euclidean. The y's will be called the components of a Euclidean coordinate system. Notice that $g_{\alpha\beta} = \delta_{\alpha\beta}$. Any coordinate system for which the g_{ij} are constants is called a cartesian coordinate system.

We can choose the metric tensor symmetric, for

$$g_{ij} \equiv \tfrac{1}{2}(g_{ij} + g_{ji}) + \tfrac{1}{2}(g_{ij} - g_{ji})$$

and the terms $\frac{1}{2}(g_{ij} - g_{ji})\, dx^i\, dx^j$ contribute nothing to the sum ds^2. The terms $\frac{1}{2}(g_{ij} + g_{ji})$ are symmetric in i and j.

Example 129. In a three-dimensional Euclidean space $ds^2 = (dx^1)^2 + (dx^2)^2 + (dx^3)^2$ for an orthogonal coordinate system, so that

$$\|g\| = \begin{Vmatrix} 1 & 0 & 0 \\ 0 & 1 & 0 \\ 0 & 0 & 1 \end{Vmatrix}$$

Let

$$\begin{aligned} x^1 &= r \sin\theta \cos\varphi = y^1 \sin y^2 \cos y^3 \\ x^2 &= r \sin\theta \sin\varphi = y^1 \sin y^2 \sin y^3 \\ x^3 &= r \cos\theta = y^1 \cos y^2 \end{aligned}$$

Now

$$\begin{aligned} \bar{g}_{ij}(r, \theta, \varphi) &= g_{\alpha\beta} \frac{\partial x^\alpha}{\partial y^i} \frac{\partial x^\beta}{\partial y^j} \\ &= \frac{\partial x^1}{\partial y^i} \frac{\partial x^1}{\partial y^j} + \frac{\partial x^2}{\partial y^i} \frac{\partial x^2}{\partial y^j} + \frac{\partial x^3}{\partial y^i} \frac{\partial x^3}{\partial y^j} \end{aligned}$$

Hence

$$\begin{aligned}\bar{g}_{11} &= (\sin y^2 \cos y^3)^2 + (\sin y^2 \sin y^3)^2 + (\cos y^2)^2 \\ &\equiv 1\end{aligned}$$

Similarly

$$\bar{g}_{22} = (y^1)^2, \qquad \bar{g}_{33} = (y^1)^2(\sin y^2)^2, \qquad \bar{g}_{ij} = 0 \quad \text{for } i \neq j$$

so that

$$\begin{aligned}ds^2 &= (dy^1)^2 + (y^1)^2(dy^2)^2 + (y^1 \sin y^2)^2(dy^3)^2 \\ &= dr^2 + r^2\, d\theta^2 + r^2 \sin^2 \theta\, d\varphi^2\end{aligned}$$

is the line element in spherical coordinates. Since the g's are not constants, a spherical coordinate system is not a cartesian coordinate system.

Example 130. We define g^{ij} as the reciprocal tensor to g_{ij}, that is, $g^{i\alpha}g_{\alpha j} = \delta^i_j$ (see Prob. 4, Sec. 126). The g^{ij} are the signed minors of the g_{ji} divided by the determinant of the g_{ij}. For spherical coordinates in a Euclidean space

$$\|g_{ij}\| = \begin{Vmatrix} 1 & 0 & 0 \\ 0 & r^2 & 0 \\ 0 & 0 & r^2 \sin^2 \theta \end{Vmatrix}, \qquad \|g^{ij}\| = \begin{Vmatrix} 1 & 0 & 0 \\ 0 & \dfrac{1}{r^2} & 0 \\ 0 & 0 & \dfrac{1}{r^2 \sin^2 \theta} \end{Vmatrix}$$

Example 131. We define the length L of a vector A^i in a Riemannian space by the quadratic form

$$\boxed{L^2 = g_{\alpha\beta}A^\alpha A^\beta} \tag{499}$$

The *associated vector* of A^i is the covariant vector

$$A_i = g_{i\alpha}A^\alpha$$

It is easily seen that $A^i = g^{i\beta}A_\beta$, so that

$$L^2 = g_{\alpha\beta}g^{\alpha\mu}A_\mu g^{\beta\nu}A_\nu = g^{\mu\nu}A_\mu A_\nu$$

We see that a vector and its associate have the same length. If $L^2 \equiv 1$, the vector is a unit vector.

Example 132. *Angle between two vectors.* Let A^i and B_j be unit vectors. We define the cosine of the angle between these two vectors by

$$\cos\theta = A^iB_i = A^ig_{ij}B^j = g_{ij}A^iB^j$$
$$= g^{ij}A_jB_i = g^{ij}A_iB_j \tag{500}$$

If the vectors are not unit vectors,

$$\cos\theta = \frac{g_{ij}A^iB^j}{(g_{ij}A^iA^j)^{\frac{1}{2}}(g_{ij}B^iB^j)^{\frac{1}{2}}} \tag{500a}$$

If $g_{ij}A^iB^j = 0$, the vectors are orthogonal. We must show that $|\cos\theta| \leqq 1$. Consider the vector $\lambda A^i + \mu B^i$. Assuming a positive definite form, that is, $g_{\alpha\beta}z^\alpha z^\beta > 0$ unless $z^i \equiv 0$, we have

$$g_{\alpha\beta}(\lambda A^\alpha + \mu B^\alpha)(\lambda A^\alpha + \mu B^\alpha) > 0$$

or

$$y = \lambda^2(g_{\alpha\beta}A^\alpha A^\beta) + 2\lambda\mu(g_{\alpha\beta}A^\alpha B^\beta) + \mu^2(g_{\alpha\beta}B^\alpha B^\beta) > 0$$

This is a quadratic form in λ^2/μ^2, so that the discriminant must be negative, for if it were nonnegative, y would vanish for some value of λ/μ or μ/λ. Hence

$$g_{\alpha\beta}A^\alpha B^\beta < (g_{\alpha\beta}A^\alpha A^\beta)^{\frac{1}{2}}(g_{\alpha\beta}B^\alpha B^\beta)^{\frac{1}{2}}$$

or $|\cos\theta| < 1$. Moreover, if $A^i = kB^i$, it is easy to see that $\cos\theta = \pm 1$. Hence $|\cos\theta| \leqq 1$.

Example 133. A hypersurface in a Riemannian space is given by $x^i = x^i(u^1, u^2)$. If we keep u^1 fixed, $u^1 = u_0^1$, we obtain the space curve $x^i = x^i(u_0^1, u^2)$, called the u^2 curve. Similarly, $x^i = x^i(u^1, u_0^2)$ represents a u^1 curve on the surface. These curves are called the coordinate curves of the surface. We have at once that on the surface

$$ds^2 = g_{\alpha\beta}\,dx^\alpha\,dx^\beta = \sum_{i,j=1}^{2} g_{\alpha\beta}\frac{\partial x^\alpha}{\partial u^i}\frac{\partial x^\beta}{\partial u^j}\,du^i\,du^j$$
$$= g_{\alpha\beta}\frac{\partial x^\alpha}{\partial u^i}\frac{\partial x^\beta}{\partial u^j}\,du^i\,du^j$$

$$\underline{\overline{ds^2 = h_{ij}\,du^i\,du^j}} \tag{501}$$

where $h_{ij} = g_{\alpha\beta}\dfrac{\partial x^\alpha}{\partial u^i}\dfrac{\partial x^\beta}{\partial u^j}$.

Example 134. *The special theory of relativity.* Let us consider the one-parameter group of transformations

$$\begin{aligned} x &= \beta(\bar{x} - V\bar{t}) \\ y &= \bar{y} \\ z &= \bar{z} \\ t &= \beta\left(\bar{t} - \frac{V}{c^2}\bar{x}\right) \end{aligned} \tag{502}$$

where $\beta = [1 - (V^2/c^2)]^{-\frac{1}{2}}$ and V is the parameter. c is the speed of light. These are the Einstein-Lorentz transformations (see Prob. 11, Sec. 24). The transformations form a group because (1) if we set $V = 0$, we obtain the identity transformations $x = \bar{x}$, $y = \bar{y}$, $z = \bar{z}$, $t = \bar{t}$; (2) the inverse transformation exists since $\bar{x} = \beta(x + Vt)$, $\bar{y} = y$, $\bar{z} = z$, $\bar{t} = \beta[t + (V/c^2)x]$, the inverse transformation obtained by replacing the parameter V by $-V$; (3) the result of applying two such transformations yields a new Lorentz transformation, for if

$$\begin{aligned} \bar{x} &= \bar{\beta}(\bar{\bar{x}} - W\bar{\bar{t}}) \\ \bar{y} &= \bar{\bar{y}} \\ \bar{z} &= \bar{\bar{z}} \\ \bar{t} &= \bar{\beta}\left(\bar{\bar{t}} - \frac{W}{c^2}\bar{\bar{x}}\right) \end{aligned}$$

where $\bar{\beta} = [1 - (W^2/c^2)]^{-\frac{1}{2}}$, then

$$\begin{aligned} x &= \bar{\bar{\beta}}(\bar{\bar{x}} - U\bar{\bar{t}}) \\ y &= \bar{\bar{y}} \\ z &= \bar{\bar{z}} \\ t &= \bar{\bar{\beta}}\left(\bar{\bar{t}} - \frac{U}{c^2}\bar{\bar{x}}\right) \end{aligned}$$

where

$$U = \frac{V + W}{1 + (VW/c^2)}, \qquad \bar{\bar{\beta}} = \left(1 - \frac{U^2}{c^2}\right)^{-\frac{1}{2}}$$

We now assume that (x, y, z, t) represents an event in space and time as observed by S and that $(\bar{x}, \bar{y}, \bar{z}, \bar{t})$ represents the same event observed by $\bar{S}$ (see Fig. 101).

The origin $\bar{O}$ has $\bar{x} = \bar{y} = \bar{z} = 0$, so that from (502) $\frac{dx}{dt} = -V$, showing that $\bar{S}$ moves with a constant speed $-V$ relative to S. Similarly S moves with speed $+V$ relative to $\bar{S}$.

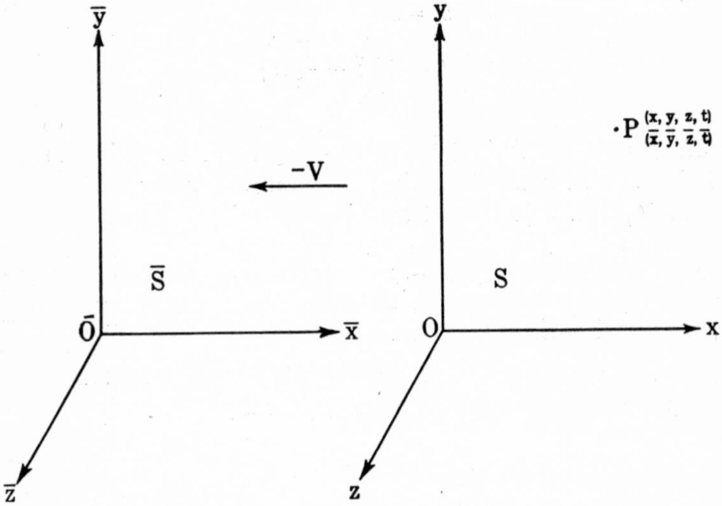

FIG. 101.

From (502) we see that O and $\bar{O}$ coincide at $t = 0$. At this instant assume that an event is the sending forth of a light wave. The results of Prob. 11, Sec. 24 show that

$$\frac{x^2 + y^2 + z^2}{t^2} = \frac{\bar{x}^2 + \bar{y}^2 + \bar{z}^2}{\bar{t}^2} = c^2$$

so that the speed of light is the same for both observers. This is one of the postulates of the special theory of relativity. Starting with this postulate and desiring the group property, we could have shown that the transformations (502) are the only transformations which keep $dx^2 + dy^2 + dz^2 - c^2\,dt^2 = 0$ an invariant.

Let us now consider a clock fixed in the S frame. We have $x = \text{constant}$, so that $dx = 0$, and from (502) $dt = \beta\,d\bar{t}$. Hence a unit of time as observed by $\bar{S}$ is not a unit of time as observed by S because of the factor $\beta \neq 1$. S remarks that $\bar{S}$'s clock is running slowly. The same is true for clocks fixed in the S frame.

We choose for the interval of our four-dimensional space the invariant

$$ds^2 = c^2\,dt^2 - dx^2 - dy^2 - dz^2 = (dx^4)^2 - (dx^1)^2 - (dx^2)^2 - (dx^3)^2$$

where $x^1 = x$, $x^2 = y$, $x^3 = z$, $x^4 = ct$. The interval ds^2 yields two types of measurements, length and time, but takes care to distinguish between them. If we keep a clock fixed in the S frame, then $dx = dy = dz = 0$, so that $ds^2 = c^2\,dt^2$, and the measurement of interval ds is real and proportional to the time dt. Now if we keep t fixed, $dt = 0$, and

$$ds^2 = -(dx^2 + dy^2 + dz^2)$$

so that ds is a pure imaginary, its absolute value denoting length as measured by meter sticks in a Euclidean space.

We shall describe the laws of physics by tensor equations, the components of the tensors subject to the transformations (502). This will guarantee the invariance of our laws of physics.

The momentum of a particle of mass m_0 will be defined by $p^\alpha = m_0 \dfrac{dx^\alpha}{ds}$. If the speed of the particle is u,

$$u^2 = \left(\frac{dx}{dt}\right)^2 + \left(\frac{dy}{dt}\right)^2 + \left(\frac{dz}{dt}\right)^2$$

as measured by S, then

$$ds^2 = c^2\,dt^2 - (dx^2 + dy^2 + dz^2) = (c^2 - u^2)\,dt^2$$

so that

$$p^\alpha = \frac{m_0}{(c^2 - u^2)^{\frac{1}{2}}}\frac{dx^\alpha}{dt} = \frac{1}{c}\frac{m_0}{[1 - (u^2/c^2)]^{\frac{1}{2}}}\frac{dx^\alpha}{dt}$$

and

$$p^4 = \frac{m_0}{[1 - (u^2/c^2)]^{\frac{1}{2}}} = m$$

We define the Minkowski force by the equations

$$\begin{aligned} f^\alpha &= c^2 \frac{d}{ds}\left(m_0 \frac{dx^\alpha}{ds}\right) \\ &= \frac{1}{[1 - (u^2/c^2)]^{\frac{1}{2}}}\frac{d}{dt}\left(m \frac{dx^\alpha}{dt}\right), \qquad \alpha = 1, 2, 3, 4 \end{aligned}$$

The Minkowski force differs from the Newtonian force by the factor $[1 - (u^2/c^2)]^{-\frac{1}{2}}$. The work done by the Newtonian force $F^\alpha = \frac{d}{dt}\left(m\,\frac{dx^\alpha}{dt}\right)$ for a displacement dx^α is

$$
\begin{aligned}
dE &= \sum_{\alpha=1}^{3} \frac{d}{dt}(m\dot{x}^\alpha)\,dx^\alpha \\
&= \sum_{\alpha=1}^{3} \left(m\ddot{x}^\alpha\,dx^\alpha + \frac{dm}{dt}\,\dot{x}^\alpha\,dx^\alpha\right) \\
&= \frac{m_0 u\,du}{[1 - (u^2/c^2)]^{\frac{3}{2}}}
\end{aligned}
$$

and integrating, $E = [1 - (u^2/c^2)]^{-\frac{1}{2}}m_0c^2 - m_0c^2 = (m - m_0)c^2$, with $E = 0$ for $u = 0$. Expanding $[1 - (u^2/c^2)]^{-\frac{1}{2}}$ in a Maclaurin series, we have $E \approx \frac{1}{2}m_0u^2$ for $(u^2/c^2) \ll 1$.

The reader is referred to Probs. 1, 2, and 3 of Sec. 82 for the application of special relativity theory to electromagnetic theory. Let the reader derive (285) by use of this theory, choosing the frame S so that at a particular instant the charge ρ is fixed in this frame. The force on the charge as measured by $\bar{S}$ is given by (285).

Problems

1. For paraboloidal coordinates

$$
\begin{aligned}
x^1 &= y^1y^2 \cos y^3 \\
x^2 &= y^1y^2 \sin y^3 \\
x^3 &= \tfrac{1}{2}[(y^1)^2 - (y^2)^2]
\end{aligned}
$$

show that $ds^2 = [(y^1)^2 + (y^2)^2][(dy^1)^2 + (dy^2)^2] + (y^1y^2)^2(dy^3)^2$.

2. Show that for a hypersurface in a three-dimensional Euclidean space, $h_{ij} = \sum_{\alpha=1}^{3} \frac{\partial x^\alpha}{\partial u^i}\frac{\partial x^\alpha}{\partial u^j}$.

3. Show that the unit vectors tangent to the u^1 and u^2 curves are given by $\frac{1}{\sqrt{h_{11}}}\frac{\partial x^i}{\partial u^1}$ and $\frac{1}{\sqrt{h_{22}}}\frac{\partial x^i}{\partial u^2}$.

4. If ω is the angle between the coordinate curves, show that $\cos\omega = h_{12}/\sqrt{h_{11}h_{22}}$.

5. $\varphi(x^1, x^2, \ldots, x^n) =$ constant determines a hypersurface of a V_n. If dx^i is any infinitesimal displacement on the hypersurface, we have $\frac{\partial \varphi}{\partial x^\alpha} dx^\alpha = 0$. Why does this show that the $\frac{\partial \varphi}{\partial x^i}$ are the components of a covariant vector that is normal to the hypersurface?

6. If $\varphi(x^1, x^2, \ldots, x^n) =$ constant, show that a unit vector normal to the surface is given by $\left(g^{\sigma\beta} \frac{\partial \varphi}{\partial x^\sigma} \frac{\partial \varphi}{\partial x^\beta}\right)^{-\frac{1}{2}} g^{r\alpha} \frac{\partial \varphi}{\partial x^\alpha}$.

7. Consider the vector with components $(dx^1, 0, 0, \ldots, 0)$. Under a coordinate transformation the components become $\left(\frac{\partial \bar{x}^1}{\partial x^1} dx^1, \frac{\partial \bar{x}^2}{\partial x^1} dx^1, \ldots, \frac{\partial \bar{x}^n}{\partial x^1} dx^1\right)$. Consider the new components for the vectors with components $(0, dx^2, \ldots, 0), \ldots, (0, 0, \ldots, 0, dx^n)$, and interpret

$$d\bar{x}^1 \, d\bar{x}^2 \cdots d\bar{x}^n \equiv \left|\frac{\partial \bar{x}}{\partial x}\right| dx^1 \, dx^2 \cdots dx^n$$

Using the result of Prob. 7, Sec. 121, show that $\sqrt{|g|}\, dx^1 \, dx^2 \cdots dx^n$ is an invariant. We define the volume by

$$V = \int\int \cdots \int \sqrt{|g|}\, dx^1 \, dx^2 \cdots dx^n$$

8. Show that $\frac{dx^i}{ds}$ is a unit vector for a V_n.

9. The surfaces $x^i =$ constant, $i = 1, 2, \ldots, n$, are called the coordinate surfaces of Riemannian space. On these surfaces all variables but one are allowed to vary. This determines subspaces of dimensions $(n - 1)$. If we let only x^1 vary, we obtain a coordinate curve. Show that the unit vectors to the coordinate curves are given by $a_i = 1/\sqrt{g_{ii}}$, $i = 1, 2, \ldots, n$, and that the angle of intersection between two coordinate curves is given by

$$\cos \omega_{ij} = \frac{g_{ij}}{\sqrt{g_{ii} g_{jj}}}$$

10. Show that a length observed by $\bar{S}$ appears to be longer as observed by S. How does $\bar{S}$ compare lengths with S?

11. Let j_x, j_y, j_z, ρ be the components of a vector as measured by S. What are the components of the same vector as measured by $\bar{S}$?

12. Let $\frac{d^2\bar{x}}{d\bar{t}^2}, \frac{d^2\bar{y}}{d\bar{t}^2}, \frac{d^2\bar{z}}{d\bar{t}^2}$ be the components of the acceleration of a particle as measured by $\bar{S}$. Find $\frac{d^2x}{dt^2}, \frac{d^2y}{dt^2}, \frac{d^2z}{dt^2}$ from (502).

128. Geodesics in a Riemannian Space. If a space curve in a Riemannian space is given by $x^i = x^i(t)$, we can compute the distance between two points of the curve by the formula

$$s = \int_{t_0}^{t_1} \left(g_{\alpha\beta} \frac{dx^\alpha}{dt} \frac{dx^\beta}{dt} \right)^{\frac{1}{2}} dt \tag{503}$$

To find the geodesics we extremalize (503) (see Sec. 40). The differential equations of the geodesics are [see Eq. (146)]

$$\frac{d}{dt}\left(\frac{\partial f}{\partial \dot{x}^i}\right) - \frac{\partial f}{\partial x^i} = 0 \tag{504}$$

where $f = (g_{\alpha\beta}\, \dot{x}^\alpha \dot{x}^\beta)^{\frac{1}{2}} = \frac{ds}{dt}$. Now

$$\frac{\partial f}{\partial x^i} = \frac{1}{2f}\left(\frac{\partial g_{\alpha\beta}}{\partial x^i}\, \dot{x}^\alpha \dot{x}^\beta\right)$$

and

$$\begin{aligned}\frac{d}{dt}\left(\frac{\partial f}{\partial \dot{x}^i}\right) &= \frac{d}{dt}\left(\frac{g_{\alpha i}\dot{x}^\alpha + g_{i\beta}\dot{x}^\beta}{2\, ds/dt}\right) \\ &= \frac{1}{2\, ds/dt}\left(g_{\alpha i}\ddot{x}^\alpha + g_{i\beta}\ddot{x}^\beta + \frac{\partial g_{\alpha i}}{\partial x^\beta}\, \dot{x}^\alpha \dot{x}^\beta + \frac{\partial g_{i\beta}}{\partial x^\alpha}\, \dot{x}^\beta \dot{x}^\alpha\right) \\ &\quad - \frac{1}{2(ds/dt)^2}\frac{d^2s}{dt^2}\,(g_{\alpha i}\dot{x}^\alpha + g_{i\beta}\dot{x}^\beta)\end{aligned}$$

If we choose s for the parameter t, $s = t$, $\frac{ds}{dt} = 1$, $\frac{d^2s}{dt^2} = 0$, and use the fact that $g_{ij} = g_{ji}$, (504) reduces to

$$g_{i\alpha}\ddot{x}^\alpha + \frac{1}{2}\left(\frac{\partial g_{\alpha i}}{\partial x^\beta} + \frac{\partial g_{i\beta}}{\partial x^\alpha} - \frac{\partial g_{\alpha\beta}}{\partial x^i}\right) \dot{x}^\alpha \dot{x}^\beta = 0 \tag{505}$$

Multiplying (505) by $g^{\tau i}$ and summing on i, we obtain

$$\ddot{x}^\tau + \frac{g^{\tau i}}{2}\left(\frac{\partial g_{\alpha i}}{\partial x^\beta} + \frac{\partial g_{i\beta}}{\partial x^\alpha} - \frac{\partial g_{\alpha\beta}}{\partial x^i}\right)\dot{x}^\alpha \dot{x}^\beta = 0$$

or

$$\frac{d^2x^\tau}{ds^2} + \Gamma^\tau_{\alpha\beta}\frac{dx^\alpha}{ds}\frac{dx^\beta}{ds} = 0 \tag{506}$$

where

$$\Gamma^\tau_{\alpha\beta} = \frac{g^{\tau\sigma}}{2}\left(\frac{\partial g_{\alpha\sigma}}{\partial x^\beta} + \frac{\partial g_{\sigma\beta}}{\partial x^\alpha} - \frac{\partial g_{\alpha\beta}}{\partial x^\sigma}\right) \tag{507}$$

The functions $\Gamma^\tau_{\alpha\beta}$ are called the Christoffel symbols of the second kind. Equations (506) are the differential equations of the geodesics or paths.

Example 135. For a Euclidean space using orthogonal coordinates, we have $ds^2 = (dx^1)^2 + \cdots + (dx^n)^2$, so that $g_{\alpha\beta} = \delta_{\alpha\beta}$ and $\dfrac{\partial g_{ij}}{\partial x^k} = 0$. Hence the geodesics are given by $\dfrac{d^2x}{ds^2} = 0$ or $x^\tau = a^\tau s + b^\tau$, a linear path.

Example 136. Assume that we live in a space for which $ds^2 = (dx^1)^2 + [(x^1)^2 + c^2](dx^2)^2$, the surface of a right helicoid immersed in a Euclidean three-space. We have

$$\|g_{ij}\| = \begin{Vmatrix} 1 & 0 \\ 0 & (x^1)^2 + c^2 \end{Vmatrix}, \qquad \|g^{ij}\| = \begin{Vmatrix} 1 & 0 \\ 0 & \dfrac{1}{(x^1)^2 + c^2} \end{Vmatrix}$$

Thus we have

$$\begin{aligned} &\Gamma^1_{11} = 0, && \Gamma^2_{11} = 0 \\ &\Gamma^1_{21} = \Gamma^1_{12} = 0, && \Gamma^2_{12} = \Gamma^2_{21} = \frac{x^1}{(x^1)^2 + c^2} \\ &\Gamma^1_{22} = -x^1, && \Gamma^2_{22} = 0 \end{aligned}$$

so that the differential equations of the geodesics on the surface are

$$\frac{d^2x^1}{ds^2} - x^1\left(\frac{dx^2}{ds}\right)^2 = 0$$

$$\frac{d^2x^2}{ds^2} + \frac{2x^1}{(x^1)^2 + c^2}\frac{dx^1}{ds}\frac{dx^2}{ds} = 0$$

Problems

1. Derive the Γ^i_{jk} of Example 136.

2. Find the differential equations of the geodesics for the line element $ds^2 = (dx^1)^2 + (\sin x^1)^2(dx^2)^2$.

3. Show that $\Gamma^\tau_{\alpha\beta} = \Gamma^\tau_{\beta\alpha}$.

4. For a Euclidean space using a cartesian coordinate system, show that $\Gamma^\tau_{\alpha\beta} \equiv 0$.

5. Obtain the Christoffel symbols and the equations of the geodesics for the surface

$$\begin{aligned} x^1 &= u^1 \cos u^2 \\ x^2 &= u^1 \sin u^2 \\ x^3 &= 0 \end{aligned}$$

This surface is the plane $x^3 = 0$, and the coordinates are polar coordinates.

6. From (507) show that $\dfrac{\partial g_{\alpha\beta}}{\partial x^\mu} = g_{\sigma\beta}\Gamma^\sigma_{\alpha\mu} + g_{\sigma\alpha}\Gamma^\sigma_{\beta\mu}$.

7. Obtain the Christoffel symbols for a Euclidean space using cylindrical coordinates. Set up the equations of the geodesics. Do the same for spherical coordinates.

8. Write out the explicit form for the Christoffel symbols of the first kind: $\{i, jk\} \equiv g_{i\sigma}\Gamma^\sigma_{jk}$.

9. Let $ds^2 = E\,du^2 + 2F\,du\,dv + G\,dv^2$. Calculate $|g|$, g^{ij}, $i, j = 1, 2$. Write out the Γ^i_{jk}.

10. If $\bar{\Gamma}^i_{jk} = \Gamma^\alpha_{\beta\gamma} \dfrac{\partial x^\beta}{\partial \bar{x}^j} \dfrac{\partial x^\gamma}{\partial \bar{x}^k} \dfrac{\partial \bar{x}^i}{\partial x^\alpha} + \dfrac{\partial^2 x^\sigma}{\partial \bar{x}^j\, \partial \bar{x}^k} \dfrac{\partial \bar{x}^i}{\partial x^\sigma}$, show that $\Gamma^\alpha_{\beta\gamma} - \Gamma^\alpha_{\gamma\beta}$ are the components of a tensor.

129. Law of Transformation for the Christoffel Symbols. Let the equations of the geodesics be given by

$$\frac{d^2x^i}{ds^2} + \Gamma^i_{jk} \frac{dx^j}{ds} \frac{dx^k}{ds} = 0 \tag{508}$$

and

$$\frac{d^2\bar{x}^i}{ds^2} + \bar{\Gamma}^i_{jk} \frac{d\bar{x}^j}{ds} \frac{d\bar{x}^k}{ds} = 0 \tag{509}$$

for the two coordinate systems x^i, $\bar{x}^i$ in a Riemannian space. We now find the relationship between the Γ^i_{jk} and $\bar{\Gamma}^i_{jk}$. Now

$$\frac{d\bar{x}^i}{ds} = \frac{\partial \bar{x}^i}{\partial x^\alpha}\frac{dx^\alpha}{ds} \quad \text{and} \quad \frac{d^2\bar{x}^i}{ds^2} = \frac{\partial^2 \bar{x}^i}{\partial x^\beta\,\partial x^\alpha}\frac{dx^\alpha}{ds}\frac{dx^\beta}{ds} + \frac{\partial \bar{x}^i}{\partial x^\alpha}\frac{d^2x^\alpha}{ds^2}$$

Substituting into (509), we obtain

$$\frac{\partial \bar{x}^i}{\partial x^\alpha}\frac{d^2x^\alpha}{ds^2} + \frac{\partial^2 \bar{x}^i}{\partial x^\beta\,\partial x^\alpha}\frac{dx^\alpha}{ds}\frac{dx^\beta}{ds} + \bar{\Gamma}^i_{jk}\frac{\partial \bar{x}^j}{\partial x^\alpha}\frac{\partial \bar{x}^k}{\partial x^\beta}\frac{dx^\alpha}{ds}\frac{dx^\beta}{ds} = 0 \qquad (510)$$

We multiply (510) by $\dfrac{\partial x^\sigma}{\partial \bar{x}^i}$ and sum on i to obtain

$$\frac{d^2x^\sigma}{dx^2} + \left(\bar{\Gamma}^i_{jk}\frac{\partial \bar{x}^j}{\partial x^\alpha}\frac{\partial \bar{x}^k}{\partial x^\beta}\frac{\partial x^\sigma}{\partial \bar{x}^i} + \frac{\partial^2 \bar{x}^i}{\partial x^\beta\,\partial x^\alpha}\frac{\partial x^\sigma}{\partial \bar{x}^i}\right)\frac{dx^\alpha}{ds}\frac{dx^\beta}{ds} = 0$$

Comparing with (508), we see that (using the fact that $\Gamma^i_{jk} = \Gamma^i_{kj}$)

$$\Gamma^i_{jk} = \bar{\Gamma}^\alpha_{\beta\gamma}\frac{\partial \bar{x}^\beta}{\partial x^j}\frac{\partial \bar{x}^\gamma}{\partial x^k}\frac{\partial x^i}{\partial \bar{x}^\alpha} + \frac{\partial^2 \bar{x}^\sigma}{\partial x^j\,\partial x^k}\frac{\partial x^i}{\partial \bar{x}^\sigma} \qquad (511)$$

This is the law of transformation for the Γ^i_{jk}. We note that the Γ^i_{jk} are not the components of a tensor, so that the Γ^i_{jk} may be zero in one coordinate system but not in all coordinate systems.

Example 137. From (481) we have

$$\frac{\partial |g|}{\partial x^\mu} = |g| g^{\alpha\beta}\frac{\partial g_{\alpha\beta}}{\partial x^\mu}$$

and from Prob. 6, Sec. 128,

$$\frac{\partial g_{\alpha\beta}}{\partial x^\mu} = g_{\sigma\beta}\Gamma^\sigma_{\alpha\mu} + g_{\sigma\alpha}\Gamma^\sigma_{\beta\mu}$$

so that

$$\begin{aligned}\frac{\partial \log |g|}{\partial x^\mu} &= g^{\alpha\beta}g_{\sigma\beta}\Gamma^\sigma_{\alpha\mu} + g^{\alpha\beta}g_{\sigma\alpha}\Gamma^\sigma_{\beta\mu} \\ &= \delta^\alpha_\sigma\Gamma^\sigma_{\alpha\mu} + \delta^\beta_\sigma\Gamma^\sigma_{\beta\mu} \\ &= \Gamma^\alpha_{\alpha\mu} + \Gamma^\beta_{\beta\mu} = 2\Gamma^\alpha_{\alpha\mu}\end{aligned}$$

or

$$\frac{\partial \log \sqrt{|g|}}{\partial x^\mu} = \Gamma^\alpha_{\alpha\mu} \qquad (512)$$

Example 138. We may arrive at the Christoffel symbols and their law of transformation by another method. Differentiating the law of transformation

$$\bar{g}_{ij} = g_{\alpha\beta} \frac{\partial x^\alpha}{\partial \bar{x}^i} \frac{\partial x^\beta}{\partial \bar{x}^j}$$

with respect to $\bar{x}^k$, we have

$$\frac{\partial \bar{g}_{ij}}{\partial \bar{x}^k} = \frac{\partial g_{\alpha\beta}}{\partial x^\gamma} \frac{\partial x^\gamma}{\partial \bar{x}^k} \frac{\partial x^\alpha}{\partial \bar{x}^i} \frac{\partial x^\beta}{\partial \bar{x}^j} + g_{\alpha\beta} \left(\frac{\partial x^\alpha}{\partial \bar{x}^i} \frac{\partial^2 x^\beta}{\partial \bar{x}^k \partial \bar{x}^j} + \frac{\partial x^\beta}{\partial \bar{x}^j} \frac{\partial^2 x^\alpha}{\partial \bar{x}^k \partial \bar{x}^i} \right) \tag{513}$$

If we now subtract (513) from the two equations obtained from it by cyclic permutations of the indices i, j, k, we obtain

$$\bar{\Gamma}^i_{jk} = \Gamma^\alpha_{\beta\gamma} \frac{\partial x^\beta}{\partial \bar{x}^j} \frac{\partial x^\gamma}{\partial \bar{x}^k} \frac{\partial \bar{x}^i}{\partial x^\alpha} + \frac{\partial^2 x^\alpha}{\partial \bar{x}^j \partial \bar{x}^k} \frac{\partial \bar{x}^i}{\partial x^\alpha} \tag{511a}$$

where

$$\Gamma^i_{jk} = \frac{1}{2} g^{\sigma i} \left(\frac{\partial g_{k\sigma}}{\partial x^j} + \frac{\partial g_{j\sigma}}{\partial x^k} - \frac{\partial g_{jk}}{\partial x^\sigma} \right)$$

Example 139. Let us consider a Euclidean space for which

$$ds^2 = (dx^1)^2 + (dx^2)^2 + \cdots + (dx^n)^2$$

In this case the $\Gamma^i_{jk}(x) = 0$. In any other coordinate system, we have

$$\bar{\Gamma}^i_{jk}(\bar{x}) = \frac{\partial^2 x^\sigma}{\partial \bar{x}^j \partial \bar{x}^k} \frac{\partial \bar{x}^i}{\partial x^\sigma}$$

If the new coordinate system is also cartesian (the g_{ij} = constants), then $\bar{\Gamma}^i_{jk} = 0$, or

$$\frac{\partial^2 x^\sigma}{\partial \bar{x}^j \partial \bar{x}^k} = 0$$

$$x^\sigma = a^\sigma_\alpha \bar{x}^\alpha + b^\sigma \tag{514}$$

where a^σ_α, b^σ are constants of integration.

Hence the coordinate transformation between two cartesian coordinate systems is linear. If, furthermore, we desire the distance between two points to be an invariant, we must have

$$\sum_{\sigma=1}^{n} dx^\sigma \, dx^\sigma = \sum_{\sigma=1}^{n} d\bar{x}^\sigma \, d\bar{x}^\sigma = \sum_{\sigma=1}^{n} a^\sigma_\alpha a^\sigma_\beta \, dx^\alpha \, dx^\beta$$

so that

$$\sum_{\sigma=1}^{n} a_\alpha^\sigma a_\beta^\sigma = \delta_{\alpha\beta} \tag{515}$$

A linear transformation such that (515) holds is called an orthogonal transformation.

For orthogonal transformations,

$$\bar{g}_{ij} = g_{\alpha\beta} \frac{\partial x^\alpha}{\partial \bar{x}^i} \frac{\partial x^\beta}{\partial \bar{x}^j}$$

reduces to $\delta_{ij} = \delta_{\alpha\beta} \dfrac{\partial x^\alpha}{\partial \bar{x}^i} \dfrac{\partial x^\beta}{\partial \bar{x}^j}$. We multiply both sides by $\dfrac{\partial \bar{x}^i}{\partial x^\mu}$ and sum on i, so that

$$\frac{\partial \bar{x}^j}{\partial x^\mu} = \delta_{\alpha\beta}\, \delta_\mu^\alpha \frac{\partial x^\beta}{\partial \bar{x}^j} = \delta_{\mu\beta} \frac{\partial x^\beta}{\partial \bar{x}^j} = \frac{\partial x^\mu}{\partial \bar{x}^j} \tag{516}$$

Now let us compare the laws of transformation for covariant and contravariant vectors. We have

$$\bar{A}^i = A^\alpha \frac{\partial \bar{x}^i}{\partial x^\alpha}, \qquad \bar{A}_i = A_\alpha \frac{\partial x^\alpha}{\partial \bar{x}^i} \tag{517}$$

Replacing $\dfrac{\partial \bar{x}^i}{\partial x^\alpha}$ by $\dfrac{\partial x^\alpha}{\partial \bar{x}^i}$ from (516), we see that

$$\bar{A}^i = \sum_{\alpha=1}^{n} A^\alpha \frac{\partial x^\alpha}{\partial \bar{x}^i} \tag{518}$$

so that orthogonal transformations affect contravariant vectors in exactly the same way that covariant vectors are affected [compare (517) and (518)]. This is why there was no distinction made between covariant and contravariant vectors in the elementary treatment of vectors.

Problems

1. From (511) show that

$$\bar{\Gamma}_{jk}^i = \Gamma_{\beta\gamma}^\alpha \frac{\partial x^\beta}{\partial \bar{x}^j} \frac{\partial x^\gamma}{\partial \bar{x}^k} \frac{\partial \bar{x}^i}{\partial x^\alpha} + \frac{\partial^2 x^\sigma}{\partial \bar{x}^j\, \partial \bar{x}^k} \frac{\partial \bar{x}^i}{\partial x^\sigma}$$

2. By differentiating the identity $g^{i\alpha}g_{\alpha j} \equiv \delta^i_j$, show that

$$\frac{\partial g^{ik}}{\partial x^j} = -g^{hk}\Gamma^i_{hj} - g^{hi}\Gamma^k_{hj}$$

3. Derive (513) by performing the permutations.

4. If $\dfrac{\partial^2 x^\sigma}{\partial \bar{x}^j \, \partial \bar{x}^k} \dfrac{\partial \bar{x}^i}{\partial x^\sigma} = 0$, show that $\dfrac{\partial^2 x^\sigma}{\partial \bar{x}^j \, \partial \bar{x}^k} = 0$.

5. If

$$\bar{\Gamma}^i_{jk} = \bar{\Gamma}^\alpha_{\beta\gamma} \frac{\partial \bar{x}^\beta}{\partial \bar{\bar{x}}^j} \frac{\partial \bar{x}^\gamma}{\partial \bar{\bar{x}}^k} \frac{\partial \bar{\bar{x}}^i}{\partial \bar{x}^\alpha} + \frac{\partial^2 \bar{x}^\sigma}{\partial \bar{\bar{x}}^j \, \partial \bar{\bar{x}}^k} \frac{\partial \bar{\bar{x}}^i}{\partial \bar{x}^\sigma}$$

show that

$$\bar{\bar{\Gamma}}^i_{jk} = \Gamma^\alpha_{\beta\gamma} \frac{\partial x^\beta}{\partial \bar{\bar{x}}^j} \frac{\partial x^\gamma}{\partial \bar{\bar{x}}^k} \frac{\partial \bar{\bar{x}}^i}{\partial x^\alpha} + \frac{\partial^2 x^\sigma}{\partial \bar{\bar{x}}^j \, \partial \bar{\bar{x}}^k} \frac{\partial \bar{\bar{x}}^i}{\partial x^\sigma}$$

6. If $x^\alpha = x^\alpha(u^1, u^2, \ldots, u^r)$, $\alpha = 1, 2, \ldots, n$, $r < n$, and if $h_{ij} = g_{\alpha\beta} \dfrac{\partial x^\alpha}{\partial u^i} \dfrac{\partial x^\beta}{\partial u^j}$, and if

$$(\Gamma^i_{jk})_h = \frac{1}{2} h^{i\sigma} \left(\frac{\partial h_{\sigma j}}{\partial u^k} + \frac{\partial h_{k\sigma}}{\partial u^j} - \frac{\partial h_{jk}}{\partial u^\sigma} \right)$$

show that

$$h_{\lambda i}(\Gamma^\lambda_{jk})_h = g_{\alpha\mu}(\Gamma^\mu_{\beta\gamma})_g \frac{\partial x^\alpha}{\partial u^i} \frac{\partial x^\beta}{\partial u^j} \frac{\partial x^\gamma}{\partial u^k} + g_{\alpha\beta} \frac{\partial x^\alpha}{\partial u^i} \frac{\partial^2 x^\beta}{\partial u^j \, \partial u^k}$$

7. Define $\bar{g}_{\alpha\beta}(x)$ by the equation $\bar{g}_{\alpha\beta}(x) \equiv \mu(x) g_{\alpha\beta}(x)$. We see that the metric tensor $g_{\alpha\beta}(x)$ is determined only up to a factor of multiplication $\mu(x)$. In this space (conformal) we do not compare lengths at two different points, or, in other words, the unit of length changes from point to point. Show that

$$\bar{\Gamma}^\alpha_{\beta\gamma}(x) = \Gamma^\alpha_{\beta\gamma}(x) + \varphi_\gamma \delta^\alpha_\beta + \varphi_\beta \delta^\alpha_\gamma - g^{\alpha\sigma} g_{\beta\gamma} \varphi_\sigma$$

where $\varphi_\sigma = \dfrac{1}{2} \dfrac{\partial \log \mu}{\partial x^\sigma}$. $\bar{\Gamma}^\alpha_{\beta\gamma}$ and $\Gamma^\alpha_{\beta\gamma}$ are defined by (507) using $\bar{g}_{\alpha\beta}$ and $g_{\alpha\beta}$.

8. Prove that a geodesic of zero length (minimal geodesic) [that is, $x^\alpha(s)$ satisfies (506) and $g_{\alpha\beta} \dfrac{dx^\alpha}{ds} \dfrac{dx^\beta}{ds} = 0$] remains a minimal geodesic under a conformal transformation.

130. Covariant Differentiation. Let us differentiate the absolute covariant vector given by the transformation

$$\bar{A}_i = A_\alpha \frac{\partial x^\alpha}{\partial \bar{x}^i}$$

We obtain

$$\frac{\partial \bar{A}_i}{\partial \bar{x}^j} = \frac{\partial A_\alpha}{\partial x^\beta}\frac{\partial x^\alpha}{\partial \bar{x}^i}\frac{\partial x^\beta}{\partial \bar{x}^j} + A_\alpha \frac{\partial^2 x^\alpha}{\partial \bar{x}^j\,\partial \bar{x}^i} \tag{519}$$

It is at once apparent that $\frac{\partial A_i}{\partial x^j}$ are not the components of a tensor. However, we can construct a tensor by the following device: From (511a) (see page 292)

$$\bar{\Gamma}^\alpha_{ij} = \Gamma^\rho_{\sigma\tau}\frac{\partial x^\sigma}{\partial \bar{x}^i}\frac{\partial x^\tau}{\partial \bar{x}^j}\frac{\partial \bar{x}^\alpha}{\partial x^\rho} + \frac{\partial^2 x^\sigma}{\partial \bar{x}^i\,\partial \bar{x}^j}\frac{\partial \bar{x}^\alpha}{\partial x^\sigma} \tag{520}$$

Multiplying (520) by $\bar{A}_\alpha$ and subtracting from (519), we obtain

$$\frac{\partial \bar{A}_i}{\partial \bar{x}^j} - \bar{A}_\alpha \bar{\Gamma}^\alpha_{ij} = \left(\frac{\partial A_\alpha}{\partial x^\beta} - A_\rho \Gamma^\rho_{\alpha\beta}\right)\frac{\partial x^\alpha}{\partial \bar{x}^i}\frac{\partial x^\beta}{\partial \bar{x}^j} \tag{521}$$

so that if we define

$$\bar{A}_{i,j} \equiv \frac{\partial \bar{A}_i}{\partial \bar{x}^j} - \bar{A}_\alpha \bar{\Gamma}^\alpha_{ij} \tag{522}$$

we have that

$$\bar{A}_{i,j} = A_{\alpha,\beta}\frac{\partial x^\alpha}{\partial \bar{x}^i}\frac{\partial x^\beta}{\partial \bar{x}^j}$$

and $A_{i,j}$ is a covariant tensor of rank 2. The tensor is called the covariant derivative of A_i with respect to x^j. The comma will denote covariant differentiation. For a cartesian coordinate system, $\Gamma^i_{jk} = 0$, so that $A_{i,j} = \frac{\partial A_i}{\partial x^j}$, our ordinary derivative.

For a scalar of weight N we have

$$\bar{A} = \left|\frac{\partial x}{\partial \bar{x}}\right|^N A$$

so that

$$\frac{\partial \bar{A}}{\partial \bar{x}^j} = \left|\frac{\partial x}{\partial \bar{x}}\right|^N \frac{\partial A}{\partial x^\alpha}\frac{\partial x^\alpha}{\partial \bar{x}^j} + N\left|\frac{\partial x}{\partial \bar{x}}\right|^{N-1} \frac{\partial \left|\frac{\partial x}{\partial \bar{x}}\right|}{\partial \bar{x}^j} A$$

and from (482), $\dfrac{\partial \left|\frac{\partial x}{\partial \bar{x}}\right|}{\partial \bar{x}^j} = \left|\dfrac{\partial x}{\partial \bar{x}}\right| \dfrac{\partial \bar{x}^\alpha}{\partial x^\beta}\dfrac{\partial^2 x^\beta}{\partial \bar{x}^j\,\partial \bar{x}^\alpha}$. Hence

$$\frac{\partial \bar{A}}{\partial \bar{x}^j} = \left|\frac{\partial x}{\partial \bar{x}}\right|^N \frac{\partial A}{\partial x^\alpha}\frac{\partial x^\alpha}{\partial \bar{x}^j} + N\left|\frac{\partial x}{\partial \bar{x}}\right|^{N} \frac{\partial \bar{x}^\alpha}{\partial x^\beta}\frac{\partial^2 x^\beta}{\partial \bar{x}^j\,\partial \bar{x}^\alpha} A \tag{523}$$

Multiplying $\bar{\Gamma}^\alpha_{j\alpha} = \Gamma^\sigma_{\alpha\sigma}\dfrac{\partial x^\alpha}{\partial \bar{x}^j} + \dfrac{\partial^2 x^\sigma}{\partial \bar{x}^\alpha\,\partial \bar{x}^j}\dfrac{\partial \bar{x}^\alpha}{\partial x^\sigma}$ by $N\bar{A}$ and subtracting from (523), we have

$$\frac{\partial \bar{A}}{\partial \bar{x}^j} - N\bar{A}\bar{\Gamma}^\alpha_{j\alpha} = \left|\frac{\partial x}{\partial \bar{x}}\right|^N \left(\frac{\partial A}{\partial x^\alpha} - NA\Gamma^\sigma_{\alpha\sigma}\right)\frac{\partial x^\alpha}{\partial \bar{x}^j} \tag{524}$$

Hence $A_{,j} \equiv \dfrac{\partial A}{\partial x^j} - NA\Gamma^\sigma_{j\sigma}$ is a relative covariant vector of weight N. It is called the covariant derivative of the relative scalar A. For a cartesian coordinate system it reduces to the ordinary derivative.

In general, it can be proved that if $T^{\alpha_1\alpha_2\cdots\alpha_r}_{\beta_1\beta_2\cdots\beta_s}$ is a relative tensor of weight N, then

$$\begin{aligned} T^{\alpha_1\alpha_2\cdots\alpha_r}_{\beta_1\beta_2\cdots\beta_s,m} \equiv \frac{\partial T^{\alpha_1\alpha_2\cdots\alpha_r}_{\beta_1\beta_2\cdots\beta_s}}{\partial x^m} &+ T^{\mu\alpha_2\cdots\alpha_r}_{\beta_1\beta_2\cdots\beta_s}\Gamma^{\alpha_1}_{\mu m} + \cdots + T^{\alpha_1\alpha_2\cdots\mu}_{\beta_1\beta_2\cdots\beta_s}\Gamma^{\alpha_r}_{\mu m} \\ &- T^{\alpha_1\alpha_2\cdots\alpha_r}_{\mu\beta_2\cdots\beta_s}\Gamma^{\mu}_{\beta_1 m} - \cdots - T^{\alpha_1\alpha_2\cdots\alpha_r}_{\beta_1\beta_2\cdots\mu}\Gamma^{\mu}_{\beta_s m} \\ &- NT^{\alpha_1\alpha_2\cdots\alpha_r}_{\beta_2\beta_2\cdots\beta_s}\Gamma^{\mu}_{\mu m} \end{aligned} \tag{525}$$

is a relative tensor of weight N, of covariant order one greater than $T^{\alpha_1\alpha_2\cdots\alpha_r}_{\beta_1\beta_2\cdots\beta_s}$, and it is called the covariant derivative of $T^{\alpha_1\alpha_2\cdots\alpha_r}_{\beta_1\beta_2\cdots\beta_s}$.

Example 140. We have

$$g_{ij,k} = \frac{\partial g_{ij}}{\partial x^k} - g_{\mu j}\Gamma^\mu_{ik} - g_{i\mu}\Gamma^\mu_{jk}$$

so that from Prob. 6, Sec. 128,

$$g_{ij,k} = 0 \tag{526}$$

Example 141. If φ is an absolute scalar, $\varphi = \bar{\varphi}$, we call $\varphi_{,j} = \dfrac{\partial \varphi}{\partial x^j}$ the gradient of φ.

Example 142. *Curl of a vector.* Let A_i be an absolute covariant vector. We have $A_{i,j} = \dfrac{\partial A_i}{\partial x^j} - A_\alpha \Gamma^\alpha_{ij}$. Similarly,

$$A_{j,i} = \frac{\partial A_j}{\partial x^i} - A_\alpha \Gamma^\alpha_{ji}$$

Hence $A_{i,j} - A_{j,i} \equiv \dfrac{\partial A_i}{\partial x_j} - \dfrac{\partial A_j}{\partial x^i}$ is a covariant tensor of rank 2. It is called the curl of the vector A_i. If the A_i are the components of the gradient of a scalar, $A_i = \dfrac{\partial \varphi}{\partial x^i}$, then

$$\text{curl } A_i = \frac{\partial^2 \varphi}{\partial x^j \, \partial x^i} - \frac{\partial^2 \varphi}{\partial x^j \, \partial x^i} = 0$$

so that the curl of a gradient is zero. It can be shown that the converse holds. If the curl is identically zero, the covariant vector is the gradient of a scalar.

Example 143. *Intrinsic derivatives.* Since $A_{i,j}$ and $\dfrac{dx^j}{ds}$ are tensors, we know that $A_{i,j} \dfrac{dx^j}{ds}$ is a covariant vector. We call it the intrinsic derivative of A_i. We have

$$A_{i,j} \frac{dx^j}{ds} = \frac{\partial A_i}{\partial x^j} \frac{dx^j}{ds} - A_\alpha \Gamma^\alpha_{ij} \frac{dx^j}{ds}$$

$$\frac{\delta A_i}{\delta s} \equiv \frac{dA_i}{ds} - A_\alpha \Gamma^\alpha_{ij} \frac{dx^j}{ds} \tag{527}$$

and write the intrinsic derivative of A_i as $\dfrac{\delta A_i}{\delta s}$.

Example 144. The divergence of an absolute contravariant vector is defined as the contraction of its covariant derivative.

Hence

$$\text{div } A^i = A^\alpha_{,\alpha} = \frac{\partial A^\alpha}{\partial x^\alpha} + A^\alpha \Gamma^i_{\alpha i}$$

Now $\Gamma^i_{\alpha i} = \dfrac{\partial \log \sqrt{|g|}}{\partial x^\alpha}$ from (512), so that

$$\text{div } A^i = \frac{\partial A^\alpha}{\partial x^\alpha} + \frac{1}{\sqrt{|g|}} \frac{\partial \sqrt{|g|}}{\partial x^\alpha} A^\alpha$$

$$\text{div } A^i = \frac{1}{\sqrt{|g|}} \frac{\partial}{\partial x^\alpha} (\sqrt{|g|}\, A^\alpha) \tag{528}$$

In spherical coordinates, we have

$$\sqrt{|g|} = \begin{vmatrix} 1 & 0 & 0 \\ 0 & r^2 & 0 \\ 0 & 0 & r^2 \sin^2 \theta \end{vmatrix}^{\frac{1}{2}} = r^2 \sin \theta$$

so that

$$\text{div } A^i = \frac{1}{r^2 \sin \theta} \left[\frac{\partial}{\partial r} (r^2 \sin \theta \, A^r) + \frac{\partial}{\partial \theta} (r^2 \sin \theta \, A^\theta) + \frac{\partial}{\partial \varphi} (r^2 \sin \theta \, A^\varphi) \right]$$

and changing A^θ and A^φ into physical components having the dimensions of A^r (see Example 121), we have

$$\text{div } A^i = \frac{1}{r^2 \sin \theta} \left[\frac{\partial}{\partial r} (r^2 \sin \theta \, A^r) + \frac{\partial}{\partial \theta} (r \sin \theta \, A^\theta) + \frac{\partial}{\partial \varphi} (r A^\varphi) \right]$$

Example 145. *The Laplacian of a scalar invariant.* If φ is a scalar invariant, $\varphi_{,j}$ is the gradient of φ, and the div $(\varphi_{,j})$ is called the Laplacian of φ.

$$\text{Lap } \varphi = \nabla^2 \varphi = \text{div } (\varphi_{,j}) = \text{div } \frac{\partial \varphi}{\partial x^j}$$

Thus

$$\nabla^2 \varphi = \frac{1}{\sqrt{|g|}} \frac{\partial}{\partial x^\alpha} \left(\sqrt{|g|}\, g^{\alpha j} \frac{\partial \varphi}{\partial x^j} \right) \tag{529}$$

We changed $\dfrac{\partial\varphi}{\partial x^j}$ into a contravariant vector so that we could apply (528). The associate of $\dfrac{\partial\varphi}{\partial x^j}$ is $g^{\alpha j}\dfrac{\partial\varphi}{\partial x^j}$.

In spherical coordinates

$$|g_{ij}|^{\frac{1}{2}} = \begin{vmatrix} 1 & 0 & 0 \\ 0 & r^2 & 0 \\ 0 & 0 & r^2\sin^2\theta \end{vmatrix}^{\frac{1}{2}}, \qquad |g^{ij}| = \begin{vmatrix} 1 & 0 & 0 \\ 0 & \dfrac{1}{r^2} & 0 \\ 0 & 0 & \dfrac{1}{r^2\sin\theta} \end{vmatrix}$$

so that

$$\nabla^2 F = \frac{1}{r^2\sin\theta}\left[\frac{\partial}{\partial r}\left(r^2\sin\theta\,\frac{\partial F}{\partial r}\right) + \frac{\partial}{\partial\theta}\left(\sin\theta\,\frac{\partial F}{\partial\theta}\right) + \frac{\partial}{\partial\varphi}\left(\frac{1}{\sin\theta}\frac{\partial F}{\partial\varphi}\right)\right]$$

Example 146. In Example 144 we defined the divergence of the vector A^i as div $A^i = A^\alpha_{,\alpha} = \dfrac{\partial A^\alpha}{\partial x^\alpha} + A^\alpha\Gamma^i_{\alpha i}$. For a Euclidean space using cartesian coordinates, the $\Gamma^i_{jk} = 0$, so that

$$\text{div } A^i = \frac{\partial A^1}{\partial x^1} + \frac{\partial A^2}{\partial x^2} + \cdots + \frac{\partial A^n}{\partial x^n}$$

The quantity $A^\alpha_{,\alpha}$ is a scalar invariant. If we let N_i be the components of the unit normal vector to the surface $d\sigma$, then $A^\alpha N_\alpha$ is also an invariant. In cartesian coordinates the divergence theorem is

$$\iiint_R \text{div } \mathbf{A}\, d\tau = \iint_S \mathbf{A}\cdot d\boldsymbol{\sigma} = \iint_S \mathbf{A}\cdot\mathbf{N}\, d\sigma$$

In tensor form it becomes

$$\iiint_R A^\alpha_{,\alpha}\, d\tau = \iint_S A^\alpha N_\alpha\, d\sigma \tag{530}$$

We can obtain Green's formula by considering the covariant vectors $\varphi_{,i}$ and $\psi_{,i}$. Now let

$$A_i = \psi\varphi_{,i} - \varphi\psi_{,i}$$

The associated vector of A_i is $A^\alpha = g^{\alpha i}A_i = g^{\alpha i}(\psi\varphi_{,i} - \varphi\psi_{,i})$. We easily see that $A^\alpha_{,\alpha} = g^{\alpha i}(\psi\varphi_{,i\alpha} - \varphi\psi_{,i\alpha})$. Now $g^{\alpha i}\varphi_{,i\alpha}$ is an invariant and in cartesian coordinates reduces to

$$\frac{\partial^2\varphi}{(\partial x^1)^2} + \frac{\partial^2\varphi}{(\partial x^2)^2} + \frac{\partial^2\varphi}{(\partial x^3)^2} = \text{Lap}\ \varphi$$

Hence, using (530), we obtain

$$\iiint\limits_R (\psi\ \text{Lap}\ \varphi - \varphi\ \text{Lap}\ \psi)\, d\tau = \iint\limits_S g^{\alpha i}A_i N_\alpha\, d\sigma$$
$$= \iint\limits_S (\psi\varphi_{,i} - \varphi\psi_{,i})N^i\, d\sigma$$

Example 147. Let us consider the covariant vector F_α. We multiply it by the contravariant vector dx^α and sum on α to obtain the invariant $F_\alpha\, dx^\alpha = F_\alpha \dfrac{dx^\alpha}{ds}\, ds$, which reduces to $\mathbf{f} \cdot d\mathbf{r}$ in cartesian coordinates. In Example 142 we constructed the curl of a vector, which turned out to be a tensor of rank 2. We now construct a vector whose components will also be those of the curl of a vector. We know that $F_{\alpha,\beta}$ is a tensor. Now define $\epsilon^{\alpha\beta\gamma} = \dfrac{1}{\sqrt{|g|}}$ if α, β, γ is an even permutation of 1, 2, 3; $\epsilon^{\alpha\beta\gamma} = -\dfrac{1}{\sqrt{|g|}}$ if α, β, γ, is an odd permutation of 1, 2, 3; $\epsilon^{\alpha\beta\gamma} = 0$ otherwise. Thus

$$\epsilon^{123} = \frac{1}{\sqrt{|g|}}, \qquad \epsilon^{231} = \frac{1}{\sqrt{|g|}}, \qquad \epsilon^{132} = -\frac{1}{\sqrt{|g|}}, \qquad \epsilon^{112} = 0$$

We obtain a new invariant

$$G^\gamma = \epsilon^{\alpha\beta\gamma}F_{\alpha,\beta}$$

In cartesian coordinates $F_{\alpha,\beta} = \dfrac{\partial F_\alpha}{\partial x^\beta}$, and

$$G^1 = \epsilon^{\alpha\beta 1}F_{\alpha,\beta} = \epsilon^{231}F_{2,3} + \epsilon^{321}F_{3,2} = \frac{\partial F_2}{\partial x^3} - \frac{\partial F_3}{\partial x^2}$$

and similarly for G^2 and G^3. Hence Stokes's theorem in tensor form reads

$$\int_\Gamma F_\alpha \frac{dx^\alpha}{ds}\, ds = \iint_S \epsilon^{\alpha\beta\gamma} F_{\alpha,\beta} N_\gamma \, d\sigma \tag{531}$$

Problems

1. By starting with $\bar{A}^i = A^\alpha \dfrac{\partial \bar{x}^i}{\partial x^\alpha}$, show that

$$A^i_{,j} \equiv \frac{\partial A^i}{\partial x^j} + A^\alpha \Gamma^i_{\alpha j}$$

is a mixed tensor.

2. Prove that $(g_{i\alpha} A^\alpha)_{,j} = g_{i\alpha} A^\alpha_{,j}$.
3. Prove that $(A^\alpha B_\alpha)_{,j} = A^\alpha B_{\alpha,j} + A^\alpha_{,j} B_\alpha$.
4. Prove that $|g|_{,j} = 0$.
5. Use (529) to find the Laplacian of F in cylindrical coordinates.
6. Prove that $\delta^i_{j,k} = 0$.
7. Prove that $\dfrac{1}{\sqrt{|g|}} \dfrac{\partial}{\partial x^\alpha} (\sqrt{|g|}\, g^{i\alpha}) + \Gamma^i_{\alpha\beta} g^{\alpha\beta} = 0$.
8. As in Example 143, show that the intrinsic derivative $\dfrac{\delta A^i}{\delta s} \equiv \dfrac{dA^i}{ds} + A^\alpha \Gamma^i_{\alpha\beta} \dfrac{dx^\beta}{ds}$ is a contravariant vector.
9. Show that the intrinsic derivative of a scalar of weight N is $\dfrac{\delta A}{\delta s} = \dfrac{dA}{ds} - NA\Gamma^\sigma_{j\sigma} \dfrac{dx^j}{ds}$, so that if A is an absolute constant, $\dfrac{\delta A}{\delta s} = 0$.
10. Show that $(g_{\alpha\beta} A^\alpha A^\beta)_{,j} = g_{\alpha\beta} A^\alpha_{,j} A^\beta + g_{\alpha\beta} A^\alpha A^\beta_{,j}$.
11. Show that $A^r_{s,t} \equiv \dfrac{\partial A^r_s}{\partial x^t} + \Gamma^r_{\mu t} A^\mu_s - \Gamma^\mu_{st} A^r_\mu$ for an absolute mixed tensor A^r_s.
12. Show that $\nabla^2(\varphi\psi) = \varphi\, \nabla^2\psi + 2\, \nabla\varphi \cdot \nabla\psi + \psi\, \nabla^2\varphi$.
13. Show that $A^\alpha_{i,\alpha} = \dfrac{1}{\sqrt{|g|}} \dfrac{\partial}{\partial x^\alpha} (\sqrt{|g|}\, A^\alpha_i) - A^\alpha_\beta \Gamma^\beta_{i\alpha}$.

14. If $A_i = A_i(x, t)$ is a covariant vector, show that

$$\frac{\delta A_i}{\delta t} = \frac{\partial A_i}{\partial t} + A_{i,j}\frac{dx^j}{dt}$$

and hence that the acceleration $f_i \equiv \dfrac{\delta v_i}{\delta t} = \dfrac{\partial v_i}{\partial t} + v_{i,j}v^j$.

15. Let λ_α be an arbitrary vector whose covariant derivative vanishes; that is, $\lambda_{\alpha,\beta} = 0$. Consider $\iint\limits_S T^{\alpha\beta}\,\lambda_\alpha N_\beta\,d\sigma$, and apply the divergence theorem to the vector $T^{\alpha\beta}\lambda_\alpha$. Hence show that

$$\iint\limits_S T^{\alpha\beta}N_\beta\,d\sigma = \iiint\limits_R T^{\alpha\beta}_{,\beta}\,d\tau.$$

16. Let s_i be the displacement vector of any particle from its position of equilibrium (see Sec. 115). We know that $s_{i,j}$ is a covariant tensor. The relative displacements of the particles are given by

$$\begin{aligned}\delta s_i &= s_{i,j}\,dx \\ &= \tfrac{1}{2}(s_{i,j} + s_{j,i})\,dx^j + \tfrac{1}{2}(s_{i,j} - s_{j,i})\,dx^j\end{aligned}$$

Show that the term $\frac{1}{2}(s_{i,j} - s_{j,i})\,dx^j$ represents a rotation.

We define the symmetric strain tensor E_{ij} by the equation

$$E_{ij} = \tfrac{1}{2}(s_{i,j} + s_{j,i})$$

The stress tensor T_{ij} is defined by the equations

$$\Delta F_i = T_{ij}N^j\,\Delta\sigma$$

where ΔF_i is the force acting on the element of area $\Delta\sigma$ with normal vector N^j (see Sec. 116).

Let f_r be the acceleration of the volume $d\tau$ and F_r be the force per unit mass acting on the mass in question. Show that

$$\iiint\limits_R \rho F_r\,d\tau + \iint\limits_S T_{rj}N^j\,d\sigma = \iiint\limits_R \rho f_r\,d\tau$$

or using contravariant components,

$$\iiint\limits_R \rho F^r\,d\tau + \iint\limits_S T^{rj}N_j\,d\sigma = \iiint\limits_R \rho f^r\,d\tau$$

Now deduce the equations of motion

$$\rho F^r + T^{rj}_{,j} = \rho f^r$$

If $T^{rj} = pg^{rj}$, show that

$$\rho F^r - p_{,j}g^{rj} = \rho f^r$$

or

$$\rho F_r - p_{,r} = \rho f_r \qquad \text{[see (411)]}$$

131. Geodesic Coordinates. The equations of the geodesics are given by

$$\frac{d^2x^i}{ds^2} + \Gamma^i_{jk}\frac{dx^j}{ds}\frac{dx^k}{ds} = 0$$

where the Γ^i_{jk} transform according to the law

$$\bar{\Gamma}^i_{jk} = \Gamma^\alpha_{\beta\gamma}\frac{\partial x^\beta}{\partial \bar{x}^j}\frac{\partial x^\gamma}{\partial \bar{x}^k}\frac{\partial \bar{x}^i}{\partial x^\alpha} + \frac{\partial^2 x^\alpha}{\partial \bar{x}^j\,\partial \bar{x}^k}\frac{\partial \bar{x}^i}{\partial x^\alpha} \tag{532}$$

We ask ourselves the following question: If the Γ^i_{jk} are different from zero at a point $x^i = q^i$, can we find a coordinate system such that $\bar{\Gamma}^i_{jk} = 0$ at the corresponding point? The answer is "Yes"!

Let

$$\bar{x}^i = (x^i - q^i) + \tfrac{1}{2}(\Gamma^i_{\alpha\beta})_q(x^\alpha - q^\alpha)(x^\beta - q^\beta) \tag{533}$$

so that $\left.\frac{\partial \bar{x}^i}{\partial x^j}\right|_q = \delta^i_j$ and $\left|\frac{\partial \bar{x}^i}{\partial x^j}\right|_q = 1$, and moreover the transformation (533) is nonsingular. The point $x^i = q^i$ corresponds to the point $\bar{x}^i = 0$.

Now differentiating (533) with respect to $\bar{x}^j$, we obtain

$$\delta^i_j = \frac{\partial x^i}{\partial \bar{x}^j} + (\Gamma^i_{\alpha\beta})_q(x^\alpha - q^\alpha)\frac{\partial x^\beta}{\partial \bar{x}^j} \tag{534}$$

because of the symmetry of $\Gamma^i_{\alpha\beta}$. Hence $\left.\frac{\partial x^i}{\partial \bar{x}^j}\right|_q = \delta^i_j$.

Differentiating (534) with respect to $\bar{x}^k$, we obtain

$$0 = \frac{\partial^2 x^i}{\partial \bar{x}^k\,\partial \bar{x}^j} + (\Gamma^i_{\alpha\beta})_q\frac{\partial x^\alpha}{\partial \bar{x}^k}\frac{\partial x^\beta}{\partial \bar{x}^j} + (\Gamma^i_{\alpha\beta})_q(x^\alpha - q^\alpha)\frac{\partial^2 x^\beta}{\partial \bar{x}^k\,\partial \bar{x}^j}$$

so that

$$\left.\frac{\partial^2 x^i}{\partial \bar{x}^k\,\partial \bar{x}^j}\right|_q = -(\Gamma^i_{\alpha\beta})_q \left.\frac{\partial x^\alpha}{\partial \bar{x}^k}\right|_q \left.\frac{\partial x^\beta}{\partial \bar{x}^j}\right|_q$$
$$= -(\Gamma^i_{\alpha\beta})_q \delta^\alpha_k \delta^\beta_j = -(\Gamma^i_{jk})_q$$

Substituting into (532), we obtain

$$(\bar{\Gamma}^i_{jk})_0 = (\Gamma^\alpha_{\beta\gamma})_q \delta^\beta_j \delta^\gamma_k \delta^i_\alpha - (\Gamma^\alpha_{jk})_q \delta^i_\alpha$$
$$= (\Gamma^i_{jk})_q - (\Gamma^i_{jk})_q = 0, \qquad \text{Q.E.D.}$$

Any system of coordinates for which $(\Gamma^i_{jk})_P = 0$ at a point P is called a geodesic coordinate system. In such a system, the covariant derivative, when evaluated at the origin, becomes the ordinary derivative evaluated at the origin. For example,

$$(A^i_{,j})_O = \left(\frac{\partial A^i}{\partial x^j}\right)_O + (\Gamma^i_{\alpha j})_O (A^\alpha)_O = \left(\frac{\partial A^i}{\partial x^j}\right)_O$$

since $\Gamma^i_{\alpha j} = 0$ at the origin.

The covariant derivative of a sum or product of tensors must obey the same rules that hold for ordinary derivatives of the calculus, for at any point we can choose geodesic coordinates so that

$$A^i_{,j} + B^i_{,j} = \frac{\partial A^i}{\partial x^j} + \frac{\partial B^i}{\partial x^j} = \frac{\partial (A^i + B^i)}{\partial x^j} = (A^i + B^i)_{,j}$$

and $A^i_{,j} + B^i_{,j} - (A^i + B^i)_{,j}$ is a zero tensor for geodesic coordinates. Hence $A^i_{,j} + B^i_{,j} - (A^i + B^i)_{,j}$ is zero in all coordinate systems, so that

$$A^i_{,j} + B^i_{,j} \equiv (A^i + B^i)_{,j}$$

We leave it as an exercise for the reader to prove that

$$(A^i B_j)_{,k} = A^i_{,k} B_j + A^i B_{j,k}$$

Equation (533) yields one geodesic coordinate system. There are infinitely many such systems, since we could have added $\varphi_{\alpha\beta\gamma}(x)(x^\alpha - q^\alpha)(x^\beta - q^\beta)(x^\gamma - q^\gamma)$ to the right-hand side of (533) and still have obtained $(\bar{\Gamma}^i_{jk})_0 = 0$.

A special type of geodesic coordinate is the following: Let $x^i = x^i(s)$ be a geodesic passing through the point P, $x^i = x^i_0$, and let $\xi^i = \left.\dfrac{dx^i}{ds}\right|_P$. Define

$$\bar{x}^i = \xi^i s \tag{535}$$

where s is arc length along the geodesic. Each ξ^i determines a geodesic through P, and s determines a point on this geodesic. Hence every point in the neighborhood of P has the definite coordinate $\bar{x}^i$ attached to it. The equations of the geodesics in this coordinate system are

$$\frac{d^2\bar{x}^i}{ds^2} + \Gamma^i_{jk}\frac{d\bar{x}^j}{ds}\frac{d\bar{x}^k}{ds} = 0$$

But $\dfrac{d\bar{x}^i}{ds} = \xi^i$ and $\dfrac{d^2\bar{x}^i}{ds^2} = 0$, so that

$$\Gamma^i_{jk}\xi^j\xi^k = 0 \tag{536}$$

Since this equation holds at the point P for all directions ξ^i, we must have $\Gamma^i_{jk} + \Gamma^i_{kj} = 2\Gamma^i_{jk} = 0$, so that the $\bar{x}^i$ are geodesic coordinates. The $\bar{x}^i = \xi^i s$ are called Riemannian coordinates.

Example 148. If ξ^i is a unit vector, we have

$$g_{\alpha\beta}\xi^\alpha\xi^\beta \equiv 1$$

The intrinsic derivative is

$$g_{\alpha\beta}\frac{\delta\xi^\alpha}{\delta s}\xi^\beta + g_{\alpha\beta}\xi^\alpha\frac{\delta\xi^\beta}{\delta s} = 0$$

since $(g_{\alpha\beta})_{,j} = 0$ (see Example 140). Hence $g_{\alpha\beta}\xi^\alpha\dfrac{\delta\xi^\beta}{\delta s} = 0$, and $g_{\alpha\beta}\xi^\alpha\left(\dfrac{d\xi^\beta}{ds} + \xi^\mu\Gamma^\beta_{\mu\sigma}\dfrac{dx^\sigma}{ds}\right) = 0$. We see that the vector

$$\frac{d\xi^i}{ds} + \Gamma^i_{\mu\sigma}\xi^\mu\frac{dx^\sigma}{ds}$$

is normal to the vector ξ^i.

Problems

1. Show that $g_{\alpha\beta}\dfrac{dx^\alpha}{ds}\dfrac{dx^\beta}{ds}$ remains constant along a geodesic.
2. Show that for normal coordinates $\bar{x}^i$, $\Gamma^i_{jk}\bar{x}^j\bar{x}^k = 0$.

3. If s is arc length of the curve C, show that the intrinsic derivative of the unit tangent $\frac{dx^i}{ds}$ in the direction of the curve has the components

$$p^i = \frac{d^2x^i}{ds^2} + \Gamma^i_{jk}\frac{dx^j}{ds}\frac{dx^k}{ds}$$

What are the components for a geodesic?

4. Prove that $\frac{\delta}{\delta t}(X^\alpha Y_\alpha) = \frac{\delta X^\alpha}{\delta t} Y_\alpha + X^\alpha \frac{\delta Y_\alpha}{\delta t}$.

132. The Curvature Tensor. Let us consider the absolute contravariant vector V^i. Its covariant derivative yields the mixed tensor

$$V^i_{,j} = \frac{\partial V^i}{\partial x_j} + V^\alpha \Gamma^i_{\alpha j}$$

On again differentiating covariantly, we obtain

$$\begin{aligned} V^i_{,jk} &= \frac{\partial V^i_{,j}}{\partial x^k} + V^\alpha_{,j}\Gamma^i_{\alpha k} - V^i_{,\alpha}\Gamma^\alpha_{jk} \\ &= \frac{\partial^2 V^i}{\partial x^k\,\partial x^j} + \frac{\partial V^\alpha}{\partial x^k}\Gamma^i_{\alpha j} + V^\alpha \frac{\partial \Gamma^i_{\alpha j}}{\partial x^k} + \left(\frac{\partial V^\alpha}{\partial x^j} + V^\beta \Gamma^\alpha_{\beta j}\right)\Gamma^i_{\alpha k} \\ &\qquad - \left(\frac{\partial V^i}{\partial x^\alpha} + V^\beta \Gamma^i_{\beta\alpha}\right)\Gamma^\alpha_{jk} \end{aligned}$$

Interchanging k and j and subtracting, we have

$$V^i_{,jk} - V^i_{,kj} = V^\alpha B^i_{\alpha jk} \tag{537}$$

where

$$B^i_{\alpha jk} = \frac{\partial \Gamma^i_{\alpha j}}{\partial x^k} - \frac{\partial \Gamma^i_{\alpha k}}{\partial x^j} + \Gamma^\beta_{\alpha j}\Gamma^i_{\beta k} - \Gamma^\beta_{\alpha k}\Gamma^i_{\beta j} \tag{538}$$

Since $V^i_{,jk} - V^i_{,kj}$ and V^i are tensors, $V^i_{\alpha jk}$ must be the components of a tensor, from the quotient law (Sec. 126). It is called the curvature tensor. We can obtain two new tensors of the second order by contraction.

Let

$$R_{ij} = B^{\alpha}_{i\alpha j} = \frac{\partial \Gamma^{\alpha}_{i\alpha}}{\partial x^j} - \frac{\partial \Gamma^{\alpha}_{ij}}{\partial x^{\alpha}} + \Gamma^{\beta}_{i\alpha}\Gamma^{\alpha}_{\beta j} - \Gamma^{\beta}_{ij}\Gamma^{\alpha}_{\beta\alpha} \tag{539}$$

This tensor is called the Ricci tensor and plays an important role in the theory of relativity.

We obtain another tensor by defining

$$S_{ij} = B^{\alpha}_{\alpha ij} = \frac{\partial \Gamma^{\alpha}_{\alpha i}}{\partial x^j} - \frac{\partial \Gamma^{\alpha}_{\alpha j}}{\partial x^i} \tag{540}$$

Evidently $S_{ij} = -S_{ji,}$ and if we use the fact that

$$\frac{\partial \log \sqrt{|g|}}{\partial x^{\mu}} = \Gamma^{\alpha}_{\alpha\mu},$$

we have that

$$S_{ij} = \frac{\partial^2 \log \sqrt{|g|}}{\partial x^i \, \partial x^j} - \frac{\partial^2 \log \sqrt{|g|}}{\partial x^j \, \partial x^i} \equiv 0$$

Now $R_{ij} - R_{ji} = S_{ij} = 0$, so that the Ricci tensor is symmetric in its indices. We could have deduced this fact by examining (539) directly.

The invariant $R = g^{ij}R_{ij}$ is called the *scalar curvature*.

133. Riemann-Christoffel Tensor. The tensor

$$R_{hijk} = g_{h\alpha}B^{\alpha}_{ijk} \tag{541}$$

is called the Riemann-Christoffel, or covariant curvature, tensor. Let us note the following important result: Assume that the Riemannian space is Euclidean and that we are dealing with a cartesian coordinate system. Since $\Gamma^i_{jk}(x) = 0$, we have from (538)

$$B^i_{jkl} = 0 \tag{542}$$

in this coordinate system. But if $B^i_{jkl} = 0$ in one coordinate system, the components are zero in all coordinate systems. Hence if a space is Euclidean, the curvature tensor must vanish. We shall show later that if $B^i_{jkl} = 0$, the space is Euclidean.

If we differentiate (538) and evaluate at the origin of a geodesic coordinate system, we obtain

$$B^i_{\alpha jk,\sigma} = \frac{\partial^2 \Gamma^i_{\alpha j}}{\partial x^\sigma \, \partial x^k} - \frac{\partial^2 \Gamma^i_{\alpha k}}{\partial x^\sigma \, \partial x^j}$$

Permuting j, k, σ and adding, we have the Bianchi identity

$$B^i_{\alpha kj,\sigma} + B^i_{\alpha\sigma j,k} + B^i_{\alpha k\sigma,l} = 0 \tag{543}$$

134. Euclidean Space. We have seen that if a space is Euclidean, of necessity $B^i_{jkl} = 0$. We shall now prove that if the $B^i_{jkl} = 0$, the space is Euclidean. Now

$$\bar{\Gamma}^i_{jk}(y) = \Gamma^\alpha_{\beta\gamma}(x) \frac{\partial x^\beta}{\partial y^j} \frac{\partial x^\gamma}{\partial y^k} \frac{\partial y^i}{\partial x^\alpha} + \frac{\partial^2 x^\alpha}{\partial y^j \, \partial y^k} \frac{\partial y^i}{\partial x^\alpha}$$

If there is a coordinate system $(x^1, x^2, \ldots, x^n)$ for which $\Gamma^\alpha_{\beta\gamma}(x) = 0$, then

$$\bar{\Gamma}^i_{jk}(y) = \frac{\partial^2 x^\alpha}{\partial y^j \, \partial y^k} \frac{\partial y^i}{\partial x^\alpha} \tag{544}$$

and conversely, if (544) holds, the $\Gamma^\alpha_{\beta\gamma}(x) = 0$. Now let us investigate under what conditions (544) may result. We write (544) as

$$\frac{\partial^2 x^\sigma}{\partial y^j \, \partial y^k} = \frac{\partial x^\sigma}{\partial y^i} \bar{\Gamma}^i_{jk}(y) \tag{545}$$

which represents a system of second-order differential equations. Let us define

$$u^\sigma_i = \frac{\partial x^\sigma}{\partial y^i} \tag{546}$$

so that (545) becomes

$$\frac{\partial u^\sigma_k}{\partial y^j} = u^\sigma_i \bar{\Gamma}^i_{jk}(y) \tag{547}$$

For each σ we have the first-order system of differential equations given by (546) and (547), which are special cases of the more general system

$$\frac{\partial z^k}{\partial y^j} = f_j^k(z^1, z^2, \ldots z^n, z^{n+1}, y^1, \ldots, y^n) \qquad \begin{aligned} k &= 1, 2, \ldots, n+1 \\ j &= 1, 2, \ldots, n \end{aligned} \tag{548}$$

If we let $z^1 = x^\sigma$, $z^2 = u_1^\sigma$, $z^3 = u_2^\sigma$, . . . , $z^{n+1} = u_n^\sigma$, Eqs. (546) and (547) reduce to (548).

We certainly must have $\dfrac{\partial^2 z}{\partial y^l\, \partial y^j} = \dfrac{\partial^2 z}{\partial y^j\, \partial y^l}$, and this implies

$$\left.\begin{aligned} \frac{\partial f_j^k}{\partial y^l} + \frac{\partial f_j^k}{\partial z^\mu}\frac{\partial z^\mu}{\partial y^l} &= \frac{\partial f_l^k}{\partial y^j} + \frac{\partial f_l^k}{\partial z^\mu}\frac{\partial z^\mu}{\partial y^j} \\ \text{or} \qquad \frac{\partial f_j^k}{\partial y^l} + \frac{\partial f_j^k}{\partial z^\mu} f_l^\mu &= \frac{\partial f_l^k}{\partial y^j} + \frac{\partial f_l^k}{\partial z^\mu} f_j^\mu \end{aligned}\right\} \tag{549}$$

If the f_j^i are analytic, it can be shown that the integrability conditions (549) are also sufficient that (548) have a solution satisfying the initial conditions $z^k = z_0^k$ at $y^i = y_0^i$. The reader is referred to advanced texts on differential equations and especially to the elegant proof found in Gaston Darboux, "Leçons systèmes orthogonaux et les coordonnées curvilignes," pp. 325–336, Gauthier-Villars, Paris, 1910.

The integrability conditions (549), when referred to the system (546), (547), become

$$\begin{aligned} \Gamma_{jk}^\alpha u_\alpha^\sigma &= \Gamma_{kj}^\alpha u_\alpha^\sigma \\ B_{jkl}^\alpha u_\alpha^\sigma &= 0 \end{aligned} \tag{550}$$

The first equation of (550) is satisfied from the symmetry of the Γ_{jk}^i, and the second is satisfied if $B_{jkl}^\alpha = 0$. Hence, if $B_{jkl}^\alpha = 0$, we can solve (545) for $z^1 = x^\sigma$ in terms of $y^1, y^2, \ldots, y^n$. For the coordinate system $(x^1, x^2, \ldots, x^n)$, we have $\Gamma_{jk}^i(x) = 0$.

Problems

1. Show that $R_{hijk} = -R_{ihjk} = -R_{hikj}$ and that

$$R_{iijk} = R_{hikk} = 0$$

2. Show that $R_{hijk} + R_{hkij} + R_{hjki} = 0$.
3. If $R_{ij} = kg_{ij}$, show that $R = nk$.

4. For a two-dimensional space for which $g_{12} = g_{21} = 0$, show that $R_{12} = 0$, $R_{11}g_{22} = R_{22}g_{11} = R_{1221}$ and that

$$R = \frac{R_{1221}}{g_{11}g_{22}}$$

$R_{ij} = \frac{1}{2}Rg_{ij}$.

5. Show that $B^i_{jkl} \neq 0$ for a space whose line element is given by $ds^2 = (dx^1)^2 + (\sin x^1)^2(dx^2)^2$.

6. Derive (550) from (546), (547), (549).

7. If $R^i_j = g^{i\alpha}R_{\alpha j}$, show that $(R^i_j)_{,i} = \dfrac{1}{2}\dfrac{\partial R}{\partial x^j}$.

CHAPTER 9

FURTHER APPLICATIONS OF TENSOR ANALYSIS

135. Frenet-Serret Formulas. Let $\lambda^i = \dfrac{dx^i}{ds}$ be the unit tangent vector to the space curve $x^i = x^i(t)$, $i = 1, 2, 3$, in a Riemannian space. In Example 148 we saw that the contravariant vector $\dfrac{\delta\lambda^i}{\delta s}$ is normal to λ^i. Let us define the curvature as $\kappa = g_{\alpha\beta}\dfrac{\delta\lambda^\alpha}{\delta s}\dfrac{\delta\lambda^\beta}{\delta s}$ and the principal unit normal μ^i by

$$\frac{\delta\lambda^i}{\delta s} = \kappa\mu^i \tag{551}$$

Since μ^i is a unit vector, we know that $\dfrac{\delta\mu^i}{\delta s}$ is normal to μ^i. Now $g_{\alpha\beta}\lambda^\alpha\mu^\beta = 0$, so that the intrinsic derivative yields

$$g_{\alpha\beta}\lambda^\alpha\frac{\delta\mu^\beta}{\delta s} + g_{\alpha\beta}\frac{\delta\lambda^\alpha}{\delta s}\mu^\beta = 0$$

since $g_{\alpha\beta,j} = 0$ or $\dfrac{\delta g_{\alpha\beta}}{\delta s} = 0$. Hence

$$g_{\alpha\beta}\,\lambda^\alpha\frac{\delta\mu^\beta}{\delta s} + \kappa g_{\alpha\beta}\mu^\alpha\mu^\beta = 0$$

or

$$g_{\alpha\beta}\lambda^\alpha\left(\frac{\delta\mu^\beta}{\delta s} + \kappa\lambda^\beta\right) = 0 \tag{552}$$

since $g_{\alpha\beta}\mu^\alpha\mu^\beta = g_{\alpha\beta}\lambda^\alpha\lambda^\beta \equiv 1$.

Equation (552) shows that $\dfrac{\delta\mu^i}{\delta s} + \kappa\lambda^i$ is normal to λ^i, and since $\dfrac{\delta\mu^i}{\delta s}$ and λ^i are normal to μ^i, $\dfrac{\delta\mu^i}{\delta s} + \kappa\lambda^i$ is also normal to μ^i. We define the binormal ν^i by the equation

$$\nu^i = -\frac{1}{\tau}\left(\frac{\delta\mu^i}{\delta s} + \kappa\lambda^i\right) \tag{553}$$

or

$$\frac{\delta\mu^i}{\delta s} = -\kappa\lambda^i - \tau\nu^i \tag{554}$$

where τ is called the torsion and is the magnitude of

$$-\left(\frac{\delta\mu^i}{\delta s} + \kappa\lambda^i\right)$$

Since ν^i is normal to both λ^i and μ^i, we have

$$\begin{aligned} g_{\alpha\beta}\nu^\alpha\lambda^\beta &= 0 \\ g_{\alpha\beta}\nu^\alpha\mu^\beta &= 0 \end{aligned} \tag{555}$$

By differentiating (555) and using (551), (554), (555), we leave it to the reader to show that

$$g_{\alpha\beta}\mu^\beta\left(\tau\mu^\alpha - \frac{\delta\nu^\alpha}{\delta s}\right) = 0 \tag{556}$$

The vector $\tau\mu^i - \dfrac{\delta\nu^i}{\delta s}$ is thus normal to all three vectors λ^i, μ^i, ν^i. Since we are dealing in three-space, this is possible only if $\tau\mu^i - \dfrac{\delta\nu^i}{\delta s} = 0$, or

$$\frac{\delta\nu^i}{\delta s} = \tau\mu^i \tag{557}$$

Writing (551), (554), (557) in full, we have the Frenet-Serret formulas

$$\begin{aligned} \frac{d\lambda^i}{ds} + \Gamma^i_{\alpha\beta}\,\lambda^\alpha\,\frac{dx^\beta}{ds} &= \kappa\mu^i \\ \frac{d\mu^i}{ds} + \Gamma^i_{\alpha\beta}\,\mu^\alpha\,\frac{dx^\beta}{ds} &= -(\kappa\lambda^i + \tau\nu^i) \\ \frac{d\nu^i}{ds} + \Gamma^i_{\alpha\beta}\,\nu^\alpha\,\frac{dx^\beta}{ds} &= \tau\mu^i \end{aligned} \tag{558}$$

For a Euclidean space using cartesian coordinates, the $\Gamma^i_{\alpha\beta} = 0$, and (558) reduces to the formulas encountered in Sec. 24.

Problems

1. Derive (556).
2. Using cylindrical coordinates,

$$ds^2 = (dx^1)^2 + (x^1)^2(dx^2)^2 + (dx^3)^2$$

and for a circle $x^1 = a$, $x^2 = t$, $x^3 = 0$. Expand (558) for this case, and show that $\kappa = 1/a$, $\tau = 0$.

3. Show that $\dfrac{\delta^2\lambda^i}{\delta s^2} = -\kappa^2\lambda^i + \dfrac{\delta\kappa}{\delta s}\mu^i - \kappa\tau\nu^i$.

4. Since (558) is true for a Euclidean space using cartesian coordinates, why would (558) hold for all other coordinate systems in this Euclidean space?

5. Since $\lambda^i = \dfrac{dx^i}{ds}$, show that $\dfrac{d^2x^i}{ds^2} + \Gamma^i_{\alpha\beta}\dfrac{dx^\alpha}{ds}\dfrac{dx^\beta}{ds}$ are the components of a contravariant vector.

136. Parallel Displacement of Vectors. Consider an absolute contravariant vector $A^i(x^1, x^2, \ldots, x^n)$ in a cartesian coordinate system. Let us assume that the components A^i are constants. Now

$$\bar{A}^i = A^\alpha \frac{\partial \bar{x}^i}{\partial x^\alpha}, \qquad A^i = \bar{A}^\alpha \frac{\partial x^i}{\partial \bar{x}^\alpha}$$

so that

$$d\bar{A}^i = A^\alpha \frac{\partial^2 \bar{x}^i}{\partial x^\beta\, \partial x^\alpha} \frac{\partial x^\beta}{\partial \bar{x}^\gamma} d\bar{x}^\gamma$$

since $dA^i = 0$. We thus obtain

$$d\bar{A}^i = \bar{A}^\sigma \frac{\partial^2 \bar{x}^i}{\partial x^\beta\, \partial x^\alpha} \frac{\partial x^\beta}{\partial \bar{x}^\gamma} \frac{\partial x^\alpha}{\partial \bar{x}^\sigma} d\bar{x}^\gamma$$

From (511*a*) (see page 292)

$$\bar{\Gamma}^i_{\gamma\sigma} = \frac{\partial^2 x^\alpha}{\partial \bar{x}^\gamma\, \partial \bar{x}^\sigma} \frac{\partial \bar{x}^i}{\partial x^\alpha}, \qquad \text{since } \Gamma^i_{jk} = 0$$

$$= -\frac{\partial^2 \bar{x}^i}{\partial x^\beta\, \partial x^\alpha} \frac{\partial x^\beta}{\partial \bar{x}^\gamma} \frac{\partial x^\alpha}{\partial \bar{x}^\sigma} \qquad \text{from (483)}$$

so that

$$d\bar{A}^i = -\bar{A}^\sigma \bar{\Gamma}^i_{\sigma\gamma}\, d\bar{x}^\gamma \tag{559}$$

In general, a Riemannian space is not Euclidean. We generalize (559) and define parallelism of a vector field A^i with respect to a curve C given by $x^i = x^i(s)$ as follows: We say that A^i is parallelly displaced with respect to the Riemannian V_n along the curve C, if

$$\frac{dA^i}{ds} = -A^\sigma \Gamma^i_{\sigma\tau} \frac{dx^\tau}{ds}$$

or

$$\boxed{\frac{\delta A^i}{\delta s} \equiv \frac{dA^i}{ds} + \Gamma^i_{\sigma\tau} A^\sigma \frac{dx^\tau}{ds} = 0} \tag{560}$$

We say that the vector A^i suffers a parallel displacement along the curve. Notice that the intrinsic derivative of A^i along the curve $x^i(s)$ vanishes.

In particular, for a geodesic we have $\dfrac{d^2x^i}{ds^2} + \Gamma^i_{\alpha\beta} \dfrac{dx^\alpha}{ds}\dfrac{dx^\beta}{ds} = 0$, so that the unit tangent vector $\dfrac{dx^i}{ds}$ suffers a parallel displacement along the geodesic.

Example 149. Let us consider two unit vectors A^i, B^i, which undergo parallel displacements along a curve. We have

$$\cos\theta = g_{\alpha\beta} A^\alpha B^\beta$$

and

$$\frac{\delta(\cos\theta)}{\delta s} = g_{\alpha\beta} \frac{\delta A^\alpha}{\delta s} B^\beta + g_{\alpha\beta} A^\alpha \frac{\delta B^\beta}{\delta s} = 0$$

so that $\theta \equiv$ constant. Hence, if two vectors of constant magnitudes undergo parallel displacements along a given curve, they are inclined at a constant angle.

Two vectors at a point are said to be parallel if their corresponding components are proportional. If A^i is a vector of constant magnitude, the vector $B^i = \varphi A^i$, $\varphi =$ scalar, is parallel to A^i. If A^i is also parallel with respect to the V_n along a curve $x^i = x^i(s)$, we have $\dfrac{\delta A^i}{\delta s} = 0$. Now

$$\frac{\delta B^i}{\delta s} = \varphi \frac{\delta A^i}{\delta s} + \frac{d\varphi}{ds} A^i = \frac{d\varphi}{ds} A^i = \frac{1}{\varphi}\frac{d\varphi}{ds} B^i = \frac{d(\log\varphi)}{ds} B^i$$

We desire B^i to be parallel with respect to the V_n along the curve, so that a vector B^i of variable magnitude must satisfy an equation of the type

$$\frac{\delta B^i}{\delta s} = f(s)B^i \tag{561}$$

if it is to be parallelly displaced along the curve.

Problems

1. Show that if the vector A^i of constant magnitude is parallelly displaced along a geodesic, it makes a constant angle with the geodesic.

2. If a vector A^i satisfies (560), show that it is of constant magnitude.

3. If a vector B^i satisfies (561) along a curve Γ, by letting $A^i = \psi B^i$ show that it is possible to find ψ so that A^i suffers a parallel displacement along Γ.

4. Let $x^i(t)$, $0 \leqq t \leqq 1$, be an infinitesimal closed path. The change in the components of a contravariant vector on being parallelly displaced along this closed path is $\Delta A^i = -\oint \Gamma^i_{\alpha\beta} A^\alpha \, dx^\beta$, from (560). Expand $A^\alpha(x)$, $\Gamma^i_{\alpha\beta}(x)$ in Taylor series about $x^i_0 = x^i(0)$, and neglecting infinitesimals of higher order, show that

$$\Delta A^i = \tfrac{1}{4} R^i_{\alpha\beta\gamma} A^\alpha \oint x^\gamma \, dx^\beta - x^\beta \, dx^\gamma$$

where $R^i_{\alpha\beta\gamma}$ is the curvature tensor (see Secs. 132, 133, 134).

137. Parallelism in a Subspace. We start with the Riemannian space, V_n, $ds^2 = g_{\alpha\beta} \, dx^\alpha \, dx^\beta$. If we consider the transformation

$$x^\alpha = x^\alpha(u^1, u^2, \ldots, u^m), \qquad m < n \tag{562}$$

we see that a point with coordinates $u^1, u^2, \ldots, u^m$ is a point of V_m and also a point of V_n. The converse is not true, for given the point with coordinates $x^1, x^2, \ldots, x^n$, there may not exist $u^1, u^2, \ldots, u^m$ which satisfy $x^\alpha = x^\alpha(u^1, u^2, \ldots, u^m)$, since $m < n$. Now

$$\begin{aligned} ds^2 &= g_{\alpha\beta} \, dx^\alpha \, dx^\beta = g_{\alpha\beta} \frac{\partial x^\alpha}{\partial u^i} \frac{\partial x^\beta}{\partial u^j} \, du^i \, du^j \\ &= h_{ij} \, du^i \, du^j \end{aligned}$$

so that the fundamental metric tensor in the subspace, V_m, is given by

$$h_{ij} = g_{\alpha\beta} \frac{\partial x^\alpha}{\partial u^i} \frac{\partial x^\beta}{\partial u^j}$$

Now $dx^\alpha = \frac{\partial x^\alpha}{\partial u^i} du^i$, so that if du^i are the components of a contravariant vector in the V_m, dx^α are the components of the same vector in the V_n. In general, if $a^i(u^1, \ldots, u^m)$, $i = 1, 2, \ldots, m$, are the components of a contravariant vector in the V_m, we say that

$$A^\alpha \equiv \frac{\partial x^\alpha}{\partial u^i} a^i \quad \alpha = 1, 2, \ldots, n \tag{563}$$

are the components of the same vector in the V_n.

We now find a relationship between $\frac{\delta A^\alpha}{\delta s}$ and $\frac{\delta a^i}{\delta s}$, where s is arc length along the curve $u^i = u^i(s)$ or the space curve

$$x^\alpha = x^\alpha[u^i(s)]$$

Differentiating (563), we have

$$\frac{dA^\alpha}{ds} = \frac{\partial x^\alpha}{\partial u^i} \frac{da^i}{ds} + \frac{\partial^2 x^\alpha}{\partial u^j \, \partial u^i} \frac{du^j}{ds} a^i$$

and

$$\frac{\delta A^\alpha}{\delta s} \equiv \frac{dA^\alpha}{ds} + (\Gamma^\alpha_{\beta\gamma})_g A^\beta \frac{dx^\gamma}{ds} = \frac{\partial x^\alpha}{\partial u^i} \frac{da^i}{ds} + \frac{\partial^2 x^\alpha}{\partial u^j \, \partial u^i} \frac{du^j}{ds} a^i + (\Gamma^\alpha_{\beta\gamma})_g a^i \frac{\partial x^\beta}{\partial u^i} \frac{\partial x^\gamma}{\partial u^j} \frac{du^j}{ds}$$

Hence

$$g_{\sigma\alpha} \frac{\partial x^\sigma}{\partial u^k} \frac{\delta A^\alpha}{\delta s} = h_{ik} \frac{da^i}{ds} + a^i \frac{du^j}{ds} \left[g_{\sigma\alpha} (\Gamma^\alpha_{\beta\gamma})_g \frac{\partial x^\beta}{\partial u^i} \frac{\partial x^\gamma}{\partial u^j} \frac{\partial x^\sigma}{\partial u^k} + g_{\sigma\alpha} \frac{\partial^2 x^\alpha}{\partial u^j \, \partial u^i} \frac{\partial x^\sigma}{\partial u^k} \right]$$

$$g_{\sigma\alpha} \frac{\partial x^\sigma}{\partial u^k} \frac{\delta A^\alpha}{\delta s} = h_{lk} \frac{da^l}{ds} + a^i \frac{du^j}{ds} h_{lk} (\Gamma^l_{ji})_h, \quad \text{(see Prob. 6, Sec. 129)}$$

Hence

$$g_{\sigma\alpha}\frac{\partial x^\sigma}{\partial u^k}\frac{\delta A^\alpha}{\delta s} = h_{lk}\left[\frac{da^l}{ds} + (\Gamma^l_{ji})_h a^i \frac{du^j}{ds}\right]$$

and

$$g_{\sigma\alpha}\frac{\partial x^\sigma}{\partial u^k}\frac{\delta A^\alpha}{\delta s} = h_{lk}\frac{\delta a^l}{\delta s} \tag{564}$$

From (564) we see that if a^i is parallelly displaced along $x^\alpha[u^i(s)]$, that is, if $\frac{\delta A^i}{\delta s} = 0$, then $\frac{\delta a^i}{\delta s} = 0$. Thus the theorem: If a curve C lies in a subspace V_m of V_n, and a vector field in V_m is parallel along C with respect to V_n, then it is also parallel along C with respect to V_m.

Problems

1. Prove that if a curve is a geodesic in a V_n, it is a geodesic in any subspace V_m of V_n. Consider the unit tangents to the geodesics.

2. By considering k fixed, show that $\frac{\partial x^\sigma}{\partial u^k}$ is a contravariant vector of V_n tangent to the u^k curve, obtained by considering $u^1, u^2, \ldots u^{k-1}, u^{k+1}, \ldots u^m$ fixed in the equations

$$x^\alpha = x^\alpha(u^1, \ldots, u^m)$$

3. If a^i is parallel along C with respect to the V_m, show that $\frac{\delta A^\alpha}{\delta s}$ is normal to the space V_m, that is, normal to the u^i curves, $i = 1, 2, \ldots, m$.

4. Under a coordinate transformation $\bar{u}^i = \bar{u}^i(u^1, \ldots, u^m)$, $i = 1, 2, \ldots, m$, the x^α remain invariant. Hence show that

$$x^\alpha_{,j} = \frac{\partial x^\alpha}{\partial u^j}, \qquad g_{\alpha\beta,j} = \frac{\partial g_{\alpha\beta}}{\partial x^\sigma}x^\sigma_{,j}$$

where the covariant derivatives are performed relative to the metric h_{ij}, that is, $a^i_{,j} = \frac{\partial a^i}{\partial u^j} + a^k(\Gamma^i_{jk})_h$.

5. Show that $g_{\alpha\beta}(x^\alpha_{,ik}x^\beta_{,j} + x^\alpha_{,i}x^\beta_{,jk}) + x^\alpha_{,i}x^\beta_{,j}x^\sigma_{,k}\dfrac{\partial g_{\alpha\beta}}{\partial x^\sigma} = 0$, where covariant differentiation is with respect to u^i and h_{ij}.

6. Show that $A^\alpha_{i,j} = x^\alpha_{,ij}a^i + x^\alpha_{,i}a^i_{,j}$, for each α.

138. Generalized Covariant Differentiation. The quantities $\dfrac{\partial x^\alpha}{\partial u^i}$ are contravariant vectors if we consider i fixed; for if

$$x^\alpha = x^\alpha(u^1, \ldots, u^m)$$

and $y^\alpha = y^\alpha(x^1, \ldots, x^n)$, $\alpha = 1, 2, \ldots, n$, then

$$\frac{\partial y^\alpha}{\partial u^i} = \frac{\partial y^\alpha}{\partial x^\beta}\frac{\partial x^\beta}{\partial u^i}$$

showing that the $\dfrac{\partial x^\alpha}{\partial u^i}$ transform like a contravariant vector. However, if we consider α as fixed, the $\dfrac{\partial x^\alpha}{\partial u^i}$, $i = 1, 2, \ldots, m$, are covariant vectors in the V_m; for if $\bar{u}^i = \bar{u}^i(u^1, \ldots, u^m)$, we have $\dfrac{\partial x^\alpha}{\partial \bar{u}^i} = \dfrac{\partial x^\alpha}{\partial u^j}\dfrac{\partial u^j}{\partial \bar{u}^i}$. We propose to consider tensors of this type, Latin indices indicating tensors of the V_m, and Greek indices indicating tensors of the V_n.

Let us consider the tensor A^α_i. We wish to derive a new tensor which will be a tensor in the V_n for Greek indices and a tensor in V_m for Latin indices. We consider a curve C in V_m given by $u^i = u^i(s)$ and by $x^\alpha = x^\alpha(s)$ in V_n. Let b^i be the components of a vector field in V_m parallel along C with respect to V_m, and let c_α be the components of a vector field in V_n parallel along C with respect to V_n. We have

$$\begin{aligned} \frac{db^i}{ds} + \Gamma^i_{jk}b^j\frac{du^k}{ds} &= 0 \\ \frac{dc_\alpha}{ds} - \Gamma^\mu_{\alpha\beta}c_\mu\frac{dx^\beta}{ds} &= 0 \end{aligned} \tag{565}$$

We now consider the product $b^ic_\alpha A^\alpha_i$. In V_n this product is an invariant (scalar product) for each i, and in V_m it is a scalar invariant and is a function of arc length s along C. Its derivative is

$$\frac{d}{ds}(b^i c_\alpha A_i^\alpha) = b^i c_\alpha \frac{dA_i^\alpha}{ds} + b^i A_i^\alpha \frac{dc_\alpha}{ds} + \frac{db^i}{ds} c_\alpha A_i^\alpha$$
$$= b^j c_\sigma \left(\frac{dA_j^\sigma}{ds} + A_j^\alpha \Gamma_{\alpha\beta}^\sigma \frac{dx^\beta}{ds} - A_i^\sigma \Gamma_{jk}^i \frac{du^k}{ds} \right)$$

making use of (565). Since b^j and c_σ are arbitrary vectors, and since $\frac{d}{ds}(b^i c_\alpha A_i^\alpha)$ is a scalar invariant, it follows from the quotient law that

$$\frac{dA_j^\sigma}{ds} + A_j^\alpha \Gamma_{\alpha\beta}^\sigma \frac{dx^\beta}{ds} - A_i^\sigma \Gamma_{jk}^i \frac{du^k}{ds} \tag{566}$$

is a tensor of the same type as A_j^σ. We call it the intrinsic derivative of A_j^σ with respect to s.

We may write (566) as

$$\left(\frac{\partial A_j^\sigma}{\partial u^k} + A_j^\alpha \Gamma_{\alpha\beta}^\sigma \frac{\partial x^\beta}{\partial u^k} - A_i^\sigma \Gamma_{jk}^i \right) \frac{du^k}{ds}$$

and since this is a tensor for all directions $\frac{du^k}{ds}$ (the directions $\frac{du^k}{ds}$ of C are arbitrary), it follows from the quotient law that

$$A_{j;k}^\sigma \equiv \frac{\partial A_j^\sigma}{\partial u^k} + A_j^\alpha \Gamma_{\alpha\beta}^\sigma \frac{\partial x^\beta}{\partial u^k} - A_i^\sigma \Gamma_{jk}^i \tag{567}$$

is the generalized covariant derivative of A_j^σ with respect to the V_m.

Problems

1. Why is $A_{j;k}^\sigma$ a contravariant vector in V_n?
2. Show that

$$A_{\beta i;j}^\alpha \equiv \frac{\partial A_{\beta i}^\alpha}{\partial u^j} + \Gamma_{\sigma\tau}^\alpha A_{\beta i}^\sigma \frac{\partial x^\tau}{\partial u^j} - \Gamma_{\beta\tau}^\sigma A_{\sigma i}^\alpha \frac{\partial x^\tau}{\partial u^j} - \Gamma_{ij}^k A_{\beta k}^\alpha$$

is a mixed tensor, by considering the scalar invariant $b^\beta c^i d_\alpha A_{\beta i}^\alpha$.

3. Show that $x_{;i}^\alpha = x_{,i}^\alpha = \frac{\partial x^\alpha}{\partial u^i}$, and that

$$x^\alpha_{;ij} = \frac{\partial^2 x^\alpha}{\partial u^i \, \partial u^j} - \Gamma^h_{ij} x^\alpha_{,h} + \Gamma^\alpha_{\beta\gamma} x^\beta_{,i} x^\gamma_{,j}$$
$$= x^\alpha_{,ij} + \Gamma^\alpha_{\beta\gamma} x^\beta_{,i} x^\gamma_{,j}$$

4. Show that $g_{\alpha\beta} x^\alpha_{;ik} x^\beta_{,j} + g_{\alpha\beta} x^\alpha_{,i} x^\beta_{;jk} = 0$, and show by cyclic permutations that $g_{\alpha\beta} x^\alpha_{;ij} x^\beta_{,k} = 0$.

5. The $x^\alpha_{;ij}$ of Prob. 4 are normal to the vectors $x^\beta_{,k}$, the tangent vectors to the surface. Hence the $x^\alpha_{;ij}$ are components of a vector normal to the subspace V_m. If N^i are the components of a unit normal to V_m, we must have $x^\alpha_{;ij} = b_{ij} N^\alpha$. We call $B = b_{ij} \, du^i \, du^j$ the second fundamental form. Show that $b_{ij} = g_{\alpha\beta} x^\alpha_{;ij} N^\beta$. If the V_n is a Euclidean V_3, $g_{\alpha\beta} = \delta_{\alpha\beta}$, show that $b_{11} = e$, $b_{12} = f$, $b_{22} = g$ (see Sec. 35) for the subspace $\mathbf{r} = \mathbf{r}(u, v)$.

139. Riemannian Curvature. Schur's Theorem. Let us consider a point P of a Riemannian space. We associate with P two independent vectors λ^α_1, λ^α_2. These vectors determine a pencil of directions at P, given by

$$\xi^\alpha = a^1 \lambda^\alpha_1 + a^2 \lambda^\alpha_2 = a^j \lambda^\alpha_j$$

Every pair of numbers a^1, a^2 determines a direction ξ^α. Since the geodesics are second-order differential equations, the point P and the direction ξ^α at P determine a unique geodesic. The locus of all geodesics determined in this manner will yield a surface. In a Euclidean space the surface will be a plane, since the geodesics are straight lines and two vectors determine a plane.

We now introduce normal coordinates y^α with origin at P. The equations of the geodesics take the form $y^\alpha = \xi^\alpha s$, where $\xi^\alpha = \left(\frac{dy^\alpha}{ds}\right)_P$, and the geodesic surface is given by

$$y^\alpha = a^1 s \lambda^\alpha_1 + a^2 s \lambda^\alpha_2$$
$$= u^1 \lambda^\alpha_1 + u^2 \lambda^\alpha_2 = u^j \lambda^\alpha_j \tag{568}$$

j summed from 1 to 2 and $u^1 = a^1 s$, $u^2 = a^2 s$.

The element of distance on the surface is given by

$$ds^2 = h_{ij} \, du^i \, du^j \tag{569}$$

and if $ds^2 = g_{\alpha\beta} \, dy^\alpha \, dy^\beta$ for the V_n, then

$$h_{ij} = g_{\alpha\beta} \frac{\partial y^\alpha}{\partial u^i} \frac{\partial y^\beta}{\partial u^j} = g_{\alpha\beta} \lambda^\alpha_i \lambda^\beta_j \tag{569a}$$

where the $g_{\alpha\beta}$ represent the components of the fundamental metric tensor in the system of normal coordinates.

Now let R_{ijkl} be the components of the curvature tensor for the surface S with coordinates u^1, u^2. Let us note the following: The $g_{\alpha\beta}$ of a Riemannian space completely determine the Christoffel symbols $\Gamma^{\alpha}_{\beta\gamma}$, which in turn specify completely the Riemann-Christoffel tensor $R_{\alpha\beta\gamma\delta}$. Once the metric of a surface embedded in a V_n is determined, we can determine the Γ^i_{jk} for this surface, and the R_{ijkl} can then be determined. We need not make any reference to the embedding space, V_n, to determine the R_{ijkl}. It is apparent that the h_{ij} can be determined without leaving the surface, so that all results and formulas derived from the h_{ij} are intrinsic properties of the surface. All we are trying to say is that $ds^2 = h_{ij}\,du^i\,du^j$ is the fundamental metric tensor for a Riemannian space which happens to be a surface embedded in a Riemannian V_n. We shall use Latin indices for the space determined by the metric h_{ij} and Greek letters for the V_n.

The indices of R_{ijkl} take on the values 1 or 2, and from Prob. 1, Sec. 134, we have that

$$\begin{aligned} R_{1212} = R_{2121} = -R_{1221} = -R_{2112} \\ R_{1111} = R_{1122} = R_{1122} = \cdots = R_{1121} = \cdots = R_{2222} \\ = 0 \end{aligned} \tag{570}$$

If we make an analytic transformation, $\bar{u}^i = \bar{u}^i(u^1, u^2)$, $i = 1, 2$, then

$$\bar{R}_{ijkl}(\bar{u}) = R_{abcd}(u)\,\frac{\partial u^a}{\partial \bar{u}^i}\,\frac{\partial u^b}{\partial \bar{u}^j}\,\frac{\partial u^c}{\partial \bar{u}^k}\,\frac{\partial u^d}{\partial \bar{u}^l}$$

and

$$\begin{aligned} \bar{R}_{1212} &= R_{abcd}\,\frac{\partial u^a}{\partial \bar{u}^1}\,\frac{\partial u^b}{\partial \bar{u}^2}\,\frac{\partial u^c}{\partial \bar{u}^1}\,\frac{\partial u^d}{\partial \bar{u}^2} \\ &= R_{1212}\left(\frac{\partial u^1}{\partial \bar{u}^1}\,\frac{\partial u^2}{\partial \bar{u}^2} - \frac{\partial u^1}{\partial \bar{u}^2}\,\frac{\partial u^2}{\partial \bar{u}^1}\right)^2 \end{aligned} \tag{571}$$

by making use of (570). Thus

$$\bar{R}_{1212} = R_{1212}\left[J\left(\frac{u^1, u^2}{\bar{u}^1, \bar{u}^2}\right)\right]^2 \tag{572}$$

Moreover, $d\bar{s}^2 = \bar{h}_{ij}\,d\bar{u}^i\,d\bar{u}^j$, and $\bar{h}_{ij} = h_{ab}\,\dfrac{\partial u^a}{\partial \bar{u}^i}\,\dfrac{\partial u^b}{\partial \bar{u}^j}$, so that

$|\bar{h}| = |h|J^2$. We rewrite (572) in the form

$$K = \frac{R_{1212}}{|h|} = \frac{\bar{R}_{1212}}{|\bar{h}|} \tag{573}$$

Equation (573) shows that K is an invariant, and it is called the Gaussian curvature. It is an intrinsic property of the surface.

We now determine an alternative form for K in terms of the directions λ_1^α and λ_2^α and the curvature tensor for the V_n at the point P. The coordinate transformation between the Christoffel symbols is given by (see Prob. 6, Sec. 129)

$$h_{il}\Gamma_{jk}^l(u) = g_{\alpha\mu}\Gamma_{\beta\gamma}^\mu(y)\frac{\partial y^\alpha}{\partial u^i}\frac{\partial y^\beta}{\partial u^j}\frac{\partial y^\gamma}{\partial u^k} + g_{\alpha\beta}\frac{\partial^2 y^\beta}{\partial u^j\,\partial u^k}\frac{\partial y^\alpha}{\partial u^i}$$

which reduces to

$$h_{il}\Gamma_{jk}^l = g_{\alpha\mu}\Gamma_{\beta\gamma}^\mu\lambda_i^\alpha\lambda_j^\beta\lambda_k^\gamma \tag{574}$$

since $\dfrac{\partial y^\alpha}{\partial u^j} = \lambda_j^\alpha$, $\dfrac{\partial^2 y^\alpha}{\partial u^k\,\partial u^j} = 0$, from (568).

At the point P, $\Gamma_{\beta\gamma}^\mu(y) = 0$ (see Sec. 131), so that $h_{il}\Gamma_{jk}^l = 0$ or $h^{im}h_{il}\Gamma_{jk}^l = \Gamma_{jk}^m = 0$. Hence the curvature tensor can be written

$$\begin{aligned} R_{1212}(P) &= h_{1l}R_{212}^l(P) \\ &= h_{1l}\left[\frac{\partial\Gamma_{21}^l(P)}{\partial u^2} - \frac{\partial\Gamma_{22}^l(P)}{\partial u^1}\right] \end{aligned} \tag{575}$$

from (538) and (541).

From (574), $h_{1l}\Gamma_{21}^l(u) = g_{\alpha\mu}\Gamma_{\beta\gamma}^\mu(y)\lambda_1^\alpha\lambda_2^\beta\lambda_1^\gamma$, so that at the origin of Riemannian coordinates

$$h_{1l}\frac{\partial\Gamma_{21}^l}{\partial u^2} = g_{\alpha\mu}\frac{\partial\Gamma_{\beta\gamma}^\mu}{\partial y^\tau}\lambda_1^\alpha\lambda_2^\beta\lambda_1^\gamma\lambda_2^\tau \tag{576}$$

since $\dfrac{\partial g_{\alpha\beta}}{\partial y^\mu} = g_{\sigma\beta}\Gamma_{\alpha\mu}^\sigma + g_{\sigma\alpha}\Gamma_{\beta\mu}^\sigma = 0$ at the origin (Prob. 6, Sec. 128), and from (569a) $\dfrac{\partial h_{ij}}{\partial u^k} = \lambda_i^\alpha\lambda_j^\beta\lambda_k^\gamma\dfrac{\partial g_{\alpha\beta}}{\partial y^\gamma} = 0$. Similarly

$$h_{1l}\frac{\partial\Gamma_{22}^l}{\partial u^1} = g_{\alpha\mu}\frac{\partial\Gamma_{\beta\gamma}^\mu}{\partial y^\tau}\lambda_1^\alpha\lambda_2^\beta\lambda_2^\gamma\lambda_1^\tau \tag{577}$$

Using (538), (541), (576), (577), it is easy to show that

$$R_{1212} = \lambda_1^\alpha \lambda_2^\beta \lambda_1^\gamma \lambda_2^\tau R_{\alpha\beta\gamma\tau} \tag{578}$$

Finally,

$$|h| = \begin{vmatrix} h_{11} & h_{12} \\ h_{21} & h_{22} \end{vmatrix} = h_{11}h_{22} - h_{12}^2$$

and

$$h_{11} = g_{\alpha\beta} \frac{\partial y^\alpha}{\partial u^1} \frac{\partial y^\beta}{\partial u^1} = g_{\alpha\beta}\lambda_1^\alpha \lambda_1^\beta$$
$$h_{12} = g_{\alpha\beta}\lambda_1^\alpha \lambda_2^\beta$$
$$h_{22} = g_{\alpha\beta}\lambda_2^\alpha \lambda_2^\beta$$

so that

$$|h| = \lambda_1^\alpha \lambda_2^\beta \lambda_1^\sigma \lambda_2^\tau (g_{\alpha\sigma} g_{\beta\tau} - g_{\alpha\tau} g_{\beta\sigma}) \tag{579}$$

Thus

$$K = \frac{R_{\alpha\beta\sigma\tau}\lambda_1^\alpha \lambda_2^\beta \lambda_1^\sigma \lambda_2^\tau}{\lambda_1^\alpha \lambda_2^\beta \lambda_1^\sigma \lambda_2^\tau (g_{\alpha\sigma} g_{\beta\tau} - g_{\alpha\tau} g_{\beta\sigma})} \tag{580}$$

We are now in a position to prove Schur's theorem. If at each point of a Riemannian space, K is independent of the orientation $(\lambda_1^\alpha, \lambda_2^\alpha)$, K is constant throughout the space.

It follows at once if K is independent of λ_1^α, λ_2^α, that

$$R_{\alpha\beta\sigma\tau} = K(g_{\alpha\sigma} g_{\beta\tau} - g_{\alpha\tau} g_{\beta\sigma})$$

and so

$$R_{\alpha\beta\sigma\tau,\mu} = K_{,\mu}(g_{\alpha\sigma} g_{\beta\tau} - g_{\alpha\tau} g_{\beta\sigma})$$
$$R_{\alpha\beta\mu\sigma,\tau} = K_{,\tau}(g_{\alpha\mu} g_{\beta\sigma} - g_{\alpha\sigma} g_{\beta\mu})$$
$$R_{\alpha\beta\tau\mu,\sigma} = K_{,\sigma}(g_{\alpha\tau} g_{\beta\mu} - g_{\alpha\mu} g_{\beta\tau})$$

Adding and using Bianchi's identity, (543), we have

$$K_{,\mu}(g_{\alpha\sigma} g_{\beta\tau} - g_{\alpha\tau} g_{\beta\sigma}) + K_{,\tau}(g_{\alpha\mu} g_{\beta\sigma} - g_{\alpha\sigma} g_{\beta\mu}) + K_{,\sigma}(g_{\alpha\tau} g_{\beta\mu} - g_{\alpha\mu} g_{\beta\tau}) = 0$$

Multiplying the above equation by $g^{\alpha\sigma}$ and summing, we have

$$K_{,\mu}(n g_{\beta\tau} - g_{\beta\tau}) + K_{,\tau}(g_{\beta\mu} - n g_{\beta\mu}) + K_{,\tau} g_{\beta\mu} - K_{,\mu} g_{\beta\tau} = 0$$

or

$$(n - 2) g_{\beta\tau} K_{,\mu} = (n - 2) K_{,\tau} g_{\beta\mu}$$

and $g_{\beta\tau} K_{,\mu} = g_{\beta\mu} K_{,\tau}$, or $\delta_\tau^\sigma K_{,\mu} = \delta_\mu^\sigma K_{,\tau}$, if $n > 2$. For $n = 2$ there is no arbitrary orientation.

If we choose $\sigma = \tau \neq \mu$, $K_{,\mu} = 0$. This is true for all μ since μ can be chosen arbitrarily from 1 to n. Hence $K \equiv$ constant throughout all of space. Such a space is said to be of constant curvature.

Problems

1. Derive (571).
2. Derive (575).
3. Derive (578).
4. For a V_3 for which $g_{ij} = 0$, $i \neq j$, show that if h, i, j are unequal,

$$R_{ij} = \frac{1}{g_{hh}} R_{ihhj}$$
$$R_{hh} = \frac{1}{g_{ii}} R_{hiih} + \frac{1}{g_{jj}} R_{hjjh}$$
$$R = \sum_{i,j=1}^{3} \frac{1}{g_{ii}g_{jj}} R_{ijji}$$

5. If $R_i^\alpha \equiv g^{\alpha j} R_{ij}$, show that $R_{i,\alpha}^\alpha = \dfrac{1}{2} \dfrac{\partial R}{\partial x^i}$.
6. If $R_{ij} = kg_{ij}$ (an Einstein space), show that $R = g^{ij}R_{ij} = nk$, or $R_{ij} = (R/n)g_{ij}$.
7. Show that a space of constant curvature K is an Einstein space and that $R = Kn(1 - n)$.

140. Lagrange's Equations. Let L be any scalar invariant function of the coordinates $q^1, q^2, \ldots, q^n$, their time derivatives $\dot{q}^1, \dot{q}^2, \ldots, \dot{q}^n$, and the time t:

$$L = L(q^i, \dot{q}^i, t) = \bar{L}(\bar{q}^i, \dot{\bar{q}}^i, t)$$

If we perform a transformation of coordinates,

$$q^\alpha = q^\alpha(\bar{q}^1, \bar{q}^2, \ldots, \bar{q}^n)$$

$\alpha = 1, 2, \ldots, n$, then $\dot{q}^\alpha = \dfrac{\partial q^\alpha}{\partial \bar{q}^\beta} \dot{\bar{q}}^\beta$, so that $\dot{q}^\alpha$ is a function of the $\bar{q}^i$, $\dot{\bar{q}}^i$. Now

$$\begin{aligned} \frac{\partial \bar{L}}{\partial \bar{q}^i} &= \frac{\partial L}{\partial q^\alpha} \frac{\partial q^\alpha}{\partial \bar{q}^i} + \frac{\partial L}{\partial \dot{q}^\alpha} \frac{\partial \dot{q}^\alpha}{\partial \bar{q}^i} \\ &= \frac{\partial L}{\partial q^\alpha} \frac{\partial q^\alpha}{\partial \bar{q}^i} + \frac{\partial L}{\partial \dot{q}^\alpha} \frac{\partial^2 q^\alpha}{\partial \bar{q}^i \, \partial \bar{q}^\beta} \dot{\bar{q}}^\beta \end{aligned} \tag{581}$$

where we consider $\bar{q}^i$ and $\dot{\bar{q}}$ as independent variables in $\bar{L}$. Now also

$$\frac{\partial \bar{L}}{\partial \dot{\bar{q}}^i} = \frac{\partial L}{\partial \dot{q}^\alpha}\frac{\partial \dot{q}^\alpha}{\partial \dot{\bar{q}}^i} = \frac{\partial L}{\partial \dot{q}^\alpha}\frac{\partial q^\alpha}{\partial \bar{q}^i}$$

so that

$$\frac{d}{dt}\left(\frac{\partial \bar{L}}{\partial \dot{\bar{q}}^i}\right) = \frac{\partial q^\alpha}{\partial \bar{q}^i}\frac{d}{dt}\left(\frac{\partial L}{\partial \dot{q}^\alpha}\right) + \frac{\partial L}{\partial \dot{q}^\alpha}\frac{\partial^2 q^\alpha}{\partial \bar{q}^\beta\,\partial \dot{\bar{q}}^i}\dot{\bar{q}}^\beta \qquad (582)$$

Subtracting (581) from (582), we obtain

$$\frac{d}{dt}\left(\frac{\partial \bar{L}}{\partial \dot{\bar{q}}^i}\right) - \frac{\partial \bar{L}}{\partial \bar{q}^i} = \frac{\partial q^\alpha}{\partial \bar{q}^i}\left[\frac{d}{dt}\left(\frac{\partial L}{\partial \dot{q}^\alpha}\right) - \frac{\partial L}{\partial q^\alpha}\right]$$

which shows that the $\frac{d}{dt}\left(\frac{\partial L}{\partial \dot{q}^\alpha}\right) - \frac{\partial L}{\partial q^\alpha}$ are the components of a covariant vector.

For a system of particles, let $L = T - V$, where T is the kinetic energy; $T = \sum_{i=1}^{n} \frac{1}{2} m_i \left(\frac{ds_i}{dt}\right)^2 = \sum_{i=1}^{n} \frac{1}{2} m_i g_{\alpha\beta}\dot{x}_i^\alpha \dot{x}_i^\beta$.

$$V(x_1^1, x_1^2, x_1^3, x_2^1, x_2^2, x_2^3, \ldots, x_n^3)$$

is the potential function, $-\frac{\partial V}{\partial x_s^r} = (F^r)_s$. Then

$$\frac{d}{dt}\left(\frac{\partial L}{\partial \dot{x}_s^r}\right) - \frac{\partial L}{\partial x_s^r} = \frac{d}{dt}(m_s g_{\alpha r}\dot{x}_s^\alpha) - \sum_{i=1}^{n} \frac{1}{2} m_i \frac{\partial g_{\alpha\beta}}{\partial x_s^r}\dot{x}_i^\alpha \dot{x}_i^\beta + \frac{\partial V}{\partial x_s^r}$$
$$= m_s \ddot{x}_s^r - (F^r)_s, \quad \text{if } g_{\alpha\beta} = \delta_{\alpha\beta}$$

and $m_s\ddot{x}_s^r - (F^r)_s = 0$ for a Euclidean space and Newtonian mechanics. Hence $\frac{d}{dt}\left(\frac{\partial L}{\partial \dot{x}_s^r}\right) - \frac{\partial L}{\partial x_s^r}$ vanishes in all coordinate systems. We replace the x_s^r by any system of coordinates q^1, q^2, . . . , q^n which completely specify the configuration of the system of particles, and Lagrange's equations of motion are

$$\boxed{\frac{d}{dt}\left(\frac{\partial L}{\partial \dot{q}^r}\right) - \frac{\partial L}{\partial q^r} = 0, \quad r = 1, 2, \ldots, n} \qquad (583)$$

Example 150. In spherical coordinates, a particle has the square of the velocity $v^2 = \dot{r}^2 + r^2\dot{\theta}^2 + r^2 \sin^2\theta\, \dot{\varphi}^2$, so that

$$L = T - V = \frac{m}{2}(\dot{r}^2 + r^2\dot{\theta}^2 + r^2 \sin^2 \theta \, \dot{\varphi}^2) - V$$

$$\frac{\partial L}{\partial r} = m(r\dot{\theta}^2 + r \sin^2 \theta \, \dot{\varphi}^2) - \frac{\partial V}{\partial r}$$

$$\frac{d}{dt}\left(\frac{\partial L}{\partial \dot{r}}\right) = m\ddot{r}$$

and one of Lagrange's equations of motion is

$$m\ddot{r} - m(r\dot{\theta}^2 + r \sin^2 \theta \, \dot{\varphi}^2) + \frac{\partial V}{\partial r} = 0$$

Since $-\dfrac{\partial V}{\partial r}$ represents the radial force, the quantity

$$\ddot{r} - (r\dot{\theta}^2 + r \sin^2 \theta \, \dot{\varphi}^2)$$

must be the radial acceleration.

If no potential function exists, we can modify Lagrange's equations as follows: We know that $T = (m/2)g_{\alpha\beta}\dot{x}^\alpha\dot{x}^\beta$ is a scalar invariant, so that

$$Q_r = \frac{d}{dt}\left(\frac{\partial T}{\partial \dot{x}^r}\right) - \frac{\partial T}{\partial x^r} \tag{584}$$

are the components of a covariant vector. In cartesian coordinates, the Q_r are the components of the Newtonian force, so that Q_r is the generalized force vector. If f_r are the components of the force vector in a y^1-y^2- · · · -y^n coordinate system, then

$$Q_r = f_\alpha \frac{\partial y^\alpha}{\partial x^r}$$

and

$$Q_r \, dx^r = f_\alpha \frac{\partial y^\alpha}{\partial x^r} dx^r = f_\alpha \, dy^\alpha$$

is a scalar invariant. The reader will immediately realize that $f_\alpha \, dy^\alpha$ represents the differential of work, dW, so that

$$Q_i = \frac{\partial W}{\partial x^i} \tag{585}$$

We obtain Q_i by allowing x^i to vary, keeping $x^1, x^2, \ldots, x^{i-1}, x^{i+1}, \ldots, x^n$ fixed, calculate the work ΔW_i done by the forces, and compute

$$Q_i = \lim_{\Delta x^i \to 0} \frac{\Delta W_i}{\Delta x^i}, \qquad i \text{ not summed}$$

Example 151. A particle slides in a frictionless tube which rotates in a horizontal plane with constant angular speed ω. The only horizontal force is the reaction R of the tube on the particle. We have $T = (m/2)(\dot{r}^2 + r^2\dot{\theta}^2)$, so that (584) becomes

$$\begin{aligned} m\ddot{r} - mr\dot{\theta}^2 &= Q_r \\ \frac{d}{dt}(mr^2\dot{\theta}) &= Q_\theta \end{aligned} \tag{586}$$

with $Q_r = 0$, $Q_\theta = \dfrac{R(r\,d\theta)}{d\theta} = rR$. The solution to (586) is $r = Ae^{\omega t} + Be^{-\omega t}$, $R = 2m\omega \dfrac{dr}{dt}$, since $\dot{\theta} = \omega$.

Problems

1. A particle slides in a frictionless tube which rotates in a vertical plane with constant angular speed ω. Set up the equations of motion.

2. For a rigid body with one point fixed,

$$T = \tfrac{1}{2}A(\omega_x^2 + \omega_y^2) + \tfrac{1}{2}C\omega_z^2$$

Using Eulerian angles, show that if $Q_\varphi = Q_\psi = 0$, then

$$\begin{aligned} C(\dot{\varphi} + \cos\theta\,\dot{\psi}) &= R \\ A\dot{\psi}\sin^2\theta + R\cos\theta &= S \\ A\ddot{\theta} - A\dot{\psi}^2\sin\theta\cos\theta + R\dot{\psi}\sin\theta &= Q_\theta \end{aligned}$$

where R, S are constants of integration.

3. If $T = a_{\alpha\beta}(q^1, \ldots, q^n)\dot{q}^\alpha\dot{q}^\beta$, show that $2T = \dfrac{\partial T}{\partial \dot{q}^\alpha}\dot{q}^\alpha$.

4. Define $p_r = \dfrac{\partial L(q, \dot{q})}{\partial \dot{q}^r}$, and assuming we can solve for $\dot{q}^r = \dot{q}^r(q^1, \ldots, q^n, p_1, \ldots, p_n)$, show that the Hamiltonian

H defined by $H = p_\alpha \dot{q}^\alpha - L$ satisfies

$$H = T + V = h \text{ (a constant)}$$

where $V = V(q^1, \ldots, q^n)$, $T = a_{\alpha\beta}(q^1, \ldots, q^n)\dot{q}^\alpha \dot{q}^\beta$. Also show that

$$\left\{\begin{aligned} \frac{\partial H}{\partial q^r} &= -\dot{p}_r \\ \frac{\partial H}{\partial p_r} &= \dot{q}^r \end{aligned}\right\}$$

These are Hamilton's equations of motion; the p_r are called the generalized momentum coordinates. Show that they are the components of a covariant vector.

5. By extremalizing the integral $\int_{t_0}^{t_1} L(x^i, \dot{x}^i, t)\, dt$, show that Lagrange's equations result.

6. If the action integral

$$A = \sqrt{2M} \int_A^B \left[(h - V) g_{\alpha\beta} \frac{dx^\alpha}{d\lambda} \frac{dx^\beta}{d\lambda} \right]^{\frac{1}{2}} d\lambda$$

is extremalized, show that the result yields

$$\frac{d}{dt}\left(\frac{\partial T}{\partial \dot{x}^r}\right) - \frac{\partial T}{\partial x^r} = -\frac{\partial V}{\partial x^r}$$

where $T + V = h$, the constant of energy. The path of the particle is the same as the geodesic of a space having the metric

$$ds^2 = 2M(h - V) g_{\alpha\beta}\, dx^\alpha\, dx^\beta$$

141. Einstein's Law of Gravitation. We look for a law of motion, which will be independent of the coordinate system used, describing the gravitational field of a single particle. In the special theory of relativity, the line element for the space-time coordinates is given by

$$\begin{aligned} ds^2 &= -dx^2 - dy^2 - dz^2 + c^2\, dt^2 \\ &= -dr^2 - r^2\, d\theta^2 - r^2 \sin^2\theta\, d\varphi^2 + c^2\, dt^2 \end{aligned} \tag{587}$$

In the space of x, y, z, t, the $g_{\alpha\beta}$ are constants and the space is flat (Euclidean), so that $B^i_{jkl} = 0$. For a gravitating particle we postulate that the Ricci tensor R_{ij} vanish (see Probs. 5 and 6 of this section). Since $R_{ij} = R_{ji}$, a four-dimensional space yields $n(n+1)/2 = 10$ equations involving the g_{ij} and their derivatives. From Prob. 7, Sec. 134, we have $R^i_{j,i} = \dfrac{1}{2}\dfrac{\partial R}{\partial x^j}$, where $R^i_j = g^{i\alpha}R_{\alpha j}$, $R = g^{\alpha\beta}R_{\alpha\beta}$, and for $j = 1, 2, 3, 4$ the 10 equations are essentially reduced to 6 equations.

We assume the line element (due to Schwarzchild) to be of the form

$$ds^2 = -e^{\lambda(r)}\,dr^2 - r^2\,d\theta^2 - r^2\sin^2\theta\,d\varphi^2 + e^{\nu(r)}\,dt^2 \tag{588}$$

so that our space is non-Euclidean. We do not include terms of the form $dr\,d\theta$, etc., because we expect our space to be homogeneous and isotropic. We have

$$g_{11} = -e^{\lambda}, \qquad g_{22} = -r^2, \qquad g_{33} = -r^2\sin^2\theta, \qquad g_{44} = e^{\nu}$$
$$g_{ij} = 0, \qquad i \neq j \tag{589}$$

and

$$g^{11} = -e^{-\lambda}, \qquad g^{22} = -r^{-2}, \qquad g^{33} = -(r^2\sin^2\theta)^{-1}$$
$$g^{44} = e^{-\nu}, \qquad g^{ij} = 0, \qquad i \neq j$$

Now $\Gamma^i_{jk} = \dfrac{1}{2}g^{i\sigma}\left(\dfrac{\partial g_{\sigma j}}{\partial x^k} + \dfrac{\partial g_{k\sigma}}{\partial x^j} - \dfrac{\partial g_{jk}}{\partial x^\sigma}\right)$, and since $g^{i\sigma} = 0$ for $i \neq \sigma$, we have

$$\Gamma^i_{jk} = \frac{1}{2}g^{ii}\left(\frac{\partial g_{ij}}{\partial x^k} + \frac{\partial g_{ki}}{\partial x^j} - \frac{\partial g_{jk}}{\partial x^i}\right) \qquad i \text{ not summed}$$

If i, j, k are different, then $\Gamma^i_{jk} \equiv 0$. We also see that

$$\begin{aligned}
\Gamma^i_{ii} &= \frac{1}{2}g^{ii}\frac{\partial g_{ii}}{\partial x^i} \\
\Gamma^i_{ik} &= \frac{1}{2}g^{ii}\frac{\partial g_{ii}}{\partial x^k} \\
\Gamma^i_{kk} &= -\frac{1}{2}g^{ii}\frac{\partial g_{kk}}{\partial x^i}
\end{aligned} \tag{590}$$

Applying (590), we have

$$
\begin{aligned}
\Gamma^1_{11} &= \frac{1}{2} g^{11} \frac{\partial g_{11}}{\partial r} = \frac{1}{2} \frac{d\lambda}{dr} \\
\Gamma^1_{22} &= -\frac{1}{2} g^{11} \frac{\partial g_{22}}{\partial r} = -re^{-\lambda} \\
\Gamma^1_{33} &= -\frac{1}{2} g^{11} \frac{\partial g_{33}}{\partial r} = -r \sin^2 \theta \, e^{-\lambda} \\
\Gamma^1_{44} &= -\frac{1}{2} g^{11} \frac{\partial g_{44}}{\partial r} = \frac{1}{2} e^{\nu-\lambda} \frac{d\nu}{dr} \\
\Gamma^2_{12} &= \frac{1}{2} g^{22} \frac{\partial g_{22}}{\partial r} = \frac{1}{r} \\
\Gamma^2_{33} &= -\frac{1}{2} g^{22} \frac{\partial g_{33}}{\partial \theta} = -\sin \theta \cos \theta \\
\Gamma^3_{13} &= \frac{1}{2} g^{33} \frac{\partial g_{33}}{\partial r} = \frac{1}{r} \\
\Gamma^3_{23} &= \frac{1}{2} g^{33} \frac{\partial g_{33}}{\partial \theta} = \cot \theta \\
\Gamma^4_{14} &= \frac{1}{2} g^{44} \frac{\partial g_{44}}{\partial r} = \frac{1}{2} \frac{d\nu}{dr}
\end{aligned}
\tag{591}
$$

and all other Γ^i_{jk} vanish.

From (539)

$$
R_{ij} = \frac{\partial \Gamma^\alpha_{i\alpha}}{\partial x^j} - \frac{\partial \Gamma^\alpha_{ij}}{\partial x^\alpha} + \Gamma^\beta_{i\alpha}\Gamma^\alpha_{\beta j} - \Gamma^\beta_{ij}\Gamma^\alpha_{\beta\alpha}
$$

so that

$$
\begin{aligned}
R_{11} &= \frac{\partial \Gamma^2_{12}}{\partial r} + \frac{\partial \Gamma^3_{13}}{\partial r} + \frac{\partial \Gamma^4_{14}}{\partial r} + \Gamma^1_{11}\Gamma^1_{11} + \Gamma^2_{12}\Gamma^2_{21} + \Gamma^3_{13}\Gamma^3_{31} \\
&\qquad + \Gamma^4_{14}\Gamma^4_{41} - \Gamma^1_{11}(\Gamma^1_{11} + \Gamma^2_{12} + \Gamma^3_{13} + \Gamma^4_{14}) \\
&= -\frac{1}{r^2} - \frac{1}{r^2} + \frac{1}{2}\frac{d^2\nu}{dr^2} + \frac{1}{r^2} + \frac{1}{r^2} + \frac{1}{4}\left(\frac{d\nu}{dr}\right)^2 - \frac{1}{2r}\frac{d\lambda}{dr} \\
&\qquad - \frac{1}{2r}\frac{d\lambda}{dr} - \frac{1}{4}\frac{d\lambda}{dr}\frac{d\nu}{dr}
\end{aligned}
\tag{592}
$$

by making use of (591). Hence Einstein's law $R_{ij} = 0$ yields

$$
R_{11} = \frac{1}{2}\frac{d^2\nu}{dr^2} + \frac{1}{4}\left(\frac{d\nu}{dr}\right)^2 - \frac{1}{4}\frac{d\lambda}{dr}\frac{d\nu}{dr} - \frac{1}{r}\frac{d\lambda}{dr} = 0
$$

Similarly

$$
\begin{aligned}
R_{22} &= e^{-\lambda}\left[1 + \frac{1}{2} r\left(\frac{d\nu}{dr} - \frac{d\lambda}{dr}\right)\right] - 1 = 0 \\
R_{33} &= \sin^2\theta\, e^{-\lambda}\left[1 + \frac{1}{2} r\left(\frac{d\nu}{dr} - \frac{d\lambda}{dr}\right)\right] - \sin^2\theta = 0 \qquad (593) \\
R_{44} &= e^{\nu-\lambda}\left[-\frac{1}{2}\frac{d^2\nu}{dr^2} - \frac{1}{4}\left(\frac{d\nu}{dr}\right)^2 + \frac{1}{4}\frac{d\lambda}{dr}\frac{d\nu}{dr} - \frac{1}{r}\frac{d\nu}{dr}\right] = 0
\end{aligned}
$$

Dividing R_{44} by $e^{\nu-\lambda}$ and adding to R_{11}, we obtain

$$\frac{d\lambda}{dr} + \frac{d\nu}{dr} = 0$$

or

$$\lambda + \nu = \text{constant} = c_0$$

We desire the form of (588), as $r \to \infty$, to approach that of (587). This requires that λ and ν approach zero as r approaches ∞. Hence $\lambda + \nu \equiv 0$ or $\lambda = -\nu$. From $R_{22} = 0$ we have $e^{\nu}\left(1 + r\frac{d\nu}{dr}\right) = 1$. Let $\gamma = e^{\nu}$, so that $\frac{d\nu}{dr} = \frac{1}{\gamma}\frac{d\gamma}{dr}$ and

$$\gamma\left(1 + \frac{r}{\gamma}\frac{d\gamma}{dr}\right) = 1$$

or

$$\frac{d\gamma}{1-\gamma} = \frac{dr}{r}$$

and

$$\gamma = 1 - \frac{2m}{r} \qquad (594)$$

where $2m$ is a constant of integration.

The equations of the geodesics are

$$\frac{d^2x^i}{ds^2} + \Gamma^i_{jk}\frac{dx^j}{ds}\frac{dx^k}{ds} = 0$$

which yield

$$\frac{d^2\theta}{ds^2} + 2\Gamma^2_{12}\frac{dr}{ds}\frac{d\theta}{ds} + \Gamma^2_{33}\left(\frac{d\varphi}{ds}\right)^2 = 0$$

or

$$\frac{d^2\theta}{ds^2} + \frac{2}{r}\frac{dr}{ds}\frac{d\theta}{ds} - \sin\theta\cos\theta\left(\frac{d\varphi}{ds}\right)^2 = 0 \tag{595}$$

If $\theta = \frac{\pi}{2}, \frac{d\theta}{ds} = 0$ initially, then $\theta \equiv \frac{\pi}{2}$ satisfies (595) and the boundary conditions.

We also obtain

$$\frac{d^2r}{ds^2} + \Gamma^1_{11}\left(\frac{dr}{ds}\right)^2 + \Gamma^1_{33}\left(\frac{d\varphi}{ds}\right)^2 + \Gamma^1_{44}\left(\frac{dt}{ds}\right)^2 = 0$$

or

$$\frac{d^2r}{ds^2} + \frac{1}{2}\frac{d\lambda}{dr}\left(\frac{dr}{ds}\right)^2 - re^{-\lambda}\left(\frac{d\varphi}{ds}\right)^2 + \frac{1}{2}e^{\nu-\lambda}\frac{d\nu}{dr}\left(\frac{dt}{ds}\right)^2 = 0 \tag{596}$$

making use of $\theta = \pi/2$.

Also

$$\frac{d^2\varphi}{ds^2} + 2\Gamma^3_{13}\frac{dr}{ds}\frac{d\varphi}{ds} = 0 \qquad \text{or} \qquad \frac{d^2\varphi}{ds^2} + \frac{2}{r}\frac{dr}{ds}\frac{d\varphi}{ds} = 0 \tag{597}$$

$$\frac{d^2t}{ds^2} + 2\Gamma^4_{14}\frac{dt}{ds}\frac{dr}{ds} = 0 \qquad \text{or} \qquad \frac{d^2t}{ds^2} + \frac{d\nu}{dr}\frac{dt}{ds}\frac{dr}{ds} = 0 \tag{598}$$

Integrating (597) and (598), we obtain

$$r^2\frac{d\varphi}{ds} = h \tag{599}$$

$$\log\frac{dt}{ds} + \nu = \log c \qquad \text{or} \qquad \frac{dt}{ds} = \frac{c}{\gamma} \tag{600}$$

where h and c are constants of integration. Equation (588) becomes

$$ds^2 = -\frac{1}{\gamma}\,dr^2 - r^2\,d\varphi^2 + \gamma\,dt^2$$

or

$$1 = -\frac{1}{\gamma}\left(\frac{dr}{ds}\right)^2 - r^2\left(\frac{d\varphi}{ds}\right)^2 + \gamma\left(\frac{c}{\gamma}\right)^2$$

or

$$1 = -\frac{1}{\gamma}\left(\frac{h}{r^2}\frac{dr}{d\varphi}\right)^2 - r^2\frac{h^2}{r^4} + \gamma\left(\frac{c}{\gamma}\right)^2$$

or

$$\left(\frac{h}{r^2}\frac{dr}{d\varphi}\right)^2 + \frac{h^2}{r^2} = c^2 - 1 + \frac{2m}{r} + \frac{2m}{r}\frac{h^2}{r^2}$$

and writing $u = 1/r$, we obtain

$$\left(\frac{du}{d\varphi}\right)^2 + u^2 = \frac{c^2 - 1}{h^2} + \frac{2m}{h^2}u + 2mu^3$$

and differentiating, we finally obtain

$$\boxed{\frac{d^2u}{d\varphi^2} + u = \frac{m}{h^2} + 3mu^2} \tag{601}$$

We obtain an approximate solution of (601) in the following manner: We first neglect the small term $3mu^2 = 3m/r^2$, for large r. The solution of $\dfrac{d^2u}{d\varphi^2} + u = \dfrac{m}{h^2}$ is

$$\frac{1}{r} = u = \frac{m}{h^2}[1 + e\cos(\varphi - \omega)]$$

where e, ω are constants of integration. This is Newton's solution of planetary motion. We substitute this value of u in the term $3mu^2$, and we obtain

$$\frac{d^2u}{d\varphi^2} + u = \frac{m}{h^2} + \frac{3m^3}{h^4} + \frac{6m^3}{h^4}e\cos(\varphi - \omega) + \frac{3m^3e^3}{2h^4}[1 + \cos 2(\varphi - \omega)]$$

We now neglect certain terms which yield little to our solution and obtain

$$\frac{d^2u}{d\varphi^2} + u = \frac{m}{h^2} + \frac{6m^3}{h^4}e\cos(\varphi - \omega)$$

From the theory of differential equations the solution of our new equation is

$$u = \frac{m}{h^2}\left[1 + e\cos(\varphi - \omega) + \frac{3m^2}{h^2}e\varphi\sin(\varphi - \omega)\right]$$

$$= \frac{m}{h^2}[1 + e\cos(\varphi - \omega - \epsilon)], \qquad \text{approximately}$$

where $\epsilon = (3m^2/h^2)\varphi$ and ϵ^2 is neglected.

When the planet moves through one revolution, the advance of the perihelion is given by $\delta(\omega + \epsilon) = (3m^2/h^2)\,\delta\varphi = 6\pi m^2/h^2$. When numerical results are given to the constants, it is found that the discrepancy between observed and calculated results on the advance of the perihelion of Mercury is removed.

Problems

1. Derive (591).
2. Derive (593).
3. For motion with the speed of light, $ds = 0$, so that from (599), $h = \infty$, and (601) becomes

$$\frac{d^2u}{d\varphi^2} + u = 3mu^2 \tag{602}$$

Integrate $\dfrac{d^2u}{d\varphi^2} + u = 0$, replace this value of u in $3mu^2$, and obtain an approximate solution of (602) in the form

$$u = \frac{\cos\varphi}{R} + \frac{m}{R^2}(\cos^2\varphi + 2\sin^2\varphi)$$

where R is a constant of integration. Since $u = 1/r$, $x = r\cos\varphi$, $y = r\sin\varphi$, show that

$$x = R + \frac{m}{R}\frac{x^2 + 2y^2}{(x^2 + y^2)^{\frac{1}{2}}}$$

The term $(m/R)(x^2 + 2y^2)/(x^2 + y^2)^{\frac{1}{2}}$ is the small deviation of the path of a light ray from the straight line $x = R$. The asymptotes are found by taking y large compared with x. Show that they are $x = R + (m/R)(\pm 2y)$ and that the angle (in radians) between the asymptotic lines is approximately $4m/R$. This is twice the predicted value, on the basis of the Newtonian theory, for the deflection of light as it passes the sun and has been verified during the total eclipse of the sun.

4. If $R_{ij} = \alpha g_{ij}$ is taken for the Einstein law, show that if $\gamma = e^\nu$, then $\gamma = 1 - (2m/r) - \frac{1}{3}\alpha r^2$ and

$$\frac{d^2u}{d\varphi^2} + u = \frac{m}{h^2} + 3mu^2 - \frac{1}{3}\frac{\alpha}{h^2}u^3$$

5. Assume the following: $ds^2 = g_{\alpha\beta}\,dx^\alpha\,dx^\beta$, $g_{\alpha\beta} = 0$ for $\alpha \neq \beta$, $g_{\alpha\beta} \approx 1$, $\dfrac{\partial g_{\alpha\beta}}{\partial x^4} = 0$, $\dfrac{dx^\alpha}{ds} \approx 0$, $\alpha = 1, 2, 3$, $\dfrac{dx^4}{ds} \approx 1$,

$$\frac{\psi}{c^2} = \tfrac{1}{2} g_{44} + \text{constant}$$

$x^1 = x$, $x^2 = y$, $x^3 = z$, $x^4 = ct$. Show that the equations of the geodesics reduce to Newton's law of motion $\dfrac{d^2x^i}{dt^2} + \dfrac{\partial \psi}{\partial x^i} = 0$, $i = 1, 2, 3$.

6. With the assumptions of Prob. 5 show that $R_{44} = 0$ yields Laplace's equation $\nabla^2 \psi = 0$.

142. Two-point Tensors. The tensors that we have studied have been functions of one point. Let us now consider the functions $g_{\alpha,\beta}(x_1, x_2)$ which depend on the coordinates of two points. We now allow independent coordinate transformations at the two points $M_1(x_1^1, x_1^2, \ldots, x_1^n)$, $M_2(x_2^1, x_2^2, x_2^3, \ldots, x_2^n)$. If in the new coordinate systems $\bar{x}_1$, $\bar{x}_2$ we have

$$\bar{g}_{\alpha,\beta}(\bar{x}_1, \bar{x}_2) = g_{\mu,\nu}(x_1, x_2) \frac{\partial x_1^\mu}{\partial \bar{x}_1^\alpha} \frac{\partial x_2^\nu}{\partial \bar{x}_2^\beta} \tag{603}$$

then the $g_{\alpha,\beta}$ are the components of a two-point tensor, a covariant vector relative to M_1 and a covariant vector relative to M_2. Indices preceding the comma refer to the point M_1, indices following the comma refer to M_2. If we keep the coordinates of M_2 fixed, that is, if $\bar{x}_2^i \equiv x_2^i$, then (603) reduces to

$$\bar{g}_{\alpha,\beta}(\bar{x}_1, x_2) = g_{\mu,\beta}(x_1, x_2) \frac{\partial x_1^\mu}{\partial \bar{x}_1^\alpha} \tag{604}$$

so that relative to M_1, $g_{\alpha,\beta}$ behaves like a covariant vector. A similar remark applies at the point M_2.

We leave it to the reader to consider the most general type of two-point tensor fields. We could, indeed, consider a multiple tensor field depending on a finite number of points. What difficulties would one encounter for tensors depending on a countable collection of points?

We may consider a two-point tensor field as special one-point tensors of a $2n$-dimensional space subject to a special group of coordinate transformations.

The scalar invariant

$$ds^2 = g_{\alpha,\beta}(x_1, x_2)\, dx_1^\alpha\, dx_2^\beta \tag{605}$$

is an immediate generalization of the Riemann line element. Indeed, when x_1 and x_2 coincide, we obtain the Riemann line element. Assuming $ds^2 > 0$ for $\alpha \leqq t \leqq \beta$, we can extremalize

$$\int ds = \int_\alpha^\beta \left(g_{\alpha,\beta} \frac{dx_1^\alpha}{dt} \frac{dx_2^\beta}{dt} \right)^{\frac{1}{2}} dt \tag{606}$$

and obtain a system of differential equations.

$$\begin{aligned} \ddot{x}_1^i + \Gamma^{i,}_{\alpha\beta,} \dot{x}_1^\alpha \dot{x}_1^\beta + C^{i,}_{\alpha,\beta} \dot{x}_1^\alpha \dot{x}_2^\beta &= 0 \\ \ddot{x}_2^i + \Gamma^{,i}_{,\alpha\beta} \dot{x}_2^\alpha \dot{x}_2^\beta + C^{,i}_{\alpha,\beta} \dot{x}_1^\alpha \dot{x}_\beta^2 &= 0 \end{aligned} \tag{607}$$

where

$$\Gamma^{i,}_{\alpha\beta,} = g^{i,\sigma} \frac{\partial g_{\alpha,\sigma}}{\partial x_1^\beta}, \qquad \Gamma^{,i}_{,\alpha\beta} = g^{\sigma,i} \frac{\partial g_{\sigma,\beta}}{\partial x_2^\alpha}$$

$$C^{i,}_{\sigma,\nu} = g^{i,\beta} \left(\frac{\partial g_{\sigma,\beta}}{\partial x_2^\nu} - \frac{\partial g_{\sigma,\nu}}{\partial x_2^\beta} \right), \qquad C^{,i}_{\sigma,\nu} = g^{\mu,i} \left(\frac{\partial g_{\mu,\nu}}{\partial x_1^\sigma} - \frac{\partial g_{\sigma,\nu}}{\partial x_1^\mu} \right) \tag{608}$$

$$g^{\mu,i} g_{\mu,j} = \delta^{,i}_{,j}, \qquad \dot{x}_1 = \frac{dx_1^i}{ds}$$

The unique solutions of (606), $x_1^i(s)$, $x_2^i(s)$, subject to the initial conditions $x_1^i(s_0) = \alpha_0^i$, $x_2^i(s_0) = \beta_0^i$, $\dot{x}_1^i(s_0) = \alpha_1^i$, $x_2^i(s_0) = \beta_1^i$, are called dyodesics, or dyopaths.

Problems

1. Derive (607).

2. Show that the $C^{,i}_{\alpha,\beta}(x_1, x_2)$ are the components of a two-point tensor, a mixed tensor relative to M_1, and a covariant vector relative to M_2.

3. Show that the law of transformation for the linear connection $\Gamma^{i,}_{\alpha\beta,}$ is

$$\bar{\Gamma}^{i}_{\alpha\beta,}(\bar{x}_1, \bar{x}_2) = \Gamma^{\mu,}_{\sigma\tau,}(x_1, x_2) \frac{\partial x_1^\sigma}{\partial \bar{x}_1^\alpha} \frac{\partial x_1^\tau}{\partial \bar{x}_1^\beta} \frac{\partial \bar{x}_1^i}{\partial x_1^\mu} + \frac{\partial^2 x_1^\mu}{\partial \bar{x}_1^\alpha \, \partial \bar{x}_1^\beta} \frac{\partial \bar{x}_1^i}{\partial x_1^\mu}$$

4. Show that $\Gamma^{i,}_{\alpha\beta,} = \Gamma^{i,}_{\beta\alpha,}$ if and only if $g_{\alpha,\beta} = \dfrac{\partial \varphi_{,\beta}}{\partial x_1^\alpha}$, where $\varphi_{,\beta}(x_1, x_2)$ is a scalar relative to M_1 and a covariant vector relative to M_2. If also $\Gamma^{,i}_{,\alpha\beta} = \Gamma^{,i}_{,\beta\alpha}$, show that of necessity

$$g_{\alpha,\beta} = \frac{\partial^2 \psi}{\partial x_1^\alpha \, \partial x_2^\beta}$$

where ψ is a scalar relative to both M_1 and M_2.

5. If

$$
\begin{aligned}
ds^2 &= -e^\lambda\, dr_1\, dr_2 - r_1 r_2 e^\mu\, d\varphi_1\, d\varphi_2 + e^\nu\, dt_1\, dt_2 \\
e^\nu &= \left[1 - \frac{mM^2}{(m+M)^2}\frac{1}{r_1}\right]\left[1 - \frac{m^2 M}{(m+M)^2}\frac{1}{r_2}\right] \\
e^\mu &= \left(1 + \frac{M}{r_1}\right)\left(1 + \frac{m}{r_2}\right) \\
e^\lambda &= e^{2\mu - \nu}
\end{aligned}
$$

show that the two-point tensors

$$
\begin{aligned}
T_{\alpha,\beta} &= \frac{\partial C^{\sigma,}_{\alpha,\beta}}{\partial x_1^\sigma} - \frac{\partial \Gamma^{\sigma,}_{\alpha\sigma,}}{\partial x_2^\beta} + \Gamma^{\sigma,}_{\tau\sigma,} C^{\tau,}_{\alpha,\beta} - C^{\sigma,}_{\tau,\beta}\Gamma^{\tau,}_{\alpha\sigma,} \\
T'_{\alpha,\beta} &= \frac{\partial \Gamma^{,\sigma}_{,\beta\sigma}}{\partial x_1^\alpha} - \frac{\partial C^{,\sigma}_{\alpha,\beta}}{\partial x_2^\sigma} + C^{,\sigma}_{\alpha,\tau}\Gamma^{,\tau}_{,\beta\sigma} - C^{,\tau}_{\alpha,\tau}\Gamma^{,\sigma}_{,\beta\sigma}
\end{aligned}
$$

vanish identically (m, M are constants). Show that the dyodesics satisfy

$$
\begin{aligned}
r_1 r_2 \frac{d\varphi_1}{ds} &= he^{-\mu} \\
r_1 r_2 \frac{d\varphi_2}{ds} &= he^{-\mu} \\
\frac{dt_1}{ds} &= C_1 e^{-\nu} \\
\frac{dt_2}{ds} &= C_2 e^{-\nu}
\end{aligned}
$$

$$
\frac{d^2 v}{d\varphi_1^2} + v\left[1 + \frac{Mm}{[1 + (m/M)]^4 h_1^2}\right] \approx \frac{M}{[1 + (m/M)]^2 h_1^2} + 3Mv^2\left[1 + \frac{Mm}{(m+M)^2}\right]
$$

provided that $Mr_2 \equiv mr_1$, $v = 1/r_1$, $h_1 = (M/m)h$. For $m \ll M$, $Mm/h_1^2 \ll 1$, we have $\dfrac{d^2 v}{d\varphi_1^2} + v \approx \dfrac{M}{h_1^2} + 3Mv^2$, the Einstein solution for the motion of an infinitesimal particle moving in the field of a point gravitational mass M.

REFERENCES

Brand, L. "Vectorial Mechanics," John Wiley & Sons, Inc., New York. 1930.

Brillouin, L. "Les Tenseurs," Dover Publications, New York 1946.

Graustein, W. C. "Differential Geometry," The Macmillan Company, New York. 1935.

Houston, W. V. "Principles of Mathematical Physics," McGraw-Hill Book Company, New York. 1934.

Joos, G. "Theoretical Physics," G. E. Stechert & Company, New York. 1934.

Kellogg, O. D. "Foundations of Potential Theory," John Murray, London. 1929.

McConnell, A. J. "Applications of the Absolute Differential Calculus," Blackie & Son Ltd., Glasgow. 1931.

Michal, A. D. "Matrix and Tensor Calculus," John Wiley & Sons, Inc., New York. 1947.

Milne-Thomson, L. M. "Theoretical Hydrodynamics," The Macmillan Company, New York. 1938.

Page, L. "Introduction to Theoretical Physics," D. Van Nostrand Company, New York. 1935.

Phillips, H. B. "Vector Analysis," John Wiley & Sons, Inc., New York. 1933.

Smythe, W. R. "Static and Dynamic Electricity," McGraw-Hill Book Company, New York. 1939.

Thomas, T. Y. "Differential Invariants of Generalized Spaces," Cambridge University Press, London. 1934.

Tolman, R. C. "Relativity Thermodynamics and Cosmology," Oxford University Press, New York. 1934.

Veblen, O. "Invariants of Quadratic Differential Forms," Cambridge University Press, London. 1933.

Weatherburn, C. E. "Elementary Vector Analysis," George Bell & Sons, Ltd., London. 1921.

——— "Advanced Vector Analysis," George Bell & Sons, Ltd., London, 1944.

——— "Differential Geometry," Cambridge University Press, London. 1927.

——— "Riemannian Geometry," Cambridge University Press, London. 1942.

Wilson, W. "Theoretical Physics," vols. I, II, III, Methuen & Co., Ltd, London. 1931, 1933, 1940.

INDEX

P

Q

R

S